Critical mm-Wave Components for Synthetic Automatic Test Systems

Michael Hrobak

Critical mm-Wave Components for Synthetic Automatic Test Systems

Springer Vieweg

Dr.-Ing. Michael Hrobak
Berlin, Germany

Dissertation, Friedrich-Alexander University of Erlangen-Nuremberg, 2014.
Original title: „Critical Millimeter-Wave Components for Synthetic Automatic Test Systems“.

ISBN 978-3-658-09762-2 ISBN 978-3-658-09763-9 (eBook)
DOI 10.1007/978-3-658-09763-9

Library of Congress Control Number: 2015937604

Springer Vieweg

Printed on acid-free paper

Springer Vieweg is a brand of Springer Fachmedien Wiesbaden
Springer Fachmedien Wiesbaden is part of Springer Science+Business Media
(www.springer.com)

Er is koa Waidler, Sie is koa Waidler
- was brauch'an de a Waidler Mess.

Joseph Seitz, 1973

Life is what happens to you
while you are busy making other plans.

John Lennon, 1980

Simulation ist eine Mischung aus Kunst, Wissenschaft,
Glück und verschiedenen Graden an Ehrlichkeit.

Samuel D. Stearns, 1994

Preface

The present thesis has been written predominantly in the course of my activities as Research Associate (Oct. 2007 till Dec. 2013) at the Institute of Microwaves and Photonics (LHFT), Friedrich-Alexander University of Erlangen-Nuremberg, Germany.

Outline

The thesis is structured in six chapters, a summary and an appendix. The core statements of each chapter are introduced in the following. A German translation of the summary is included in an additional chapter.

Chapter 1 Synthetic Instruments introduces history and the **fundamental idea** of synthetic instruments. Block diagrams of synthetic and traditional instruments are shown and **critical millimeter-wave components** for synthetic automatic test systems are identified.

In **Chapter 2 Resistive Diode Frequency Multipliers** a complete **modelling procedure** for discrete **Schottky diodes** is illustrated, which includes the semiconductor and linear parts of the diode. One section deals with single tone large signal analysis of nonlinear circuits, as it is used by harmonic balance circuit simulators. Different diode configurations, optimum embedding impedances and the author's frequency **multiplier design flow** are explained. Experimental results are presented. These include an octave bandwidth frequency **tripler for 20 to 40 GHz**, five frequency multipliers for **50 / 60 to 110 GHz** and frequency multipliers for **D- and Y-band**. In addition, the design equations of Dolph-Chebyshev waveguide tapers and an uncertainty analysis of spectral power measurements above 50 GHz are outlined.

In the first section of **Chapter 3 Planar Directional Couplers and Filters**, the modal and nodal scattering and impedance network parameters of coupled transmission lines according to the **general waveguide circuit theory** (GWCT) are introduced. This analysis is based on results from 2D EM Eigenmode analysis. The second section deals with **backward wave directional couplers**. A synthesis procedure for nonuniform transmission line couplers is included. Theoretical and experimental investigations on backward wave couplers with dielectric overlay technique (stripline) and wiggly-line technique to equalize the even and odd mode phase velocities are presented. The third section is dedicated to **codirectional couplers**. Beside the synthesis procedure, simulated and measured results of various codirectional couplers on thin-film processed alumina are given. Couplers for planar integration require internal nonreflective impedance terminations. Several 50 Ω **terminations** based on nickel chrome (NiCr) sheet resistors are presented in section 3.4. It further includes simulation and measurement results of the author's DC to 110 GHz **attenuator series**. Equalized phase velocities are also beneficial for edge and broadside **coupled line bandpass filters (BPF)** to suppress the parasitic second passband. This is demonstrated in section 3.5 by applying wiggly-line technique to the first and last filter element of an edge coupled line BPF on 10 mil alumina.

Chapter 4 Triple Balanced Mixers presents the author's investigations on triple balanced mixers (TBM), which are the only mixers providing overlapping RF, LO and IF frequency ranges together with large IF bandwidth and enhanced RF large signal handling capability. The first section highlights major differences and similarities between frequency multipliers and mixers. This is followed by a section about analytical and numerical methods for mixer analysis. Different methods are listed together with their strengths and weaknesses. With the results of single tone large signal analysis (harmonic balance) as a starting point, **large signal / small signal (LSSS) mixer analysis** based on conversion matrices is derived. The minimum achievable conversion loss at different embedding impedances and also optimum embedding impedances are calculated for a simplified case, which includes nonzero diode series resistances. Different mixer configurations (SBM, DBM and TBM) are compared from a small signal analysis point of view and regarding the mixers' spurious tone behaviour, following a method from Henderson. A **planar TBM realization** on SiO_2 using commercial silicon crossed quad diodes is presented. The TBM operates from **1 to 45 / 50 GHz** at RF / LO and achieves at least 20 GHz IF bandwidth. Simulation and measurement results of the component are presented. Measurement results of the TBM within an automatic test system front end module are included.

The first section of **Chapter 5 Power Detectors** introduces theoretical foundations. Expressions for detector current β and voltage sensitivity γ, and the upper ΔP_{sq}, P_{USL} and lower boundary P_{NEP}, P_{TSS} of the square law dynamic range are derived. This is followed by a section which compares common detector architectures. Two planar ultra-wideband power detector designs operating from **1 to 40 GHz** and **60 to 110 GHz** are presented. The detectors are part of the signal generator frequency extension modules, described in chapter 6, to realize monitoring of the output power level and automatic level control (ALC).

Chapter 6 Integrated Front End Assemblies presents two signal generator frequency extension modules for the output frequency ranges **20 to 40 GHz** and **60 to 110 GHz**. Such source modules are required in automatic test systems to provide internal local oscillator signals and high frequency stimulus. Although many necessary functionalities are available as microwave monolithic integrated circuits (MMIC), hybrid realization is preferred to combine the advantages of both, thin-film processed alumina Al_2O_3 or quartz SiO_2 and MMIC technology. Components from the preceding sections are utilized within the source modules.

The **Appendix Filter Synthesis** includes equations for synthesis of Chebyshev lowpass (LPF) and bandpass filters (BPF) in a very compact form. The filters are either based on **g** parameters of a lowpass prototype filter or **G** parameters of a prototype quarterwave transformer.

Annotations

- The author tried to be compatible with guidelines for submission of articles to the IEEE Transactions on Microwave Theory and Techniques, regarding presentation of content and also syntax.
- Column vectors **X** and matrices **X** are shown in bold face typing.
- The subscripts $\mathbf{X}^{\mathrm{T}}, X^{\star}, \mathbf{X}^{\mathrm{H}}$ denote the matrix or vector transposition, complex conjugation and the Hermitian adjoint, respectively.
- Real and complex quantities are not distinguished.
- The letters $\alpha, \beta, \epsilon, \gamma, \sigma, P$ and p have different meanings throughout the thesis, which are believed to be clear from context.
- Per unit length network parameters are often marked with apostrophes. Within this work, the symbols R, L, C, G are used for the conventional equivalent network parameters **and** also for the distributed ones.
- The letter j is used for the imaginary unit, whereas the complex current is denoted by i.
- The frequency f and angular frequency $\omega = 2\pi f$ are both called frequency in the continuous text.
- The following frequency range abbreviations and designations for rectangular waveguides WR are used throughout this work.

band abbrev.	frequency range [GHz]	EIA band design.	internal dim. [mm]	TE_{10} cut-off freq. [GHz]
V	50 to 75	WR-15	3.759 × 1.880	40
E	60 to 90	WR-12	3.099 × 1.549	48
W	75 to 110	WR-10	2.540 × 1.270	59
F	90 to 140	WR-8	2.032 × 1.016	74
D	110 to 170	WR-6	1.651 × 0.826	91
G	140 to 220	WR-5	1.295 × 0.648	116
Y	170 to 260	WR-4	1.092 × 0.546	137
J	220 to 325	WR-3	0.864 × 0.432	174

EIA = The Electronic Industries Alliance, www.ecianow.org

An almost uniform labeling from V- to G-band is well-established in industry, which is not the case for the letters used for WR-4 to WR-3. Especially H-band is often used instead of J-band and many designations for the rarely used Y-band can be found.

- The following symbols denote DC current and voltage sources, AC current and voltage sources, lumped elements (inductors, resistors, capacitors), variable sources (arrow), nonlinear sources, nonlinear resistors and capacitors, variable resistors and capacitors.

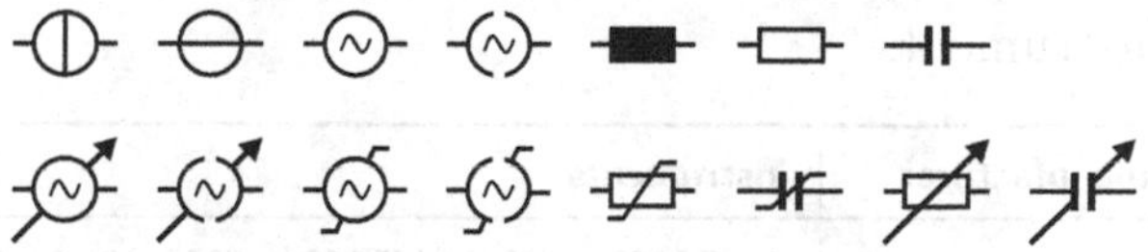

- In physics, the notational operator $[\star]$ gives the unit of the physical quantity $\star$, to which it is applied. In this sense $[f] = \text{GHz}$ would be a correct application. Anyway, it is common practice to write [GHz]. The latter is used in all labels of Figures and Table captions throughout this work.
- The complex, frequency dependent permittivity $\epsilon(\omega)$ and conductivity $\sigma(\omega)$ cover the intrinsic bulk material properties. Whereas the effective conductivity $\sigma_{\text{eff}} = \sigma(\omega) + j\omega\epsilon(\omega)$ and effective permittivity $\epsilon_{\text{eff}}(\omega) = \sigma_{\text{eff}}/j\omega$ include both effects. The utilized definition of $\epsilon(\omega)$ includes the permittivity constant ϵ_0. This leads to the

following derived expressions of the intrinsic wave impedance Z_w and propagation coefficient γ.

$$Z_\mathrm{w} = \sqrt{\frac{\mu(\omega)}{\epsilon_\mathrm{eff}(\omega)}} = \sqrt{\frac{j\omega\mu(\omega)}{\sigma_\mathrm{eff}}} = \frac{j\omega\mu(\omega)}{\gamma}$$

$$\gamma = \frac{j\omega\mu(\omega)}{Z_\mathrm{w}} = \sqrt{j\omega\mu_0} \cdot \sqrt{\sigma_\mathrm{eff}}$$

- The letters v, i are used for voltage and current time domain waveforms, whereas the letters V, I are used for the corresponding DC and spectral components in the frequency domain. Small- and large-signal quantities are not distinguished.
- In general, frequency domain variables belong to the complex double sided Fourier series. Some derivations are expressed more suitable by complex single sided Fourier series. In both cases, the same symbol $\circ\!\!-\!\!\bullet$ for time to frequency domain mapping is used.
- Presented simulation data is based on the software tools Agilent ADS, Ansys HFSS, Ansys Designer and CST Microwave Studio.
- A complete list of utilized measurement instruments is given in the following. The presented measurement data underlies the accuracy of these instruments.

manufacturer	**instruments**
signal generators	Ag E8257D, R&S SMF100A, SMZ-90, An MG3694C
spectrum analyzers	Ag N9030A PXA, R&S FSW43, FSEK30
harmonic mixers	Ag M1970V, M1970W, R&S FS-Z75, FS-Z90, SAM-110, SAM-140, SAM-170, SAM-220
power detectors	Ag V8486A, W8486A, R&S NRP-Z55, NRP-Z57, NRP-Z58, VDI Erickson PM4
network analyzers	Ag N5247A, N5251A, R&S ZVA50, ZVA110

Ag = Agilent Technologies, www.home.agilent.com,
An = Anritsu GmbH, www.anritsu.com,
R&S = Rohde & Schwarz GmbH & Co. KG, www.rohde-schwarz.com,
VDI = Virginia Diodes Inc., www.vadiodes.com.

Acknowledgements

I would like to thank my supervisor, examiner and former employer **Lorenz-Peter Schmidt** for giving me the opportunity to pursue my scientific work at the Institute of Microwaves and Photonics (LHFT) and for his professional guidance throughout the time.

My gratitude also goes to **Jan Hesselbarth** for careful reading of the thesis and his work as reviewer. Thanks to **Manfred Albach** and **Peter Wellmann** for their contribution to the oral examination.

I express my special thanks to my office colleague **Marcus Schramm** for many technical discussions and for helping me to take the ups and downs during the time at the Institute with a smile. I am very grateful to my former colleague and good friend **Michael Sterns** for his qualified advice on many critical engineering problems, his enthusiastic support and ongoing interest in cooperation and sharing engineering experiences.

Thanks to **Jochen Weinzierl** and **Jan Schür** for smooth operation of all financial aspects, the easy ones and the not so easy ones ☺. It has always been a pleasure to experience the contagious enthusiasm with respect to engineering phenomena of **Siegfried Martius**. Many thanks to all student contributions to my thesis, especially to **Wadim Stein** and **Ernst Seler** for their particularly valuable inputs. Successful prototypes would not have been possible without the help from the LHFT technology team, **Günter Bauer, Jürgen Popp, Lothar Höpfel, Ottmar Wick,** and **Johannes Ringel**. I am especially thankful to **Jürgen Popp** for the high number of accurately manufactured mechanical housings.

Berlin Michael Hrobak

Contents

Abbreviations, Constants and Symbols

Abbreviations, constants and symbols which are not mentioned in the following are explained in the continuous text.

Abbreviations

Abbrev.	Description
2D / 3D	two / three dimensional
AC	alternating current
ADC	analog-digital converter
ADS	Agilent advanced design system
ALC	automatic level control (loop)
ATCA	advanced telecommunications computing architecture
ATS	automatic test system
AWG	arbitrary waveform generator
AWR	Applied Wave Research
AXIe	ATCA extensions for instrumentation
BES	buried epitaxial layer
bFIN	bilateral finline
BPF	bandpass filter
bSTL	broadside coupled stripline
cMSL	coupled microstrip line
COT	cost of test
COTS	commercial off the shelf
CPW	coplanar waveguide

Abbrev.	Description
CTRL	control
DAC	digital-analog converter
DBM	double balanced mixer
DC	direct current
DDB	double double balanced mixer
DDS	direct digital synthesis
DGS	defected ground structure
DOD	US department of defense
DRS	design rule set
DUT	device under test
EDA	electronic design automation
eLRRM	enhanced line-reflect-reflect-match
eSTL	edge coupled stripline
FBW	fractional bandwidth
FCB	finite conductivity boundary
FEM	finite element method
FFT	fast Fourier transform
FR4	glass-reinforced epoxy laminate
gCPW	grounded coplanar waveguide
GMRES	generalized minimal residual method
GSG	ground signal ground
GWCT	general waveguide circuit theory
HBT	heterojunction bipolar transistor
HFSS	Ansys high frequency structure simulator
HPF	highpass filter
IF	intermediate frequency
LAN	local area network
LIB	layered impedance boundary
LO	local oscillator
LPF	lowpass filter

Abbrev.	Description
LSSS	large signal small signal
LU	lower upper factorization
LXI	LAN extensions for instrumentation
MAG	maximum available gain
MATLAB	Mathworks matrix laboratory
m/pHEMT	meta- / pseudomorphic high electron mobility transistor
MIC	monolithic integrated circuit
MIM	metal insulator metal
MMIC	microwave / millimeter-wave monolithic integrated circuit
MSL	microstrip line
MWS	CST microwave studio
NEP	noise equivalent power
NLTL	nonlinear transmission line
NTL	nonuniform transmission line
NxTEST	next generation automatic test systems
UV	ultraviolet
PCB	printed circuit board
PCI	peripheral component interconnect
PEC	perfect electric conductor
PLC	power line cycle
PLL	phase locked loop
PM4	Erickson power meter 4
PMC	perfect magnetic conductor
PP	parallel plate
PSD	power spectral density
PXI	PCI extensions for instrumentation
QTEM	quasi transversal electromagnetic
RF	radio frequency
RX	receiver
SBM	single balanced mixer

Abbrev.	Description
SDM	single device mixer
SIWG	synthetic instrument working group
SLC	single layer capacitor
SMA	subminiature version A
SMD	surface mounted device
SNR	signal to noise ratio
SOL	short-open-load
SOLT	short-open-load-thru
SPICE	simulation program with integrated circuit emphasis
STL	stripline
TBM	triple balanced mixer
TE	transversal electric
TEM	transversal electromagnetic
THT	through hole technology
TM	transversal magnetic
TRL	thru-reflect-line
TSS	tangential signal sensitivity
TX	transmitter
uFIN	unilateral finline
USL	upper square law limit
VBW	video bandwidth
VGA	variable gain amplifier
VME	versa module europa bus
VNA	vector network analyzer
VXI	VME extensions for instrumentation
WR	waveguide rectangular
YIG	yttrium iron garnet
ZBD	zero bias Schottky diode

Constants

Irrational numbers are truncated to the fifth decimal place.

Symbol	Value	Unit	Description
μ_0	$4\pi \cdot 10^{-7}$	$\mathrm{VsA^{-1}m^{-1}}$	permeability constant
ϵ_0	$8.85419 \cdot 10^{-12}$	$\mathrm{AsV^{-1}m^{-1}}$	permittivity constant
c	$\sqrt{\mu_0 \epsilon_0}$	$\mathrm{ms^{-1}}$	speed of light
Z_0	$\mu_0 \cdot c$	$\mathrm{VA^{-1}}$	free space wave impedance
e	2.71828		Euler constant
$h = \hbar \cdot 2\pi$	$6.62607 \cdot 10^{-34}$	Js	Planck constant
k	$1.38065 \cdot 10^{-23}$	$\mathrm{JK^{-1}}$	Boltzmann constant
q_e	$1.60218 \cdot 10^{-19}$	As	elementary charge
π	3.14159		ratio of circumference to diameter of a circle
m_0, m_e	$9.10938 \cdot 10^{-31}$	kg	electron rest mass

Symbols

Symbol	Unit	Description
a	m	anode diameter
A^*, A^{**}	$\mathrm{A/(mK)^2}$	standard and modified Richards constant
AL	m	anode length
AW	m	anode width
b	m	buffer radius
$\mathbf{B}$	T	magnetic flux density
C_epi	F	epitaxial layer capacitance
C_j	F	junction capacitance
C_j0	F	max. junction capacitance at zero bias

Symbol	Unit	Description
$\mathbf{C}$		chain matrix
δ		Dirac delta function (distribution)
δ_{skin}	m	skin depth
d	m	buffer diameter
$\mathbf{D}$	As/m^2	electric flux density
$\epsilon(\omega)$	As/(Vm)^{-1}	freq. dependent intrinsic bulk permittivity
ϵ_{eff}	As/(Vm)^{-1}	effective permittivity
ϵ_{r}		relative permittivity
$\mathbf{E}$	V/m	electric field strength
f_{c}	1/s	diode cut-off frequency (figure of merit)
f_{c1dB}	1/s	lowpass filter 1 dB cut-off frequency
f_{c3dB}	1/s	lowpass filter 3 dB cut-off frequency
F_{C}		diode capacitance coefficient
$\mathbf{F}(\mathbf{V})$	A	current error vector
$\gamma = \alpha + j\beta$	1/m	propagation coefficient
Γ		reflection coefficient
$H(m)$		unit step function
$\mathbf{H}$	A/m	magnetic field strength
$i, i_{\text{LO}}, i_{\text{RF}}, i_{nm}$	A	current time waveforms
$I, I_{\text{LO}}, I_{\text{RF}}, I_{nm}$	A	current components of Fourier series'
I_{S}	A	reverse saturation current
$\mathbf{I}_{\text{C}}$	A	vector of displacement current vectors
$\mathbf{I}_{\text{G}}$	A	vector of conductance current vectors
$\mathbf{I}_{\text{L}}$	A	vector of linear current vectors at all circuit nodes
$\mathbf{I}_{\text{NL}}$	A	vector of nonlinear current vectors
$\mathbf{I}_{\text{s}}$	A	vector of linear source current vectors
$\mathbf{J}, \mathbf{J}_{\text{vol}}$	A/m^3	volume current density
$\mathbf{J}$	$1/\Omega$	Jacobian matrix
κ	$(\Omega\text{m})^{-1}$	electrical conductivity

Symbol	Unit	Description
L_{epi}	H	epitaxial layer inductance
μ_{e}	$\text{m}^2/(\text{Vs})^{-1}$	epitaxial layer electron mobility
$m^\star$	kg	effective electron mass
M		capacitance coefficient
η		ideality factor
n, m		RF and LO harmonic numbers
N_{Depi}	$1/\text{m}^3$	donor concentration
Φ_{b}	V	metal semicond. corrected barrier height
Φ_{b0}	V	metal semicond. uncorrected barrier height
$\Delta\Phi_{\text{b0}}$	V	metal semicond. barrier height correction
Φ_{χ}	V	potential difference between semicond. conductor and free space level
Φ_{m}	V	potential difference between metal Fermi and free space level
$p(z)$	Ω/m	reflection coefficient distribution function
$P + jR$	W	real and reactive power
$\mathbf{P}, P_{\text{e}}, P_{\text{D}}$	C/m^2	dielectric polarization
Q/S	As/m^2	charge per unit surface
$\mathbf{Q}$	C	vector of charge waveform vectors
ρ, ρ_{epi}	C/m^3	charge densities
r, r_{dB}		Chebyshev filter ripple
R_{ohmic}	Ω	ohmic contact impedance
R_{s}	Ω	series resistance
σ^2		variance value
$\sigma(\omega)$	$(\Omega\text{m})^{-1}$	frequency dependent intrinsic bulk cond.
σ_{eff}	$(\Omega\text{m})^{-1}$	effective conductivity
$\sigma, \sigma_{\text{epi}}, \sigma_{\text{DC}}, \sigma_{\text{b}}$	$(\Omega\text{m})^{-1}$	bulk conductivities
S_{j}	m^2	anode surface area
$\mathbf{S}$		scattering matrix
t_{buffer}	m	thickness of buffer layer

Symbol	Unit	Description
$t_\mathrm{d}(v)$	m	thickness of depleted epitaxial layer
t_epi	m	thickness of epitaxial layer
$t_\mathrm{u}(v)$	m	thickness of undepleted epitaxial layer
T	K	absolute temperature
T_M	°	melting point
$\hat{v}, \hat{i}$	V, A	voltage and current peak values
v_d	m/s	mean drift velocity of electrons
$v, v_\mathrm{LO}, v_\mathrm{RF}, v_{nm}$	V	voltage time waveforms
$v_\mathrm{j}, i_\mathrm{j}$	V, A	junction voltage and current
V_BR	V	reverse breakdown voltage
$V_\mathrm{d}, \Phi_\mathrm{bi}$	V	built-in voltage of the junction, diffusion potential, barrier height
$V_\mathrm{T} = kT/q_\mathrm{e}$	V	thermal voltage
$V, V_\mathrm{LO}, V_\mathrm{RF}, V_{nm}$	V	voltage components of Fourier series'
V_n	V	potential difference between lower edge of conduction band and Fermi level
$\mathbf{V}_\mathrm{L}$	V	vector of linear voltage vectors
$\mathbf{V}_\mathrm{L}$	V	at all circuit nodes
ω_bound	1/s	eigenfrequency of bound electrons within Lorentz oscillator model
ω_d	1/s	dielectric relaxation frequency
ω_p	1/s	plasma resonance frequency
ω_s	1/s	scattering frequency
ω_seff	1/s	effective scattering frequency
$\mathbf{\Omega}$	1/s	frequency matrix
W_0	VAs	electron free space energy level
$W_\mathrm{Cm/s}$	VAs	lower edge of conduction band energy level
W_χ	VAs	semiconductor electron affinity
$W_\mathrm{Fm/s}$	VAs	Fermi energy level
W_V	VAs	upper edge of valence band energy level

Symbol	Unit	Description
$\mathbf{Y}, \mathbf{Z}$		admittance, impedance matrices
$\mathbf{Y}_{\mathrm{L}}^{'}$, $\mathbf{Y}_{\mathrm{L}}$	$1/\Omega$	matrix of linear admittance matrices
Z_{c}	Ω	characteristic impedance
Z_{epi}	Ω	epilayer impedance
$Z_{\mathrm{in}}, Z_{\mathrm{out}}$	Ω	input and output impedances
$Z_{\mathrm{ref}} = Z_0$	Ω	reference imp. of network parameters
Z_{s}, Z_{ℓ}	Ω	source and load impedances
Z_{skin}	Ω	skin effect impedance
Z_{spr}	Ω	spreading impedance
$Z, Z_{\mathrm{RF}}, Z_{\mathrm{LO}},$	Ω	DC and AC impedances
$Z_{\mathrm{IF}}, Z_{\mathrm{DC}}, Z_{nm}$	Ω	DC and AC impedances

Chapter 3

Symbol	Unit	Description
a		component of $\mathbf{M}_v, \mathbf{M}_i$
$\mathbf{a}, \mathbf{b}$	$\sqrt{\mathrm{W}}$	forward and backward travelling wave amplitude vectors
c		voltage coupling factor
$c = S_{31}$		coupling coefficient
c_k^+, c_k^-		complex amplitudes of fortward (+) and backward (-) travelling waves
c_{max}		maximum voltage coupling factor
$\overline{C}$		average coupling
C_{fow}	F/m	odd mode fringing cap. after wiggling
$C_{\mathrm{p}}, C_{\mathrm{f}}, C_{\mathrm{fe}}, C_{\mathrm{fo}}$	F/m	frequency dependent capacitance values
$C_{\mathrm{spec}} = S_{31\mathrm{spec}}$		specified magnitude of coupling
$d, d(c), d(z)$	m	wiggle depths
$D, \overline{D}, D_{\mathrm{min}}$		directivity, average and min. directivity
E_0, E_1, E_2		scaling parameters
$\mathbf{E}_{\perp}$	V/m	transversal electric field strength

Symbol	Unit	Description
$\mathbf{E}_z$	V/m	longitudinal electric field strength
$G(z), \phi_1, \phi_2, \psi_1, \psi_2$		functions for determination of
K_k, Q_k, f_1, f_2		coupler reflection coefficient
$\gamma_e = \alpha_e + j\beta_e$	1/m	even mode propagation coefficient
$\gamma_o = \alpha_o + j\beta_o$	1/m	odd mode propagation coefficient
$\mathbf{H}_\perp$	A/m	transversal magnetic field strength
$\mathbf{H}_z$	A/m	longitudinal magnetic field strength
i_{0k}	A	normalizing current of mode k
i_{ck}	A	nodal / conductor based current of node k
i_{mk}	A	modal current of mode k
$\mathbf{I}$		identity matrix
ℓ_0	m	waveguide / coupler length
ℓ_{i0k}	m	current integration line of mode k
ℓ_{v0k}	m	voltage integration line of mode k
$\mathbf{M}_v, \mathbf{M}_i$		modal to nodal conversion matrices
p	W	total complex power
$p(z)$	Ω/m	reflection coefficient distribution function
r_{max}		maximum ripple
R_c, R_π		components of $\mathbf{M}_v, \mathbf{M}_i$
$R_\square = R_{sheet}$	Ω	sheet resistance
$s, s(c), s(z)$	m	strip spacings
s_{lk}	m	loose coupling spacing
s_{tk}	m	tight coupling spacing
$\mathbf{S}_{ct} = \mathbf{S}_n$		nodal scattering matrix
$\mathbf{S}_{mt} = \mathbf{S}_m$		modal scattering matrix
S_P, S_{P1}, S_{P2}	m^2	port surfaces
t	m	conductor thickness
v_{0k}	V	normalizing voltage of mode k
v_{ck}	V	nodal / conductor based voltage of node k
v_{mk}	V	modal voltage of mode k

Symbol	Unit	Description
v_{e}	m/s	phase velocity of even mode
v_{g}	m/s	phase velocity of waveguide
v_{o}	m/s	phase velocity of odd mode
$w, w(c), w(z)$	m	strip widths
$\mathbf{X}$		cross power matrix
Δz	m	wiggle length
$Z_{0\mathrm{common}}$	Ω	common mode characteristic impedance
$Z_{0\mathrm{differential}}$	Ω	differential mode characteristic impedance
$Z_{0\mathrm{e}}$	Ω	even mode characteristic impedance
$Z_{0\mathrm{o}}$	Ω	odd mode characteristic impedance
$Z_{0\mathrm{p}}$	Ω	prototype waveguide's charact. impedance
Z_{PV}	Ω	characteristic impedance of power-voltage,
Z_{PI}	Ω	power-current and
Z_{VI}	Ω	voltage-current definition
$\mathbf{Z}_0$	Ω	diagonal characteristic impedance matrix
$\mathbf{Z}_{\mathrm{ct}} = \mathbf{Z}_{\mathrm{n}}$		nodal impedance matrix
$\mathbf{Z}_{\mathrm{mt}} = \mathbf{Z}_{\mathrm{m}}$		modal impedance matrix
$\mathbf{Z}_{\mathrm{ref}}$	Ω	diagonal reference impedance matrix

Chapter 4

$\mathbf{C}$	F	small signal mixer capacitance matrix
ϵ_1, ϵ_3		mixer conversion loss parameters
$\mathbf{G}$	1/Ω	small signal mixer conductance matrix
$\mathcal{G}_{\mathrm{M}}$		maximum available power gain (MAG)
$\mathcal{G}_{\mathrm{T}}$		transducer power gain
i_{LS}	A	large signal current
i_{SS}	A	small signal current
I_1		modified Bessel functions of the first kind
$\mathbf{I}_{\mathrm{C}}$	A	vector of displ. currents at all harmonics

Symbol	Unit	Description
$\mathbf{I}_{\mathrm{G}}$	A	vector of conduct. currents at all harmo.
κ_2		mixer conversion loss parameter
L_1		conv. loss with short circuited image signal
L_2		conv. loss in the broadband case
L_3		conv. loss with open circuited image signal
θ		mixer conversion loss parameter
v_{LS}	V	large signal voltage
v_{SS}	V	small signal voltage
$\mathbf{\Omega}$	1/s	frequency matrix
y_{LS}	$1/\Omega$	large signal admittance waveform
y_{SS}	$1/\Omega$	small signal admittance waveform

Chapter 5

Symbol	Unit	Description
$\beta, \beta^{'}, \beta_0^{'}, \beta^{''}$	A/W	current sensitivity
C_{EPC}	F	equivalent parallel capacitance value
C_{ESC}	F	equivalent series capacitance value
Δ_1, Δ_2		correction terms
f_{SRF}	1/s	series resonant frequency
$\gamma, \gamma^{'}, \gamma_0^{'}, \gamma^{''}$	V/W	voltage sensitivity
$g(v) = i$	$1/\Omega$	diode current-voltage dependency
I_{DC}	A	generated DC current
I_{N}	A	equivalent root mean square noise current
R_{j}	Ω	diode's junction resistance
$P_{\mathrm{NEP}}, P_{\mathrm{NEP0}}$	W	noise equivalent power level
P_{RF}	W	total RF power
$P_{R_{\mathrm{j}}}$	W	power absorbed in junction resistance
ΔP_{sq}		square law error
PSD_i	$\mathrm{A^2sK}$	one-sided current power spectral density
PSD_v	$\mathrm{V^2sK}$	one-sided voltage power spectral density

Symbol	Unit	Description
P_{TSS}	W	tangential signal sensitivity power level
P_{USL}	W	upper limit of square law operation
$T = 1/f$	s	sinusoidal periodic time
V_0	V	applied DC bias voltage
V_{DC}	V	generated DC voltage
V_{N}	V	equivalent root mean square noise voltage

Chapter 6

I_{D}	A	drain current
$\boldsymbol{\ell}_{\mathbf{J}}$	m	J inverter lengths
$\boldsymbol{\ell}_{\mathbf{res}}$	m	resonator lengths
$P_{\mathrm{in}}, P_{\mathrm{out}}, P_{\mathrm{max}}$	W	input / output/ maximum power levels
$\tan(\delta_{\mathrm{e}}), \tan(\delta_{\mathrm{m}})$		electric / magnetic tangent delta values
V_{CTRL}	V	control voltage
V_{D}	V	drain voltage
V_{G}	V	gate voltage
$\mathbf{w}_{\mathbf{J}}$	m	J inverter widths
$\mathbf{w}_{\mathbf{res}}$	m	resonator widths
$\mathbf{Z}_{\mathbf{J}}$	Ω	J inverter impedances
$\mathbf{Z}_{\mathbf{res}}$	Ω	resonator impedances

Appendix

α, A_r		prototype $\lambda/4$ transformer parameters
β		Chebyshev filter constant
$B_{\mathrm{s}}, B_{\mathrm{p}}$	1/Ω	series, parallel susceptance
ε		Chebyshev filter ripple constant
f_0	1/s	bandpass filter center frequency
γ		Chebyshev filter constant

Symbol	Unit	Description
$\mathbf{g}$		prototype lowpass filter parameters
$\mathbf{G}$		prototype $\lambda/4$ transformer parameters
h		filter adjustment parameter
J	$1/\Omega$	admittance inverter
K	Ω	impedance inverter
$\lambda_\mathrm{e}, \lambda_\mathrm{o}$	m	even and odd mode wavelengths
$\boldsymbol{\ell}_\mathrm{int}$	m	vector of interconnection line lengths
$\boldsymbol{\ell}_\mathrm{s}$	m	vector of stub lengths
M_k		fishgrate filter parameter
η		Chebyshev filter constant
N		filter order
ϕ	$^\circ$	phase of inverter circuit
p_k		poles of Chebyshev filter transfer function
r, r_dB		Chebyshev filter passband ripple
RL_min		minimum passband return loss
$X_\mathrm{s}, X_\mathrm{p}$	Ω	series, parallel reactance
$\mathbf{Y}_\mathrm{s}$	$1/\Omega$	vector of stub admittances
$\mathbf{Y}_\mathrm{int}$	$1/\Omega$	vector of interconnection line admittances
$Z_{0\mathrm{e}}, Z_{0\mathrm{o}}$	Ω	even and odd mode charact. impedances

1 Synthetic Instruments

1.1 Fundamental Idea

In the 1990s, the United States Department of Defense (DoD) initiated the development of new automatic test system (ATS) architectures for military products in its Next Generation Automatic Test Systems (NxTest) program. The main goals of the initiative are

- reduce total cost of test (CoT) and cost of ownership of ATS,
- reduce time to develop and deploy new or upgraded ATS,
- reduce physical footprint,
- reduce logistics footprint,
- provide greater flexibility through interoperable systems,
- improve quality of test.

The Synthetic Instrument Working Group (SIWG) of the NxTest program, a group joint participation between the DoD, defense prime contractors and suppliers, introduced synthetic instruments to achieve the NxTest goals. The term synthetic instrument[1] (SI) is used for a combination of well chosen functional hardware units. These are able to perform several different measurement tasks, depending on the applied control software. Whereas the opposite, traditional[2] instruments are built up to do a fixed set of measurement tasks with preferably high accuracy, normally realized as box instruments and found in rack and stack test configurations.

[1]In literature, the expressions virtual or modular instrument are used as synonyms.
[2]Traditional instruments are also called stand alone instruments or natural instruments in [1].

The reuse of the same set of hardware blocks for different fields of application, pushes instrument synthesis into the focus of ATS for commercial semiconductor products (Fig. 1.4) and also for test systems in research laboratories.

It is important to carefully partition traditional RF instruments, like signal generators (Fig. 1.1), vector network analyzers (Fig. 1.2), spectrum analyzers (Fig. 1.3), power meters, high speed scopes or frequency counters into reusable hardware units. Apart from special add ons, four basic blocks are able to synthesize the mentioned RF instruments. These are

- signal conditioning unit,
- upconversion unit,
- downconversion unit,
- signal acquisition or baseband unit including numeric processor.

The most promising developments are based on VXI, PXI, LXI or AXIe standards and the corresponding software interfaces.

The advantages, hardware reusability, the individual reconfiguration for each measurement demand and therefore reduction of CoT, are generally achieved at an expense of measurement accuracy. Fig. 1.1, 1.2, 1.3 show exemplary block diagrams of traditional RF instruments. Fig. 1.4 illustrates a front end module with high instantaneous signal bandwidth, as it is required for semiconductor ATS and surveillance radars. These block diagrams do not belong to real systems [2], but are chosen to illustrate instrument synthesis. Recurring hardware parts are highlighted with the same color, which makes clear it is possible to synthesize these traditional instruments saving redundant hardware parts. The reader should keep in mind, realization of each traditional instrument has been improved over decades for optimum performance. Hence, what looks the same in the block diagrams of Fig. 1.1, 1.2, 1.3, 1.4 is indeed realized differently and therefore the synthetic counterpart in general achieves lower performance compared to the traditional instrument.

The step from natural to synthetic instruments however gives rise to many new and different realizations of linear and nonlinear components, whereas a single optimum realization is often already found that fulfills the needs of a certain traditional instrument.

1.2 Critical Millimeter-Wave Components for Synthetic Automatic Test Systems

Within the scope of this work several critical millimeter-wave components are developed. The author's components are intended for use in synthetic automatic test systems but are also beneficial for use in traditional instruments. The focus is on ultrawide bandwidth to gain maximum versatility of the instruments with respect to coverage of measurement scenarios.
Table 1.1 and Table 1.2 give an overview of the author's component and module developments. Each development is listed with its bandwidth, utilized materials, processing technology and the figure with measurement results. The reader should keep in mind, many different waveguide transitions are either part of the components or necessary for system integration or measurement. These include transitions from coaxial and hollow waveguide to planar waveguide and baluns. The same holds true for required bias-T circuits and broadband planar filters (Table 1.2). These subcomponents deserve particular attention.

Table 1.1: Overview of Component and Module Developments within the Scope of this Thesis

component / module	description	frequency range° [GHz]	FBW [%]	technology / material	figure
x3_2040	frequency tripler	21 to 40	62	PCB / AD600	Fig. 2.35
x3_HE1	frequency tripler	60 to 110	59	thin-film / 5 mil Al_2O_3	Fig. 2.58
x3_uFIN	frequency tripler	60 to 110	59	thin-film / 5 mil Al_2O_3	Fig. 2.66
x3_SIDE	frequency tripler	67 to 110	49	thin-film / 5 mil Al_2O_3	Fig. 2.74
x2_uFIN	frequency doubler	50 to 110	75	thin-film / 5 mil Al_2O_3	Fig. 2.80
x2_uFINring	frequency doubler	50 to 110	75	thin-film / 5 mil Al_2O_3	Fig. 2.85
x2_D	frequency doubler	100 to 160	46	PCB / FF27 (PTFE)	Fig. 2.95
x3_Y	frequency tripler	180 to 230	24	PCB / FF27 (PTFE)	Fig. 2.103
BWC_STL_50	stripline backward wave coupler	1.8 to 50	186	PCB / RO4350/4450	Fig. 3.20
BWC_STL_70_SW	stripline backward wave coupler	1.3 to 62	192	PCB / RO4350/4450	Fig. 3.25
BWC_STL_70	stripline backward wave coupler	2.2 to 60	186	PCB / RO4350/4450	Fig. 3.28
BWC_WIG_50	wiggly-line backward wave coupler	2 to 50	185	thin-film / 10 mil Al_2O_3	Fig. 3.37
FWC_67	codirectional coupler	5 to 67	172	thin-film / 10 mil Al_2O_3	Fig. 3.43
FWC_20_40•	codirectional coupler	20 to 40	67	thin-film / 5 mil Al_2O_3	Fig. 3.48
FWC_60_110•	codirectional coupler	60 to 110	59	thin-film / 5 mil Al_2O_3	Fig. 3.51
LOAD_0_110	50 Ω termination	0 to 110	200	thin-film / 5 mil Al_2O_3	Fig. 3.60
ATT6/10/15_0_110	6 dB, 10 dB, 15 dB attenuators	0 to 110	200	thin-film / 5 mil Al_2O_3	Fig. 3.67
BPF_SIDE_WIG_20_30	edge coupled halfwave BPF	20 to 40	67	thin-film / 10 mil Al_2O_3	Fig. 3.73
TBM	triple balanced mixer	LO: 1 to 50	192	thin-film / 10 mil SiO_2	Fig. 4.28
TBM	triple balanced mixer	RF: 1 to 45	191	thin-film / 10 mil SiO_2	Fig. 4.28
TBM	triple balanced mixer	IF: 0.1 to 20	198	thin-film / 10 mil SiO_2	Fig. 4.27
DET_1_40	power detector	1 to 40	190	thin-film / 5 mil Al_2O_3	Fig. 5.12
DET_60_110	power detector	60 to 110	59	thin-film / 5 mil Al_2O_3	Fig. 5.20
SRC_20_40	signal source module	20 to 40	67	hybrid multi-chip module	Fig. 6.10
SRC_60_110	signal source module	60 to 110	59	hybrid multi-chip module	Fig. 6.23

° Output frequency range in case of frequency multipliers. • Simulation results only, but operation within the systems of chapter 6 is approved.

Table 1.2: Overview of Component and Module Developments within the Scope of this Thesis

component / module	description	frequency range° [GHz]	FBW [%]	technology / material	figure
BT_10	bias-T	0.01 to 10	200	PCB / RO4450	Fig. 5.6
BT_60_110	bias-T	60 to 110	59	thin-film / 5 mil Al_2O_3	Fig. 5.15 / Fig. 5.16
SCAP_0.01_110	DC block	0.01 to 110	200	thin-film / 5 mil Al_2O_3	Fig. 5.8
SCAP_60_110	DC block	60 to 110	59	thin-film / 5 mil Al_2O_3	Fig. 5.16
HPF_60_110	highpass filter	60 to 110	59	thin-film / 5 mil Al_2O_3	Fig. 5.17
LPF_0_50	lowpass filter	0 to 53	200	thin-film / 5 mil Al_2O_3	Fig. 2.46
BPF_70_110	bandpass filter	70 to 110	44	thin-film / 5 mil Al_2O_3	Fig. 2.71
BPF_20_40	fishgrate bandpass filter	20 to 40	67	thin-film / 10 mil SiO_2	Fig. 6.8
MSL_COAX_V/1MM	1.85mm / 1.00 mm to MSL	0 to 110	200	thin-film / 5 mil Al_2O_3	Fig. 2.40
MSL_WR-10	MSL to WR-10 transition	63 to 110	54	thin-film / 5 mil Al_2O_3	Fig. 2.42

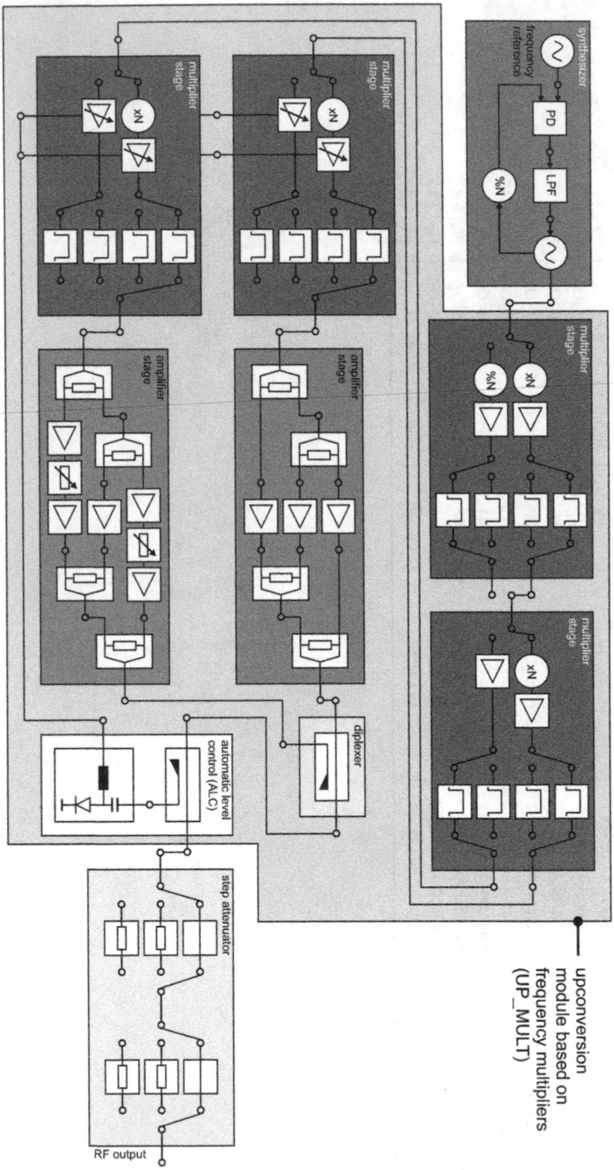

Figure 1.1: Block diagram of an exemplary signal generator front end module.

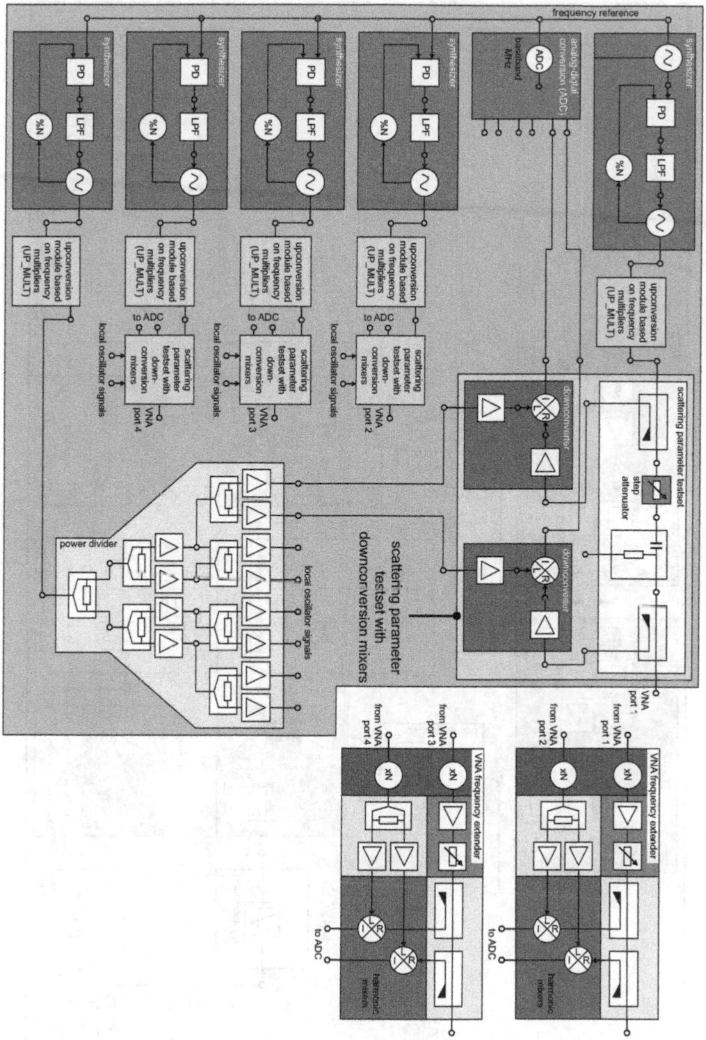

Figure 1.2: Block diagram of an exemplary vector network analyzer with frequency extenders.

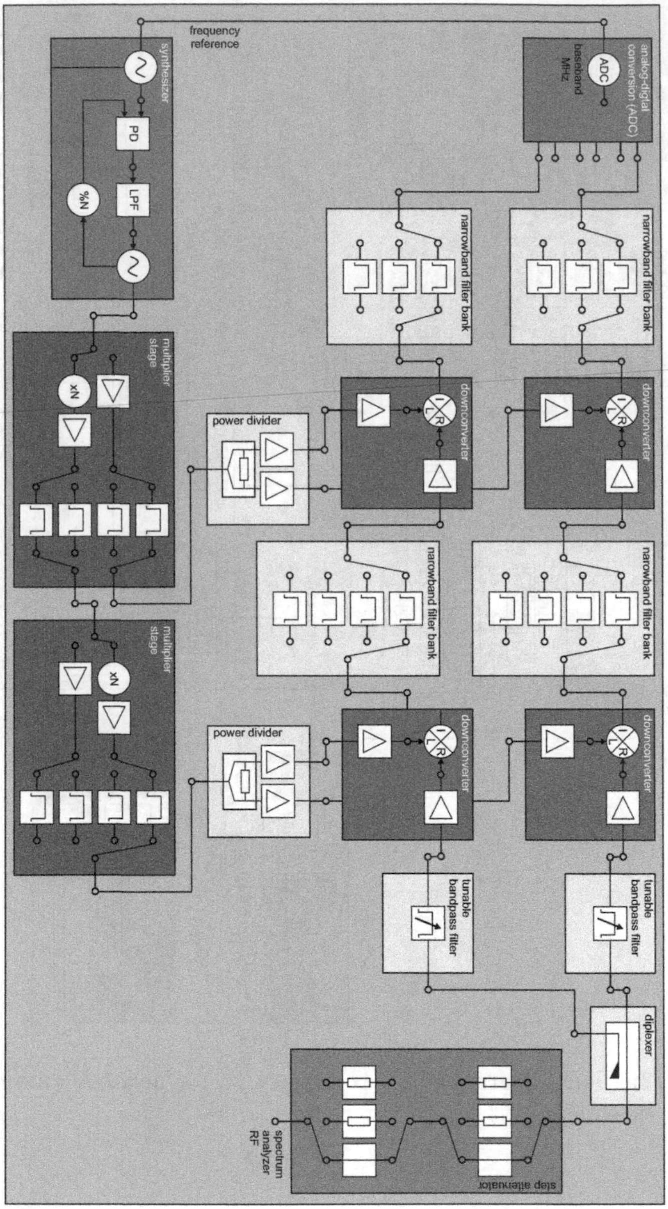

Figure 1.3: Block diagram of an exemplary spectrum analyzer front end module.

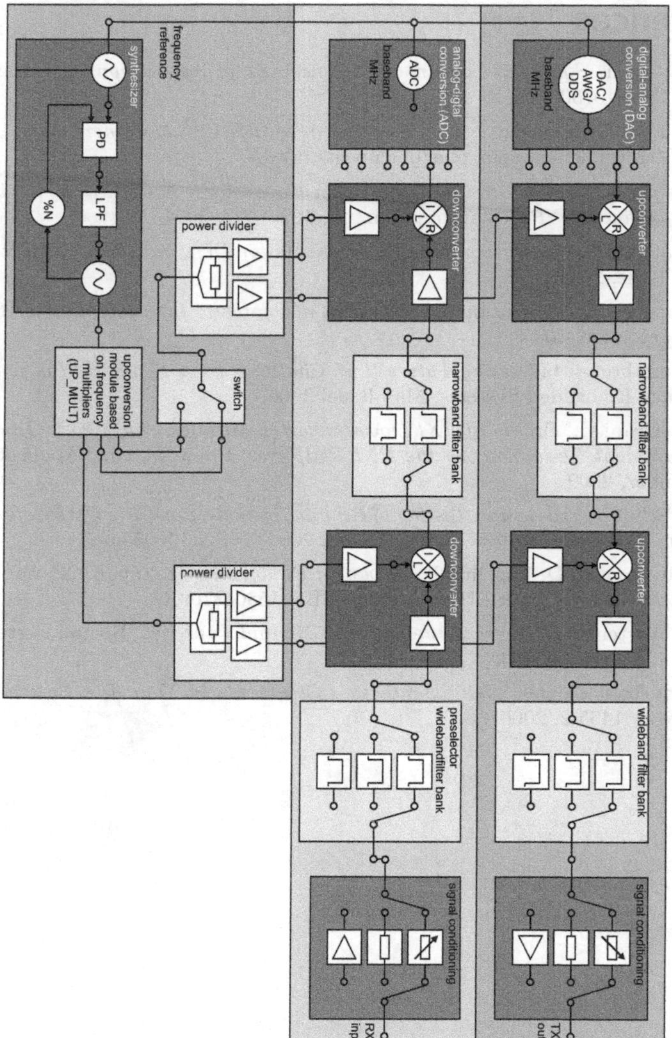

Figure 1.4: Block diagram of an exemplary broadband front end module with high instantaneous signal bandwidth for automatic test systems or surveillance radars.

References

[1] C. Nadovich. *Synthetic Instruments: Concepts and Applications.* Oxford: Newnes, 2004. ISBN 075067783X.

[2] Agilent Technologies. *Agilent PNA Documentation Catalog.* http://na.tm.agilent.com/pna/documents.html.

[3] C. Mahal and M. N. Granieri. *The Down Converter - A Critical Synthetic Instrument Technology.* Proc. Autotestcon, pp. 422 - 427, 2004.

[4] J. Stratton. *Surveying Synthetic Instruments for Defense.* Microwaves & RF, February Issue, 2005.

[5] M. Rozner. *NxTest and the Development of Synthetic Instrumentation.* RF Design, February Issue, 2005.

[6] R. Humphrey. *Addressing Future Test Challenges via Synthetic Instrumentation.* Millitary Embedded Systems, May Issue, 2006.

[7] M. N. Granieri. *Family of PXI Downconverter Modules brings 26.5 GHz RF/MW Measurement Technology to the PXI Platform.* Phase Matrix, Application Note, November, 2009.

[8] K. B. Schaub. *Production Testing of RF and System-on-a-Chip Devices for Wireless Communications.* Boston: Artech House, 2004. ISBN 1580536921.

[9] J. Kelly and M. D. Engelhardt. *Advanced Production Testing of RF, SoC, and SiP Devices.* Boston: Artech House, 2006. ISBN 158053709X.

[10] S. A. Wartenberg. *RF Measurements of Die and Packages.* Boston: Artech House, 2002. ISBN 158053273X.

[11] Agilent Technologies. *Using Synthetic Instruments in Your Test System.* Application Note 1465-2, 2006.

2 Resistive Diode Frequency Multipliers

Although there is a persistent tendency to higher operating frequencies of direct signal generation concepts, frequency multipliers are still used to provide phase locked signal stimuli at millimeter-wave frequencies. Multipliers are required for the local oscillator and signal path of synthetic automatic test systems and traditional instruments (Fig. 1.1, 1.2, 1.3, 1.4). The considerations of this chapter are restricted to resistive nonlinearities (varistor Schottky diodes). Section 2.1 outlines the theoretical conversion loss limitations of resistive diode frequency multipliers. An entire modelling procedure for hybrid Schottky diodes is given in section 2.2. Semiconductor physics of the Schottky junction is presented first in subsection 2.2.1. This is followed by a comparison of several common diode architectures. Modelling procedures for the linear part of the diode, undepleted epilayer and buffer layer, are given in subsection 2.2.3. Drude's dispersion model is applied and effective material parameters are derived, which are required for analytical or 3D EM considerations. The subsection ends with a market overview of Schottky diodes. Section 2.3 explains single tone large signal analysis of nonlinear circuits and serves as a basis for large signal / small signal analysis (LSSS) of subsection 4.2.1. Different multiplier architectures, optimum embedding impedances and the author's frequency multiplier design flow are explained in section 2.4. The author's experimental results are given in section 2.5 to section 2.7. An octave bandwidth tripler for 20 to 40 GHz utilizing bilateral finlines on PCB is presented in section 2.5. Two frequency doublers for 50 to 110 GHz and three triplers for 60 / 67 to 110 GHz on thin-film processed alumina are illustrated in section 2.6. Section 2.7 includes a D-band frequency doubler and Y-band frequency tripler. In subsection 2.7.1 the design equations of Dolph-Chebyshev waveguide tapers are presented. An uncertainty analysis of spectral power measurements above 50 GHz is given in subsection 2.7.2.

2.1 Theoretical Limitations of Resistive Diode Frequency Multipliers

Fig. 2.1 shows a shunt mounted diode, excited with DC, RF and LO voltage components $V_{\mathrm{DC}}, v_{\mathrm{RF}}(t)$ and $v_{\mathrm{LO}}(t)$.

$$\begin{aligned} v_{\mathrm{LO}}(t) &= \hat{v}_{\mathrm{LO}} \cos\left(\omega_{\mathrm{LO}} t + \varphi_{\mathrm{LO}}\right) = V_{\mathrm{LO}} e^{j\omega_{\mathrm{LO}} t} + V_{\mathrm{LO}}^{\star} e^{-j\omega_{\mathrm{LO}} t} \\ v_{\mathrm{RF}}(t) &= \hat{v}_{\mathrm{RF}} \cos\left(\omega_{\mathrm{RF}} t + \varphi_{\mathrm{RF}}\right) = V_{\mathrm{RF}} e^{j\omega_{\mathrm{RF}} t} + V_{\mathrm{RF}}^{\star} e^{-j\omega_{\mathrm{RF}} t} \end{aligned} \tag{2.1}$$

The individual signal branches are isolated by ideal LC bandpass filters, showing short circuit behaviour at the corresponding signal frequencies $\omega = 0, \omega_{\mathrm{RF}}, \omega_{\mathrm{LO}}$ and open circuit behaviour at all other frequencies. Due to the nonlinear voltage-current dependency of the diode, the diode's voltage and current spectral contents include a DC component, RF harmonics $n\,\omega_{\mathrm{RF}}$, LO harmonics $m\,\omega_{\mathrm{LO}}$ and components at the mixing frequencies $n\,\omega_{\mathrm{RF}} + m\,\omega_{\mathrm{LO}}$ (Eq. (2.2)).

$$\omega_{nm} = 2\pi f_{nm} = n\,\omega_{\mathrm{RF}} + m\,\omega_{\mathrm{LO}} \qquad n, m \in \pm\mathbb{N}_0 \tag{2.2}$$

Each spectral component ω_{nm} is filtered out in additional branches with individual load impedances Z_{nm} (Fig. 2.1), called embedding impedances or idle impedances[1]. The circuit of Fig. 2.1 and its dual circuit with series mounted diode in Fig. 2.2 constitute the most versatile mixer equivalent circuits. Choosing appropriate values of the source and embedding impedances, modelling of every practical mixer and frequency multiplier circuit is possible. If all idle impedances in Fig. 2.1 are open circuits $Z_{nm} \to \infty$, in the notation of Saleh [1]

[1] As these appear in parallel to the diode, they are called current idlers. In case of series connection with the diode, they are called voltage idlers.

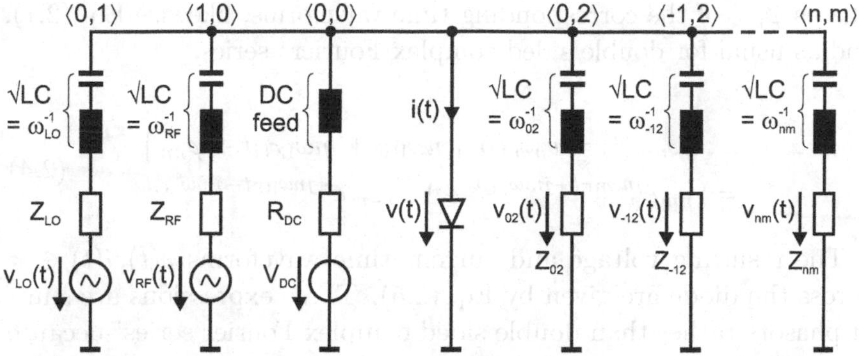

Figure 2.1: Equivalent circuit of frequency mixer consisting of shunt mounted diode, DC, RF, LO sources, load impedances and current idlers in shunt branches.

this is called a Z-mixer or impedance mixer because its small signal conversion matrix (compare subsection 4.2.1) is an impedance matrix. In this sense, the circuit of Fig. 2.2 is called a Y-mixer if all idle impedances are short circuits.

In 1958, Pantell [2] derived fundamental power relationships between source and load power of the circuit in Fig. 2.1, which are presented in the following. Within the work of Pantell, the diode is modelled as positive nonlinear resistor, which means a positive voltage derivative of current $\mathrm{d}i(t)/\mathrm{d}v(t) \geq 0$. All voltage idlers consist of resistive impedances only.

$$\begin{aligned} g : \mathbb{R} \to \mathbb{R}, v \mapsto i = g(v) \\ g \text{ infinitely differentiable,} \\ \text{and } \mathrm{d}i(t)/\mathrm{d}v(t) \geq 0 \\ Z_{\mathrm{RF}}, Z_{\mathrm{LO}}, Z_{nm} \in \mathbb{R} \end{aligned} \tag{2.3}$$

V_{nm} and I_{nm} cover the spectral content of voltage and current across the diode at each combination frequency ω_{nm}. These phasors are related to half of the peak amplitude values $\hat{v}_{nm}/2, \hat{i}_{nm}/2$ and phases φ_{nm} of the corresponding time waveforms, likewise Eq. (2.1), and as usual for double sided complex Fourier[2] series.

$$\begin{aligned} v_{nm}(t) &= \hat{v}_{nm}\cos\left[(n\omega_{\mathrm{RF}} + m\omega_{\mathrm{LO}})t + \varphi_{nm}\right] \\ &= V_{nm}e^{jn\omega_{\mathrm{RF}}t + jm\omega_{\mathrm{LO}}t} + V_{-n-m}e^{-jn\omega_{\mathrm{RF}}t - jm\omega_{\mathrm{LO}}t} \end{aligned} \tag{2.4}$$

The resulting voltage and current time waveforms $v(t), i(t) \in \mathbb{R}$ across the diode are given by Eq. (2.5). These expressions are sums of phasors rather than double sided complex Fourier series' because the individual spectral components are not only harmonics. As a consequence of $v(t), i(t)$ being real, $V_{nm} = V^{\star}_{-n-m}, I_{nm} = I^{\star}_{-n-m}$ holds true.

$$\begin{aligned} v(t) &= \sum_{n=-\infty}^{\infty}\sum_{m=-\infty}^{\infty} V_{nm}\exp\left(jn\omega_{\mathrm{RF}}t + jm\omega_{\mathrm{LO}}t\right) = \\ &= \sum_{n=-\infty}^{\infty}\sum_{m=-\infty}^{\infty} V_{nm}\exp\left(j\omega_{nm}t\right) \\ i(t) &= \sum_{n=-\infty}^{\infty}\sum_{m=-\infty}^{\infty} I_{nm}\exp\left(jn\omega_{\mathrm{RF}}t + jm\omega_{\mathrm{LO}}t\right) = \\ &= \sum_{n=-\infty}^{\infty}\sum_{m=-\infty}^{\infty} I_{nm}\exp\left(j\omega_{nm}t\right) \\ V_{nm} &= V^{\star}_{-n-m}, \quad I_{nm} = I^{\star}_{-n-m} \end{aligned} \tag{2.5}$$

Eq. (2.6) includes the average real P_{nm} and reactive power R_{nm} at the nonlinear diode and mixing frequency $\langle\mathrm{RF},\,\mathrm{LO}\rangle = \langle n, m\rangle$.

$$\begin{aligned} P_{nm} &= V_{nm}I^{\star}_{nm} + V^{\star}_{nm}I_{nm} = P_{-n-m} \\ R_{nm} &= jV^{\star}_{nm}I_{nm} - jV_{nm}I^{\star}_{nm} \end{aligned} \tag{2.6}$$

[2] Jean Baptiste Joseph Fourier (1768–1830), French mathematician.

In [2] Pantell shows the following general power relationships are valid.

$$\sum_{n=0}^{\infty}\sum_{m=-\infty}^{\infty} n^2 P_{nm} = \frac{1}{4\pi^2}\int_0^{2\pi} \mathrm{d}(\omega_{\mathrm{LO}}t) \int_0^{2\pi} \frac{\mathrm{d}i}{\mathrm{d}v}\left(\frac{\mathrm{d}v}{\mathrm{d}(\omega_{\mathrm{RF}}t)}\right)^2 \mathrm{d}(\omega_{\mathrm{RF}}t)$$
$$\sum_{n=-\infty}^{\infty}\sum_{m=0}^{\infty} m^2 P_{nm} = \frac{1}{4\pi^2}\int_0^{2\pi} \mathrm{d}(\omega_{\mathrm{RF}}t) \int_0^{2\pi} \frac{\mathrm{d}i}{\mathrm{d}v}\left(\frac{\mathrm{d}v}{\mathrm{d}(\omega_{\mathrm{LO}}t)}\right)^2 \mathrm{d}(\omega_{\mathrm{LO}}t) \quad (2.7)$$

Because $\mathrm{d}i/\mathrm{d}v \geq 0$ both expressions in Eq. (2.7) are greater than or equal to zero.

$$\sum_{n=0}^{\infty}\sum_{m=-\infty}^{\infty} n^2 P_{nm} \geq 0$$
$$\sum_{n=-\infty}^{\infty}\sum_{m=0}^{\infty} m^2 P_{nm} \geq 0 \quad (2.8)$$

The amount of power across the diode at each fundamental and mixing frequency strongly depends on the input power level and **all** the embedding impedances, including RF, LO, DC source impedances, shown in Fig. 2.1. Analytical solutions to this problem do not exist, numerical transient and balance methods (section 4.2) have to be applied. Anyway the above relations allow for an estimation of the maximum conversion efficiency or minimum conversion loss of mixers and multipliers. In case of fundamental mixing, Eq. (2.8) becomes Eq. (2.9), with P_{01} as fundamental LO power and the fundamental RF power P_{10}.

$$P_{01} + n^2 P_{nm} \geq 0$$
$$P_{10} + m^2 P_{nm} \geq 0 \quad (2.9)$$
$$R_{nm} \geq 0$$

The conversion gain is given by Eq. (2.10).

$$\frac{|P_{nm}|}{P_{01}+P_{10}} \geq \frac{1}{n^2+m^2} \tag{2.10}$$

Therefore the maximum conversion gain for fundamental mixers ($m = n = 1$) is -3 dB (**!**). The conventional definition of mixer conversion gain does not involve the local oscillator power. Hence, fundamental **and** harmonic mixers can achieve 0 dB conversion gain. Comments on the difference between commutative mixers and multipliers are given in section 4.1.

Similarly for resistive diode frequency multipliers generating the m-th harmonic, the maximum conversion gain is given by Eq. (2.11).

$$\frac{|P_{0m}|}{P_{01}} \geq \frac{1}{m^2} \tag{2.11}$$

Which is -6 dB, -9.5 dB and -12 dB for doublers, triplers and quadruplers. These results were given earlier by Page [3] in 1956.

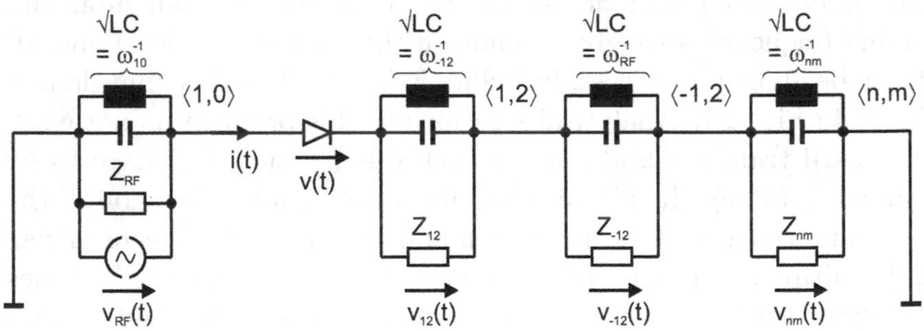

Figure 2.2: Equivalent circuit of frequency mixer consisting of series mounted diode, DC, RF, LO sources, load impedances and voltage idlers in series connection.

2.2 Schottky Barrier Diode Modelling

2.2.1 Junction Modelling

Fig. 2.3 shows energy band diagrams of metal (left side), n-type semiconductor (middle) and metal to n-type semiconductor junction (right side) with applied voltage v_{j}. The free space energy level of an electron is denoted by W_0. The necessary amount of energy to emit an electron from the lower edge of the metal conduction band W_{Cm}, which equals the Fermi[3] energy level W_{Fm}, is called the metal work function $q_{\mathrm{e}}\Phi_{\mathrm{m}}$. The n-type semiconductor's band gap, shown in Fig. 2.3, is given by $W_{\mathrm{Cs}} - W_{\mathrm{Vs}}$. Within the semiconductor, the Fermi energy level W_{Fs} differs from the conduction band level W_{Cs} by the amount of $q_{\mathrm{e}}V_{\mathrm{n}} = W_{\mathrm{Cs}} - W_{\mathrm{Fs}}$.

$$\begin{aligned} q_{\mathrm{e}}V_{\mathrm{n}} &= W_{\mathrm{Cs}} - W_{\mathrm{Fs}} \\ W_{\mathrm{Fs}} &= W_{\mathrm{Cs}} - q_{\mathrm{e}}V_{\mathrm{n}} \end{aligned} \tag{2.12}$$

Hence, the required energy to emit electrons to free space, the electron affinity $W_{\chi} = q_{\mathrm{e}}\Phi_{\chi}$, is lower than the metal work function $W_{\chi} < q_{\mathrm{e}}\Phi_{\mathrm{m}}$. Establishing infinitesimal distance (interatomic distance) between two such materials (m-n) builds the so called Schottky[4] junction (right side of Fig. 2.3). After connecting metal and semiconductor, the lower edge of the conduction band $W_{\mathrm{C}}(x)$ and the upper edge of the valence band $W_{\mathrm{V}}(x)$ become functions of location x. The Fermi energy level of the semiconductor equalizes to W_{Fm}, by emission of electrons to the metal ($v_{\mathrm{j}} = 0$). This leads to negative surface charge in the metal and positive space charge in the semiconductor. The latter is due to positively charged ionized donor atoms in the n-type material. Consequently, the arising barrier for electrons emitting

[3]Enrico Fermi (1901–1954), Italian physicist.

[4]Walter Hermann Schottky (1886–1976), German physicist.

from n-type semiconductor to metal equals the Fermi energy level difference $q_e\Phi_{bi} = q_e(\Phi_m - \Phi_s) = q_e(\Phi_m - \Phi_\chi - V_n)$.

$$q_e\Phi_{bi} = q_e(\Phi_m - \Phi_s) = q_e(\Phi_m - \Phi_\chi - V_n) \tag{2.13}$$

Φ_{bi} is called built-in potential of the Schottky junction and also known as diffusion voltage or flatband voltage. The uncorrected barrier height Φ_{b0} for electrons emitting from metal to semiconductor is greater than Φ_{bi} and given by $\Phi_{b0} = \Phi_m - \Phi_\chi$.

With an applied voltage v_j, the semiconductor's Fermi energy level shifts by the amount of $q_e v_j$ and so does the barrier. Positive voltages, from metal (anode) to semiconductor (cathode), lower the barrier and vice versa.

The so called Schottky effect lowers Φ_{b0} to the effective barrier height $\Phi_b = \Phi_{b0} - \Delta\Phi_b$. If an electric field is applied, an electron in the semiconductor at position $x = 0 + dx$ displaces electrons in the metal. This leads to an induced positive charge (image charge) at the

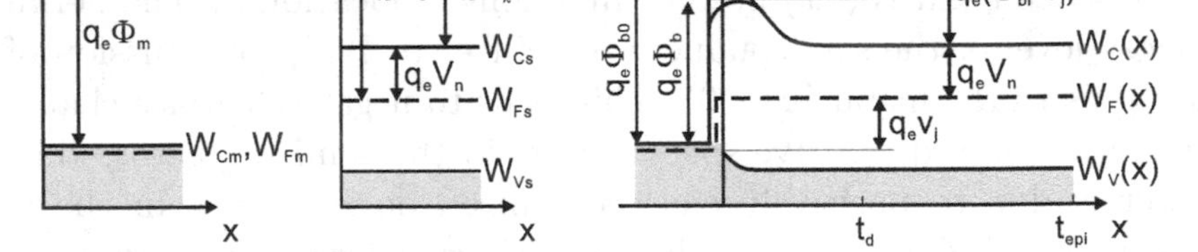

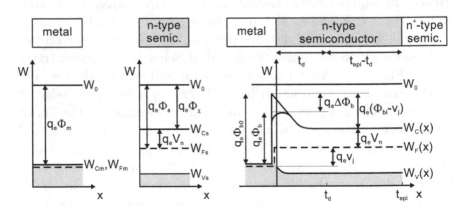

Figure 2.3: Energy band diagrams of metal (left), n-type semiconductor (midle) and metal to n-type semiconductor junction (right) with applied voltage v_j.

metal surface and an attractive force (image force) between electron and positive charge, resulting in $\Delta\Phi_\mathrm{b}$ [5].

In practice, the effective barrier heights Φ_b of m-n transitions are contrary to Eq. (2.13) almost independent of the metal work function $q_\mathrm{e}\Phi_\mathrm{m}$. At $x = 0$ the crystal lattice is disturbed, which leads to a large density of surface energy states in the semiconductor. This keeps the Fermi energy level $W_\mathrm{F}(0 + \mathrm{d}x)$ at a minimum level, greater than W_Fm, and makes the barrier height dependent on the semiconductor's surface only [5].

Derivation of Schottky Junction Capacitance From a technological point of view, the n-type semiconductor from Fig. 2.3 is an epitaxial layer (epilayer) with thickness t_epi, grown to highly doped n^+ substrate (buffer) material. In the following a derivation of the Schottky junction capacitance $C_\mathrm{j}(v_\mathrm{j})$ as a function of applied voltage v_j is presented. Without external voltage v_j, emitting electrons from n-type semiconductor to metal partly deplete the epilayer. The arising depleted epilayer with thickness t_d consists of positively charged ionized donor atoms hindering more electrons from emission to metal. The depletion layer's thickness and the corresponding junction capacitance are functions of v_j.

In the following, complete depletion is assumed, which means no electrons are left in the depletion region. This is known as depletion approximation. Furthermore, the epitaxial layer should be uniformly doped with donor concentration N_Depi, not a function of x. As a consequence, the space charge in the depletion layer does only depend on the positively charged ionized donor atoms and is constant for a given $t_\mathrm{d}(v_\mathrm{j})$. The charge per unit surface area S_j is therefore $Q_\mathrm{j}/S_\mathrm{j} = q_\mathrm{e}N_\mathrm{Depi}t_\mathrm{d}(v_\mathrm{j})$.

---

[5] In [4], the barrier height correction is given by $\Delta\Phi_\mathrm{b} \approx q_\mathrm{e}\sqrt{\frac{q_\mathrm{e}|E_\mathrm{max}|}{4\pi\epsilon_0\epsilon_\mathrm{r}}} \approx \sqrt[4]{\frac{q_\mathrm{e}^3 N_\mathrm{Depi}(\Phi_\mathrm{bi}-v_\mathrm{j})}{8\pi^2\epsilon_0^3\epsilon_\mathrm{r}^3}}$, with maximum electric field across the barrier E_max, the semiconductor's permittivity value ϵ_r and donor concentration N_Depi.

Fig. 2.4 illustrates the charge density $\rho(x)$ (top), charge per unit surface area $Q_\mathrm{j}/S_\mathrm{j}$ (middle) and electric field strength $E(x)$ (bottom) inside the m-n-n$^+$ system. On the metal surface, there is a negative surface charge, which equals the positive charge of the epilayer in magnitude. This is expressed using the Dirac[6] delta distribution in Fig. 2.4, for which the following properties hold true.

$$\delta(x) = \begin{cases} \infty & x = 0 \\ 0 & x \neq 0 \end{cases} \qquad \int\limits_{-\infty}^{+\infty} \delta(x)\mathrm{d}x = 1 \tag{2.14}$$

According to Fig. 2.3, the potential difference (voltage) from $x = 0$ to $x = t_\mathrm{d}$ equals $(\Phi_\mathrm{bi} - v_\mathrm{j})$. The differential equation in Eq. (2.15) follows directly from the definition of voltage.

$$\begin{aligned} (v_\mathrm{j} - \Phi_\mathrm{bi}) &= -\int\limits_{0}^{t_\mathrm{d}} E(x)\mathrm{d}x \\ \frac{\mathrm{d}\left(v_\mathrm{j} - \Phi_\mathrm{bi}\right)}{\mathrm{d}x} &= -E(x) \\ \frac{\mathrm{d}^2\left(v_\mathrm{j} - \Phi_\mathrm{bi}\right)}{\mathrm{d}x^2} &= -\frac{\mathrm{d}E(x)}{\mathrm{d}x} \end{aligned} \tag{2.15}$$

Gauss[7] law leads to Eq. (2.16).

$$\frac{\mathrm{d}E(x)}{\mathrm{d}x} = \frac{\rho(x)}{\epsilon_0\epsilon_\mathrm{r}} \tag{2.16}$$

Comparing Eq. (2.15) with Eq. (2.16) the dependency of junction voltage v_j from charge density $\rho(x)$ is known.

$$\frac{\mathrm{d}^2\left(v_\mathrm{j} - \Phi_\mathrm{bi}\right)}{\mathrm{d}x^2} = -\frac{\rho(x)}{\epsilon_0\epsilon_\mathrm{r}} \tag{2.17}$$

[6]Paul Adrien Maurice Dirac (1902–1984), English physicist.
[7]Johann Carl Friedrich Gauss (1777–1855), German mathematician.

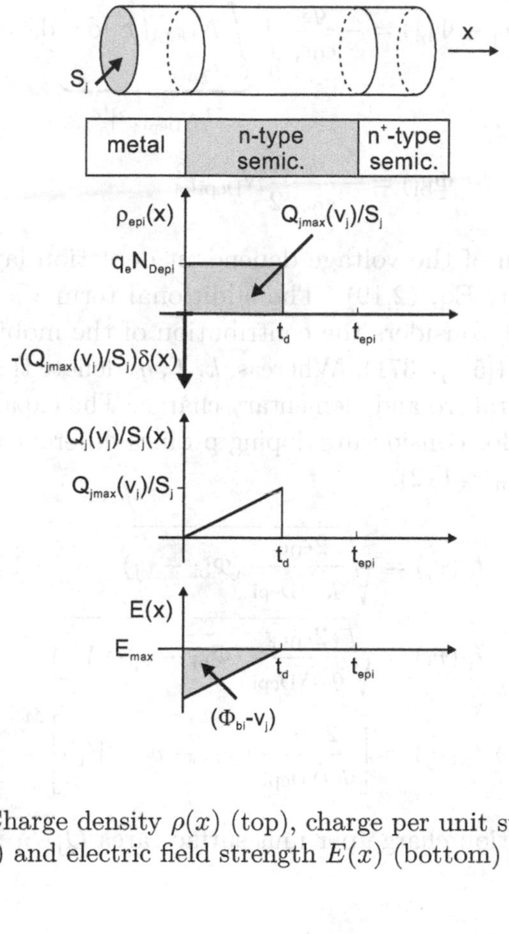

Figure 2.4: Charge density $\rho(x)$ (top), charge per unit surface area Q_j/S_j (middle) and electric field strength $E(x)$ (bottom) inside the m-n-n$^+$ system.

Inserting charge density $\rho(x) = q_\text{e}N_\text{Depi}$, known from depletion approximation, into Eq. (2.17) and integrating twice over x leads to Eq. (2.18).

$$(v_\mathrm{j} - \Phi_\mathrm{bi}) = -\frac{q_\mathrm{e}}{\epsilon_0\epsilon_\mathrm{r}} \underbrace{\int_0^{t_d}\int_0^{x} N_\mathrm{Depi}(x^{'})\mathrm{d}x^{'}\,\mathrm{d}x}_{\left[\frac{1}{2}N_\mathrm{Depi}x^2\right]_0^{t_\mathrm{d}}}$$
$$(v_\mathrm{j} - \Phi_\mathrm{bi}) = -\frac{q_\mathrm{e}}{\epsilon_0\epsilon_\mathrm{r}}\frac{1}{2}N_\mathrm{Depi}t_\mathrm{d}^2 \tag{2.18}$$

An expression of the voltage dependent depletion layer thickness $t_\mathrm{d}(v_\mathrm{j})$ is given by Eq. (2.19). The additional term $V_\mathrm{T} = \frac{kT}{q_\mathrm{e}}$, called thermal voltage, considers the contribution of the mobile carriers to the electric field ([5], p. 371). Whereas, k, T, q_e denote the Boltzmann[8] constant, temperature and elementary charge. The capacitance coefficient M allows for considering doping profiles different from uniform doping ($M_\mathrm{uniform} = 1/2$).

$$t_\mathrm{d}(v_\mathrm{j}) = \sqrt{\frac{2\epsilon_0\epsilon_\mathrm{r}}{q_\mathrm{e}N_\mathrm{Depi}}(\Phi_\mathrm{bi} - v_\mathrm{j})}$$
$$t_\mathrm{d}(v_\mathrm{j}) = \sqrt{\frac{2\epsilon_0\epsilon_\mathrm{r}}{q_\mathrm{e}N_\mathrm{Depi}}(\Phi_\mathrm{bi} - v_\mathrm{j} - V_\mathrm{T})} \tag{2.19}$$
$$\rightarrow t_\mathrm{d}(v_\mathrm{j}) = \left[\frac{2\epsilon_0\epsilon_\mathrm{r}}{q_\mathrm{e}N_\mathrm{Depi}}(\Phi_\mathrm{bi} - v_\mathrm{j} - V_\mathrm{T})\right]^M$$

Hence, the junction charge per unit surface area $Q_\mathrm{j}/S_\mathrm{j}$ is known as a function of v_j.

$$Q_\mathrm{j}(v_\mathrm{j})/S_\mathrm{j} = \int_{0-\mathrm{d}x}^{0+\mathrm{d}x} \rho(x)\mathrm{d}x = \int_{0+\mathrm{d}x}^{t_d} \rho(x)\mathrm{d}x = q_\mathrm{e}N_\mathrm{Depi}t_\mathrm{d}(v_\mathrm{j})$$
$$Q_\mathrm{j}(v_\mathrm{j})/S_\mathrm{j} = q_\mathrm{e}N_\mathrm{Depi}t_\mathrm{d}(v_\mathrm{j}) = \sqrt{2\epsilon_0\epsilon_\mathrm{r}q_\mathrm{e}N_\mathrm{Depi}(\Phi_\mathrm{bi} - v_\mathrm{j} - V_\mathrm{T})} \tag{2.20}$$

[8] Ludwig Eduard Boltzmann (1844–1906), Austrian physicist.

The capacitance per unit surface area C_j/S_j is derived using a parallel plate capacitor model (Eq. (2.21)), where ϵ_r is the semiconductor's relative permittivity. The expressions are valid, as long as the epilayer is thick enough to prevent the depletion region from punching through to the substrate at high reverse voltages.

$$\begin{aligned}
C_j/S_j &= \frac{dQ_j/S_j}{dv_j} = \epsilon_0\epsilon_r\frac{1}{t_d(v_j)} \\
C_j/S_j &= \left(\frac{q_e\epsilon_0\epsilon_r N_{Depi}}{2\left(\Phi_{bi} - v_j - V_T\right)}\right)^M \\
C_j/S_j &= \frac{C_{jmax}/S_j}{\left[1 - v_j/\left(\Phi_{bi} - V_T\right)\right]} \\
C_{jmax}/S_j &= C_j/S_j\Big|_{v_j=0} = C_{j0}/S_j = \left(\frac{q_e\epsilon_0\epsilon_r N_{Depi}}{2\left(\Phi_{bi} - V_T\right)}\right)^M
\end{aligned} \tag{2.21}$$

The simple parallel plate capacitor model predicts infinite capacitance if $t_d = 0$, instead of zero capacitance. Hence, C_j from Eq. (2.21) becomes infinite at $v_j = \Phi_{bi} - V_T$. To overcome this problem, junction capacitance is defined in a different way. In case of $v_j \leq F_C\left(\Phi_{bi} - V_T\right)$ the derived expression in Eq. (2.22) is valid.

$$C_j = C_{j0}\left(1 - \frac{v_j}{\Phi_{bi} - V_T}\right)^{-M} \tag{2.22}$$

If $v_j > F_C\left(\Phi_{bi} - V_T\right)$, Eq. (2.23) is used.

$$C_j = \frac{C_{j0}}{\left(1 - F_C^M\right)}\left\{1 + \left[\frac{M}{(\Phi_{bi} - V_T)(1 - F_C)}\right]\left[v_j - F_C\left(\Phi_{bi} - V_T\right)\right]\right\} \tag{2.23}$$

The parameter F_C usually varies between 0.4 to 0.7 and is implemented within the common SPICE model DIODE (default value $F_C = 0.5$, [6]). Alternatively, circuit simulators often prevent problems with infinite capacitance by using increments of charge to estimate displacement current $i_{\mathrm{C}}(v_{\mathrm{j}}, t)$ through the capacitor instead of the capacitance function ([7, 8]).

$$i_{\mathrm{C}}(v_{\mathrm{j}}, t) = \frac{\mathrm{d}Q_{\mathrm{j}}}{\mathrm{d}t} = \left.\frac{\mathrm{d}Q_{\mathrm{j}}}{\mathrm{d}v}\right|_{v=v_{\mathrm{j}}(t)} \frac{\mathrm{d}}{\mathrm{d}t} v_{\mathrm{j}}(t) \tag{2.24}$$

Derivation of Schottky Junction Current In 1942 Bethe[9] studied the current transport mechanism through the Schottky barrier based on pure thermionic emission [9]. It assumes all electrons with sufficiently high velocity component perpendicular to the barrier emit through the barrier. With Maxwell velocity distribution, the current from n-type semiconductor material to metal i_{jSM} and vice versa i_{jMS} is given by Eq. (2.25).

$$\begin{aligned} i_{\mathrm{jSM}} &= I_{\mathrm{S}} \left[\exp\left(\frac{v_{\mathrm{j}}}{V_{\mathrm{T}}}\right)\right] \\ i_{\mathrm{jMS}} &= -I_{\mathrm{S}} \\ I_{\mathrm{S}} &= A^{\star} T^2 S_{\mathrm{j}} \exp\left(\frac{-q_{\mathrm{e}}\Phi_{\mathrm{b}}}{kT}\right) \\ A^{\star} &= \frac{4\pi q_{\mathrm{e}} m^{\star} k^2}{h^3} \end{aligned} \tag{2.25}$$

Whereas $A^{\star}$ is the Richards constant from Richardson[10] law, $m^{\star}$ is the effective electron mass [10] and the total junction current i_{j} is the sum of i_{jSM} and i_{jMS}.

$$i_{\mathrm{j}}(v_{\mathrm{j}}) = i_{\mathrm{jSM}} + i_{\mathrm{jMS}} = I_{\mathrm{S}} \left[\exp\left(\frac{v_{\mathrm{j}}}{V_{\mathrm{T}}}\right) - 1\right] \tag{2.26}$$

Four years earlier, in 1938 Schottky[11] derived an expression for the junction current based on the diffusion theory (Fick's[12] first law,

[9] Hans Albrecht Bethe (1906–2005), German physicist.
[10] Sir Owen Willans Richardson (1879–1959), British physicist.
[11] Walter Hermann Schottky (1886–1976), German physicist.
[12] Adolf Eugen Fick (1829–1901), German physician and physiologist.

[11]). The result has the form of Eq. (2.26), which is also known as Shockley[13] equation, but suggests a different expression for the reverse saturation current.

$$I_S = \underbrace{\mu_e N_C \sqrt{\frac{2q_e(\Phi_{bi} - v_j)N_{Depi}}{\epsilon_0 \epsilon_r}}} S_j \exp\left(\frac{-q_e \Phi_b}{kT}\right) \tag{2.27}$$

The thermionic emission-diffusion theory from Crowell and Sze unifies the approaches from Bethe and Schottky and also accounts for the probability of quantum mechanical tunneling (field emission) of electrons through the barrier. It suggests the introduction of the modified Richards constant A^{**} in Eq. (2.28), which is not a true constant (compare [5] for detailed derivation).

$$I_S = A^{**} T^2 S_j \exp\left(\frac{-q_e \Phi_b}{kT}\right) \tag{2.28}$$

Supplementing Eq. (2.28) with the ideality factor $\eta = 1 \ldots 2$ allows for modelling parasitic effects, neglected in the above derivations ([12], p. 19).

$$I_S = A^{**} T^2 S_j \exp\left(\frac{-\Phi_b}{\eta V_T}\right) \quad V_T = \frac{kT}{q_e} \tag{2.29}$$

Reported values of A^{**} differ significantly[14]. Fortunately, when extracting the barrier height $\Phi_b \sim \ln(1/A^{**})$ from Eq. (2.29) the value of A^{**} has little effect on accuracy.

2.2.2 Diode Architectures

Fig. 2.5 illustrates a 3D view and cut through (BB′) a planar GaAs Schottky diode, intended for use in hybrid systems as discrete com-

[13] William Bradford Shockley Jr. (1910–1989), American physicist.

[14] Room temperature values in units of $A/(cm^2K^2)$ are [4] $A^{**} \approx 100$, [13] $A^{**}_{n-Si} \approx 7.92$, $A^{**}_{n-GaAs} \approx 230.4$, [12] $A^{**}_{Si} \approx 96$, $A^{**}_{GaAs} \approx 4.4$, [5] $A^{**}_{n-Si} \approx 252$, $A^{**}_{n-GaAs} \approx 7.2 \ldots 144$.

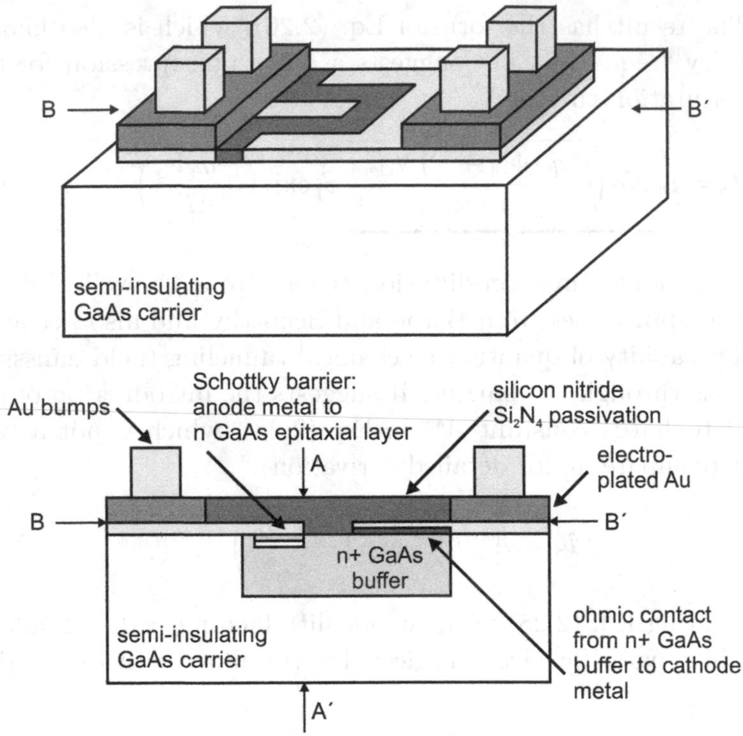

Figure 2.5: 3D view (top) of a planar buried epitaxial layer GaAs Schottky diode like UMS DBES105a and BB′ cutaway view (bottom), showing epilayer, buffer layer and ohmic contact.

ponent. The n-type epitaxial layer is grown on highly doped n^+ GaAs buffer substrate (gray), which is surrounded by semi-insulating GaAs carrier. On the left side, the anode metal contacts the epilayer, whereas on the right side an ohmic contact (red) from cathode metal to the buffer is realized. Silicon nitride (Si_2N_4) passivation is shown in green. Electroplated Au at anode and cathode, together with bumps allow for easy assembly to planar circuits.

The cross-sectional view AA′ and a topview are shown in Fig. 2.6 and Fig. 2.7, respectively. This architecture is called buried epi-

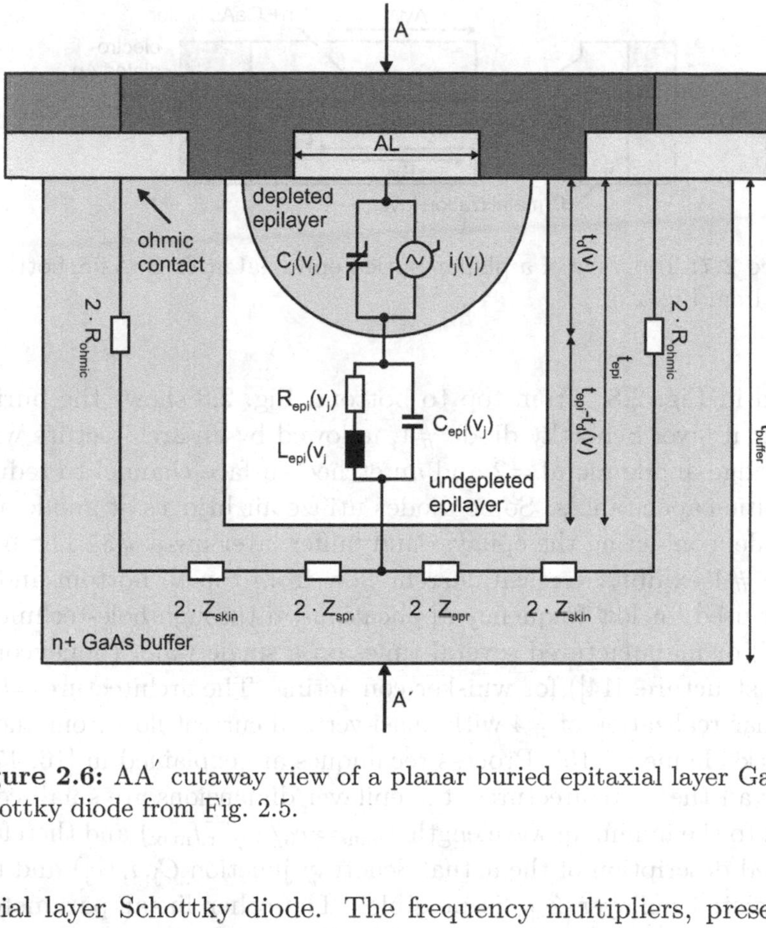

Figure 2.6: AA′ cutaway view of a planar buried epitaxial layer GaAs Schottky diode from Fig. 2.5.

taxial layer Schottky diode. The frequency multipliers, presented in sections 2.6, 2.7 utilize such a diode, the DBES105a from United Monolithic Semiconductors (UMS).

Fig. 2.6 further includes an equivalent circuit model. Beside the nonlinear epilayer junction capacitance $C_j(v_j)$ from Eq. (2.21, 2.22, 2.23) and the nonlinear diode current $i_j(v_j)$ from Eq. (2.26, 2.29), linear components of the undepleted epilayer $R_{epi}(v_j)$, $L_{epi}(v_j)$, $C_{epi}(v_j)$ and buffer layer Z_{spr}, Z_{skin}, R_{ohmic} are included.

There are several different planar diode architectures for microwave, millimeter-wave and terahertz applications. Some of which are illus-

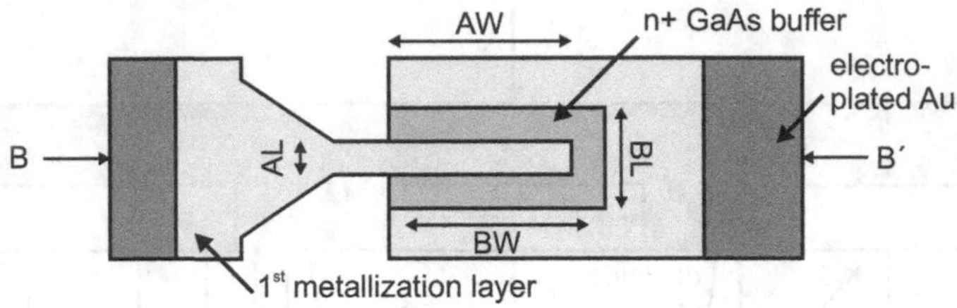

Figure 2.7: Top view of a planar buried epitaxial layer GaAs Schottky diode from Fig. 2.5.

trated in Fig. 2.8. From top to bottom, Fig. 2.8 shows the buried epitaxial layer Schottky diode #1, followed by an architecture with air bridge anode metal #2 and an etched surface channel to reduce parasitic capacitances. Some diodes utilize air bridges at anode and cathode, contacting the epilayer and buffer layer mesa #3. The bulk diode #4 exhibits vertical current flow from top to bottom and is either used for low frequency applications in through-hole-technique (THT) or manufactured several times on a single wafer (honeycomb array structure, [14]) for whisker contacting. The architecture #5 is a planar realization of #4 with quasi-vertical current flow from anode to backside metal [15]. Process techniques are explained in [16, 17].

For all these architectures, the epilayer dimensions are small compared to the minimum wavelength $\lambda_{\mathrm{min}} = c_0 / \left(\sqrt{\epsilon_{\mathrm{r}}} f_{\mathrm{max}}\right)$ and therefore lumped description of the actual Schottky junction $C_{\mathrm{j}}, i_{\mathrm{j}}(v_{\mathrm{j}})$ and the undepleted epilayer Z_{epi} is possible. The other linear parameters of the equivalent circuit of Fig. 2.9, Z_{spr}, Z_{skin}, and the influence of the first metallization layer, electroplated Au, passivation and semi-insulating GaAs carrier are not inherent diode properties but strongly depend on the interaction with the electromagnetic field. An extreme contrast is depicted in Fig. 2.10. It compares hybrid integration (left) to monolithic microwave integrated circuit (MMIC) usage (right) of a planar diode. In the hybrid case, the semi-insulating GaAs carrier constitutes a parasitic element with an electromagnetic field penetration that depends on the chosen planar waveguide, sub-

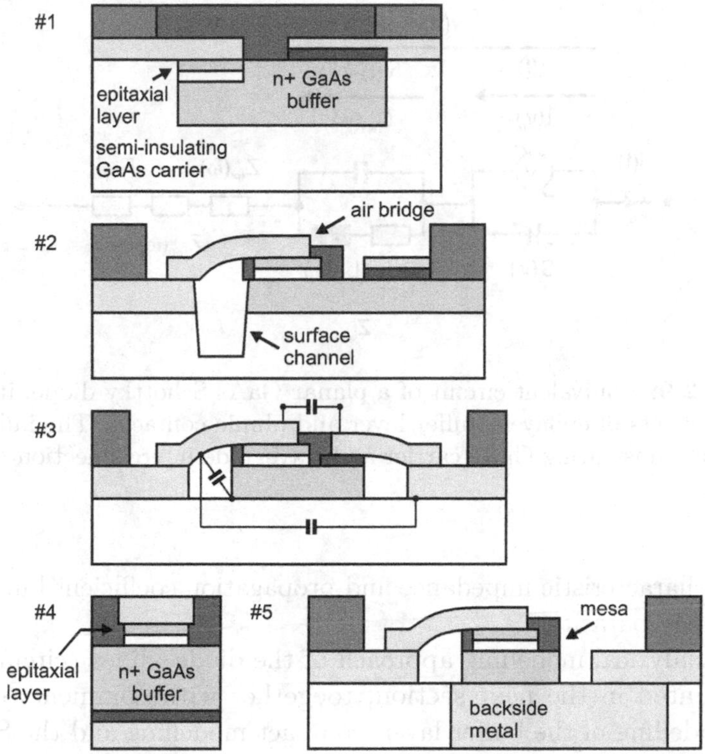

Figure 2.8: Different planar diode architectures. From top to bottom, buried epitaxial layer, anode finger bridge with surface channel, mesa diode with anode and cathode finger bridge and surface channels, bulk diode (e.g. part of whisker contacted diode array) and mesa diode with vertical current flow to cathode.

strate material and interconnect solution. In this sense, a coplanar waveguide (CPW) incorporates less interaction with the carrier than it is with microstrip line (MSL) and the capacitances, shown in the third row of Fig. 2.8, have smaller values. In case of monolithic integration, the semi-insulating GaAs carrier acts as the substrate material of the feeding waveguide and therefore has severe influence

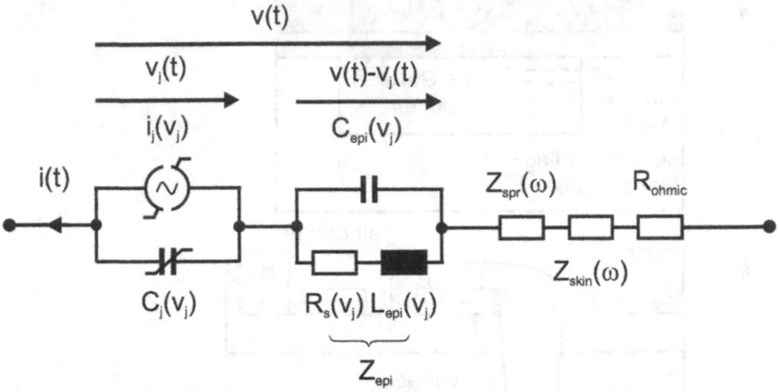

Figure 2.9: Equivalent circuit of a planar GaAs Schottky diode, including the effects of epilayer, buffer layer and ohmic contacts. The influence of the semi-insulating GaAs carrier and waveguide interconnection are not covered.

on the characteristic impedance and propagation coefficient but is not necessarily parasitic.

An analytical modelling approach of the diode's linear circuit part is presented in the next section, toegether with comments on 3D EM modelling of the buffer layer, compact modelling and the SPICE model DIODE.

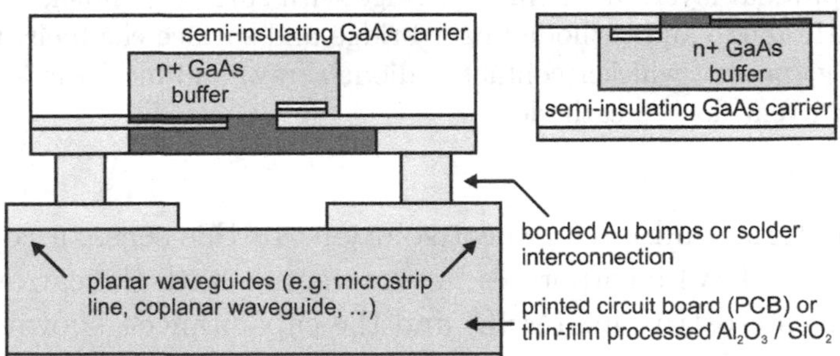

Figure 2.10: Planar diode as a discrete component for hybrid integration (left) and planar diode as part of a monolithic microwave integrated circuit (right).

2.2.3 Linear Parts of Diode Model

In the following we derive analytical equations to describe the linear parts of the diode model, the undepleted epilayer Z_{epi}, highly doped buffer layer $Z_{\text{spr}} + Z_{\text{skin}}$ and ohmic contacts R_{ohmic}. In both cases simple models are chosen as a starting point. To cover the high frequency behaviour, enhanced modelling of the material properties, electrical permittivity and conductivity, is applied. This last fact allows for covering arbitrary geometries with 3D EM simulations based on dispersive materials. Whereas, the analytical equations are valid for rather simple geometries only.

Undepleted Epitaxial Layer The modelling approach for the undepleted epilayer with thickness $t_{\text{u}}(v_{\text{j}}) = t_{\text{epi}} - t_{\text{d}}(v_{\text{j}})$ in Fig. 2.9 includes three components, covering resistive effects R_{epi}, displacement current C_{epi} and carrier inertia effects L_{epi}. The undepleted epilayer is the main contributor to the diode's series resistance.

$$Z_{\text{epi}} = (R_{\text{epi}} + j\omega L_{\text{epi}}) \parallel (j\omega C_{\text{epi}})^{-1} \tag{2.30}$$

In the following derivations, the simple cylindrical geometry of Fig. 2.11, with the anode diameter a and buffer radius b, is assumed. Utilizing the standard lumped resistor model (Pouillet's[15] law) $R = 1/\sigma \frac{\ell}{S}$ the element R_{epi} at DC is given by Eq. (2.31).

$$\begin{aligned} R_{\text{epi}}(v_{\text{j}})\Big|_{f=0} &= \frac{t_{\text{u}}(v_{\text{j}})}{\sigma_{\text{epi}} S_{\text{j}}} \\ \sigma_{\text{epi}} &= q_{\text{e}} \mu_{\text{e}} N_{\text{Depi}} \end{aligned} \tag{2.31}$$

In Eq. (2.31), the electron mobility in the epilayer μ_{e} depends on the doping concentration N_{Depi} and temperature T. Throughout this

[15] Claude Servais Mathias Pouillet (1791–1868), French physicist.

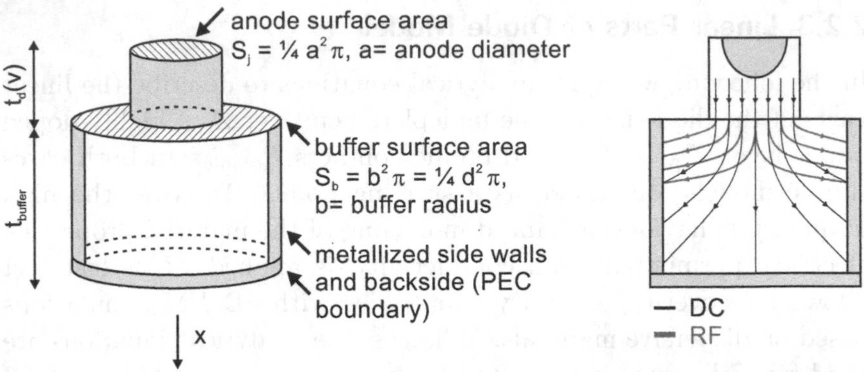

Figure 2.11: Simple cylindrical model of epilayer and buffer layer (left) for analytical analysis and current density lines at DC and RF (right).

work, the dependency is covered by the empirical low-field electron mobility model from [18], given by Eq. (2.32).

$$\mu_e\left(N_D, T\right) = 10^4\mu_{min} + \frac{10^4\mu_{max}\left(\frac{300\ K}{T}\right)^{\theta_1} - 10^4\mu_{min}}{1 + \left[\frac{10^{-6}N_D}{10^{-6}N_{ref}\left(\frac{T}{300\ K}\right)^{\theta_2}}\right]^{\theta_3}} \tag{2.32}$$

The results of Fig. 2.12 are based on the parameters for electrons in GaAs (Eq. (2.33)).

$$\begin{aligned} &\mu_{min}/\mu_{max} = 0.94/0.05\ m^2V^{-1}s^{-1} \\ &N_{ref} = 6\cdot 10^2\ m^{-3} \quad \theta_1 = 2.1 \quad \theta_2 = 3.0 \quad \theta_3 = 0.394 \end{aligned} \tag{2.33}$$

The planar junction model of Eq. (2.31) holds true as long as the effective anode diameter a is large compared to the epilayer thickness. Typical values are $t_{epi} \leq 0.1$ µm and $a \approx 1$ to 6 µm. R_{epi} has only weak dependency on v_j, due to $t_u(v_j) = t_{epi} - t_d(v_j)$. It has a maximum value at high forward bias, which advises to use $t_{epi} = \max\left(t_u(v_j)\right)$ as worst case approximation. Empirical models are available to include the increase of R_{epi} due to carrier velocity saturation at high drive

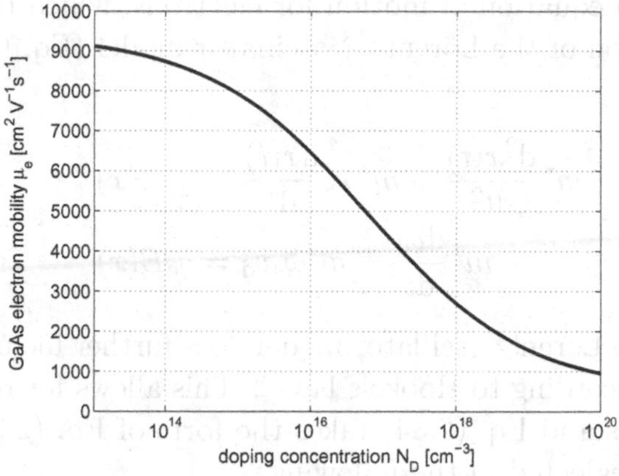

Figure 2.12: Empirical low-field electron mobility μ_e of GaAs versus doping concentration N_D from [18].

level (DC and / or RF, [19]). Beside resistive effects R_{epi}, displacement current through the undepleted epilayer and carrier inertia effects, due to nonzero effective mass $m_e^\star \sim m_e$ of the electrons [10], are covered by C_{epi} and L_{epi}.

Drude's Dispersion Model Derivation is based on Drude's[16] dispersion model [20, 21]. The undepleted epilayer is assumed to consist of a negatively charged electron gas (plasma), the valence electrons, and a positively charged background, the donor atoms. The electrons move under the influence of an applied electric field. The electron's (mean) drift velocity component ($v_d \parallel x$) in parallel with the electric field $E(x)$ increases up to a maximum value v_{dmax} before collision (scattering event) with lattice atoms leads to randomization of v_d. The average time between such scattering effects (collision time) is the reciprocal of the scattering frequency $\omega_s = \frac{q_e}{m^\star \mu_e}$, which is assumed to be constant for all involved electrons. With Newton's[17] second law,

[16] Paul Karl Ludwig Drude (1863–1906), German physicist.
[17] Sir Isaac Newton (1642–1727), English physicist.

this leads to an equation of motion for electrons, which belongs to a simplified version of the Lorentz[18] oscillator model (Eq. (2.34)).

$$\begin{aligned} m^\star \frac{\mathrm{d}^2 x(t)}{\mathrm{d}t^2} + m^\star \omega_\mathrm{s} \frac{\mathrm{d}x(t)}{\mathrm{d}t} &= q_\mathrm{e} E(x) \\ m^\star \frac{\mathrm{d}v_\mathrm{d}}{\mathrm{d}t} + m^\star \omega_\mathrm{s} v_\mathrm{d} &= q_\mathrm{e} E(x) \end{aligned} \tag{2.34}$$

The complete Lorentz oscillator model does further include a spring force F_spring, according to Hooke's law[19]. This allows for modelling of bound electrons and Eq. (2.34) takes the form of Eq. (2.35). Bound electrons are neglected in the following.

$$\underbrace{m^\star \frac{\mathrm{d}^2 x(t)}{\mathrm{d}t^2}}_{\text{inertial}} + \underbrace{m^\star \omega_\mathrm{s} \frac{\mathrm{d}x(t)}{\mathrm{d}t}}_{\text{damping}} + \underbrace{m^\star \omega_\mathrm{bound}^2 x(t)}_{\text{restoring}, F_\mathrm{spring}} = \underbrace{q_\mathrm{e} E(x)}_{\text{driving}} \tag{2.35}$$

The scattering frequency ω_s is found from the static solution of Eq. (2.34). At DC, it is $\frac{\mathrm{d}^2 x(t)}{\mathrm{d}t^2} = 0$ and $\frac{\mathrm{d}x(t)}{\mathrm{d}t} = v_\mathrm{d} = \text{constant}$. Hence, Eq. (2.34) becomes Eq. (2.36).

$$\omega_\mathrm{s} = \frac{q_\mathrm{e} E(x)}{m^\star v_\mathrm{d}} \tag{2.36}$$

The drift velocity is given by $v_\mathrm{d} = \frac{J(x)}{q_\mathrm{e} N}$, with Ohm's law[20] $J(x) = \sigma_\mathrm{DC} E(x)$ this leads to $v_\mathrm{d} = \frac{\sigma_\mathrm{DC}}{q_\mathrm{e} N} E(x)$ and ω_s.

$$\omega_\mathrm{s} = \frac{q_\mathrm{e}^2 N}{m^\star \sigma_\mathrm{DC}} \tag{2.37}$$

[18] Hendrik Antoon Lorentz (1853–1928), Dutch physicist.
[19] Robert Hooke (1635–1703), English scientist.
[20] Georg Simon Ohm (1789–1854), German physicist.

Replacing the bulk DC conductivity $\sigma_{DC} = q_e \mu_e N$, the scattering frequency becomes

$$\omega_s = \frac{q_e}{m^\star \mu_e}. \tag{2.38}$$

Inserting the trial solution $x(t) = x_0 \exp(j\omega t)$ into Eq. (2.34) leads to the electron displacement $x(t)$.

$$x(t) = \frac{q_e E(x)}{-m^\star \omega^2 + j\omega\omega_s m^\star} \tag{2.39}$$

Dielectric polarization density per unit volume $\mathbf{P}(x)$ is defined as sum of the dipole moments per unit volume. Dipole moments of the electron gas are given by the electron's charge q_e times electron displacement $x(t)$. Hence, the dielectric polarization density for the one-dimensional case is

$$\begin{aligned} P_e &= q_e x(t) N \\ P_e &= \frac{q_e^2 N E(x)}{-m^\star \omega^2 + j\omega\omega_s m^\star}. \end{aligned} \tag{2.40}$$

P_D covers the polarization density of the positively charged background (donor atoms). For GaAs, the relative permittivity ϵ_r equals 12.9 and is often referred to as ϵ_∞ in literature.

$$P_D = \epsilon_0 \chi E = \epsilon_0 (\epsilon_r - 1) E \tag{2.41}$$

Inserting Eq. (2.40) and Eq. (2.41) into Maxwell's[21] constitutive relations,[22] it is possible to derive an expression for the frequency

[21] James Clerk Maxwell (1831–1879), Scottish physicist.

[22] Eq. (2.42) includes Maxwell's equations in integral and differential form,

dependent, intrinsic bulk permittivity $\epsilon(\omega)$ of the semiconductor (electron gas and donor atoms) as a function of the polarization density.

$$\begin{aligned} D = \epsilon(\omega)E &= \epsilon_0 E + P_\mathrm{D} + P_\mathrm{e} \\ &= \epsilon_0 E + \epsilon_0 \left(\epsilon_\mathrm{r} - 1\right) E + P_\mathrm{e} \\ &= \epsilon_0 \epsilon_\mathrm{r} E + P_\mathrm{e} \\ \epsilon(\omega) &= \epsilon_0 \epsilon_\mathrm{r} \left(1 + \frac{P_\mathrm{e}}{\epsilon_0 \epsilon_\mathrm{r} E}\right) \end{aligned} \tag{2.43}$$

Inserting Eq. (2.37) and Eq. (2.40) into Eq. (2.43) leads to Eq. (2.44).

$$\epsilon(\omega) = \epsilon_0 \epsilon_\mathrm{r} \left(1 + \frac{\frac{q_\mathrm{e}^2 N}{\epsilon_0 \epsilon_\mathrm{r} m^\star}}{-\omega^2 + j\omega\omega_\mathrm{s}}\right) = \epsilon_0 \epsilon_\mathrm{r} \left(1 + \frac{\omega_\mathrm{s} \frac{\sigma_\mathrm{DC}}{\epsilon_0 \epsilon_\mathrm{r}}}{-\omega^2 + j\omega\omega_\mathrm{s}}\right) \tag{2.44}$$

$$\begin{aligned} &\text{Faraday's law:} && \oint_\ell \mathbf{E} d\boldsymbol{\ell} = -\frac{\partial}{\partial t} \int_\mathbf{S} \mathbf{B} d\mathbf{S} \\ &\text{Ampère's law:} && \oint_\ell \mathbf{H} d\boldsymbol{\ell} = \int_\mathbf{S} \left(\mathbf{J} + \frac{\partial \mathbf{D}}{\partial t}\right) d\mathbf{S} \\ &\text{Gauss' law:} && \oint_\mathbf{S} \mathbf{B} d\mathbf{S} = 0 \qquad \oint_\mathbf{S} \mathbf{D} d\mathbf{S} = Q \\ &\text{Constitutive equations:} && \mathbf{D} = \epsilon_0 \epsilon_r \mathbf{E} \qquad \mathbf{B} = \mu_0 \mu_r \mathbf{H} \\ &\text{Faraday's law:} && \mathrm{curl}\mathbf{E} = \nabla \times \mathbf{E} = -\frac{\partial \mathbf{B}}{\partial t} \\ &\text{Ampère's law:} && \mathrm{curl}\mathbf{H} = \nabla \times \mathbf{H} = \mathbf{J} + \frac{\partial \mathbf{D}}{\partial t} \\ &\text{Gauss' law:} && \mathrm{div}\mathbf{D} = \nabla \cdot \mathbf{D} = \rho \\ & && \mathrm{div}\mathbf{B} = \nabla \cdot \mathbf{B} = 0 \\ &\text{Constitutive equations:} && \mathbf{D} = \epsilon_0 \epsilon_r \mathbf{E} \qquad \mathbf{B} = \mu_0 \mu_r \mathbf{H} \end{aligned} \tag{2.42}$$

With the abbreviation $\omega_{\mathrm{d}} = \frac{\sigma_{\mathrm{DC}}}{\epsilon_0 \epsilon_{\mathrm{r}}}$, called dielectric relaxation frequency, Eq. (2.44) simplifies to the final result Eq. (2.45).

$$\epsilon(\omega) = \epsilon_0 \epsilon_{\mathrm{r}} \left(1 - \frac{\omega_{\mathrm{s}} \omega_{\mathrm{d}}}{\omega^2 - j\omega\omega_{\mathrm{s}}}\right) \tag{2.45}$$

Drude's Dispersion expressed by Effective Material Properties

$\epsilon(\omega)$ has been introduced as the frequency dependent, intrinsic bulk permittivity in Maxwell's constitutive relations. With Ampère's law[23] in the frequency domain (Eq. (2.46)), the effective permittivity $\epsilon_{\mathrm{eff}}(\omega)$ and effective conductivity σ_{eff} are introduced, which both cover conductor and displacement current phenomena.

$$\begin{aligned}
\mathrm{curl}\mathbf{H} &= \mathbf{J} + j\omega\mathbf{D} \\
\mathrm{curl}\mathbf{H} &= \left[\sigma(\omega) + j\omega\epsilon(\omega)\right]\mathbf{E} = \sigma_{\mathrm{eff}}\mathbf{E} \\
\mathrm{curl}\mathbf{H} &= j\omega\left[\epsilon(\omega) - j\frac{\sigma(\omega)}{\omega}\right]\mathbf{E} = j\omega\epsilon_{\mathrm{eff}}\mathbf{E} \\
\epsilon_{\mathrm{eff}}(\omega) &= \epsilon_0\epsilon_{\mathrm{r}}\left[1 + \frac{\sigma(\omega)}{j\omega\epsilon_0\epsilon_{\mathrm{r}}}\right] \\
\sigma_{\mathrm{eff}}(\omega) &= j\omega\epsilon_{\mathrm{eff}}
\end{aligned} \tag{2.46}$$

Assuming an intrinsic bulk conductivity $\sigma(\omega) = \frac{\sigma_{DC}}{1+j\omega/\omega_s}$ and inserting into the expression of effective permittivity from Eq. (2.46) leads to the result of Eq. (2.47), which equals the expression in Eq. (2.45). The geometric average of the dielectric relaxation frequency and the scattering frequency is called plasma frequency $\omega_{\mathrm{p}} = \sqrt{\omega_{\mathrm{d}}\omega_{\mathrm{s}}}$.

$$\epsilon_{\mathrm{eff}}(\omega) = \epsilon_0\epsilon_{\mathrm{r}}\left[1 - \frac{\omega_{\mathrm{s}}\omega_{\mathrm{d}}}{\omega^2 - j\omega\omega_{\mathrm{s}}}\right] = \epsilon_0\epsilon_{\mathrm{r}}\left[1 - \frac{\omega_{\mathrm{p}}^2}{\omega^2 - j\omega\omega_{\mathrm{s}}}\right] \tag{2.47}$$

[23] Andrè-Marie Ampère (1775–1836), French physicist.

The corresponding expression for the effective conductivity value is shown in Eq. (2.48).

$$\sigma_{\text{eff}}(\omega) = \sigma(\omega) + j\omega\epsilon(\omega) = \frac{\sigma_{\text{DC}}}{1 + j\omega/\omega_{\text{s}}} + j\omega\epsilon_0\epsilon_{\text{r}} \tag{2.48}$$

Hence, Drude's dispersion model is applied to Maxwell's equations by use of the effective permittivity $\epsilon_{\text{eff}}(\omega)$ or effective conductivity $\sigma_{\text{eff}}(\omega)$ with the intrinsic bulk conductivity $\sigma(\omega) = \frac{\sigma_{\text{DC}}}{1+j\omega/\omega_{\text{s}}}$ and intrinsic bulk permittivity $\epsilon(\omega) = \epsilon_0\epsilon_{\text{r}}$. In combination with numerical 3D electromagnetic field solvers, covering the effects of arbitrarily shaped structures following Drude's dispersion is possible [22–24].

Replacing the DC conductivity σ_{epi} in the expression of R_{epi} (Eq. (2.31)) by the effective conductivity $\sigma_{\text{eff}}(\omega)$ from Drude's dispersion model yields the complex impedance Z_{epi} (Eq. (2.49)). As Drude's model has been derived for bulk materials. Crowe [25] suggested to include the influence of the epilayer thickness on the scattering frequency. Hence, an effective scattering frequency for thin layers $\omega_{\text{seff}} = \frac{q_{\text{e}}}{m^{\star}\mu_{\text{e}}} + \frac{v_{\text{d}}}{t_{\text{u}}}$ should be used, whereas the (mean) drift velocity is approximated by its maximum value. In GaAs it is $v_{\text{dmax}} \approx 2 \cdot 10^5 \frac{\text{m}}{\text{s}}$.

$$\begin{aligned}
R_{\text{epi}} &= \frac{t_{\text{u}}(v_{\text{j}})}{\sigma_{\text{epi}} S_{\text{j}}} = \frac{t_{\text{u}}(v_{\text{j}})}{q_{\text{e}}\mu_{\text{e}} N_{\text{Depi}} S_{\text{j}}} \\
\sigma_{\text{eff}}(\omega) &= \sigma_{\text{DC}} \left[\frac{1}{1 + j\omega/\omega_{\text{seff}}} + j\frac{\omega}{\omega_{\text{d}}}\right] \\
Z_{\text{epi}} &= R_{\text{epi}} \left(\frac{1}{1 + j(\omega/\omega_{\text{seff}})} + j\omega/\omega_{\text{d}}\right)^{-1}
\end{aligned} \tag{2.49}$$

Inspecting Eq. (2.49), the equivalent circuit elements R_{epi}, C_{epi}, and L_{epi} of the epilayer are all known (Eq. (2.50)).

$$\begin{aligned}
L_{\text{epi}}(v_{\text{j}}) &= R_{\text{epi}}(v_{\text{j}})/\omega_{\text{seff}} \\
C_{\text{epi}}(v_{\text{j}}) &= \epsilon_0\epsilon_{\text{r}} \frac{S_{\text{j}}}{t_{\text{u}}(v_{\text{j}})} \qquad t_{\text{epi}} = t_{\text{d}}(v_{\text{j}}) + t_{\text{u}}(v_{\text{j}})
\end{aligned} \tag{2.50}$$

C_{epi} in Eq. (2.50) equals the parallel capacitor model. At the introduced dielectric relaxation frequency ω_d, the element values R_{epi} and $(\omega C_{epi})^{-1}$ have equal values. Hence, displacement current becomes dominant at $\omega > \omega_d$. Resonance between carrier inertia and electron displacement appears at the plasma frequency $\omega_p = \sqrt{\omega_s \omega_d}$.

Buffer Layer According to Fig. 2.11, the epilayer and buffer layer are assumed to have cylindrical shapes with anode diameter a and buffer diameter $2b$. In 1967 Dickens [26], derived expressions for the spreading resistance R_{spr} and impedance due to skin effect Z_{skin}, both dependent on the geometry of Fig. 2.11. The increase of buffer layer resistance, due to current flow spreading into the n^+ buffer layer is considered by R_{spr}.

$$R_{spr} = \frac{\operatorname{atan}\left(\frac{2b}{a}\right)}{\pi \sigma_b a} \tag{2.51}$$

The complex impedance Z_{skin} covers the current concentration near the buffer layer surface as a function of frequency and geometry (right side of Fig. 2.11).

$$Z_{skin} = \frac{(1+j)}{\sigma_b \delta_{skin}} \frac{1}{2\pi} \ln\left(\frac{2b}{a}\right) \qquad \delta_{skin} = \sqrt{\frac{2}{\omega \mu_0 \sigma_b}} \tag{2.52}$$

Dickens assumed conductive sidewalls and conductive backside together with a large buffer layer thickness $t_b >> d$, $t_b >> 2b$. These are simplifications, incorrect for both, whisker contacted Schottky diodes and planar Schottky diodes (compare Fig. 2.8). Dickens' expressions are good approximations, if appropriate effective values of anode diameter a_{eff} and buffer radius b_{eff} are chosen instead of the real physical values. If the buffer layer diameter is much larger than the anode diameter, it is

$$\lim_{(2b)/a \to \infty} \operatorname{atan}\left(\frac{2b}{a}\right) = \frac{\pi}{2} \tag{2.53}$$

and Eq. (2.51) simplifies to $R_{\text{spr}} = \frac{1}{2\sigma_{\text{b}} a}$.

Dickens' expression for Z_{skin} in Eq. (2.52) consists of the classical skin effect formula with skin depth δ_{skin} combined with a geometry dependent term.

As it is with other surface impedances, the skin effect impedance is derived from the intrinsic wave impedance for bulk materials Z_{w}. The wave impedance is defined as the phasor ratio of the electric and magnetic field strengthes perpendicular to the direction of propagation. Eq. (2.54) includes the dependency of Z_{w} from material parameters and the propagation coefficient γ.

$$\begin{aligned} Z_{\text{w}} :=& \frac{E_{\perp}(x,y,z)}{H_{\perp}(x,y,z)} \\ Z_{\text{w}} =& \sqrt{\frac{\mu(\omega)}{\epsilon_{\text{eff}}(\omega)}} = \sqrt{\frac{j\omega\mu(\omega)}{\sigma_{\text{eff}}}} = \frac{j\omega\mu(\omega)}{\gamma} \\ \gamma =& \frac{j\omega\mu(\omega)}{Z_{\text{w}}} = \sqrt{j\omega\mu_0} \cdot \sqrt{\sigma_{\text{eff}}} \end{aligned} \tag{2.54}$$

Inserting $\sigma_{\text{eff}} = \sigma(\omega) + j\omega\epsilon(\omega) = \sigma_{\text{DC}}$ into Eq. (2.54) leads to the complex skin impedance. Throughout this thesis, magnetic permeability is $\mu(\omega) = \mu_0$.

$$Z_{\text{skin}} = \sqrt{\frac{j\omega\mu_0}{\sigma_{\text{DC}}}} = \frac{j\omega\mu_0}{\sqrt{\sigma_{\text{DC}} j\omega\mu_0}} = \frac{j\omega\mu_0}{\frac{1+j}{\delta_{\text{skin}}}} = \frac{j\omega\mu_0}{\gamma} = \frac{(1+j)}{\sigma_{\text{b}}\delta_{\text{skin}}} \tag{2.55}$$

Comparing Eq. (2.54) and Eq. (2.55), we identify the term $\frac{(1+j)}{\delta_{\text{skin}}}$ as the propagation coefficient γ within the buffer layer.

Applying Drude's dispersion model to Dicken's expressions of Eq. (2.51, 2.52) is possible if the DC buffer layer conductivity $\sigma_{\text{DC}} = \sigma_{\text{b}}$ is replaced by its effective, frequency dependent value (Eq. (2.56)).

$$\sigma_{\text{eff}} = \sigma(\omega) + j\omega\epsilon(\omega) = \sigma_{\text{b}} \left(\frac{1}{1 + j(\omega/\omega_{\text{s}})} + j\omega/\omega_{\text{d}} \right) \tag{2.56}$$

In addition, the term $(1+j)/\delta_{\mathrm{skin}}$ in Eq. (2.52) is replaced by the propagation coefficient from Eq. (2.54).

$$\gamma = \sqrt{j\omega\mu_0} \cdot \sqrt{\sigma_{\mathrm{eff}}} = \sqrt{j\omega\mu_0} \cdot \left(\frac{1}{1 + j(\omega/\omega_{\mathrm{s}})} + j\omega/\omega_{\mathrm{d}} \right)^{1/2} \tag{2.57}$$

This leads to a high frequency version of Dickens' formulas [27–30].

$$\begin{aligned} Z_{\mathrm{spr}} &= \frac{\operatorname{atan}\left(\frac{2b}{a}\right)}{\pi \sigma_{\mathrm{b}} a} \left(\frac{1}{1 + j(\omega/\omega_{\mathrm{s}})} + j\omega/\omega_{\mathrm{d}} \right)^{-1} \\ Z_{\mathrm{skin}} &= \frac{\ln\left(\frac{2b}{a}\right)}{2\pi} \left(\frac{j\omega\mu_0}{\sigma_{\mathrm{b}}} \right)^{1/2} \left(\frac{1}{1 + j(\omega/\omega_{\mathrm{s}})} + j\omega/\omega_{\mathrm{d}} \right)^{-1/2} \end{aligned} \tag{2.58}$$

As already mentioned, calculated effective permittivity or effective conductivity values are readily included within 3D EM simulations and hence arbitrarily shaped three-dimensional geometries can be considered. Commercial fullwave solvers do also offer a variety of two-dimensional boundary conditions,[24] to include the influence of dispersive media without the necessity of 3D mesh grids. Anyway, in case of commercially available Schottky diodes the required process and material data are generally not available. In the author's opinion, using simple models should be preferred over assuming complicated but erroneous models.

Exemplary results of the volume current density vector $\mathbf{J}_{\mathrm{vol}}$ at $f = 85$ GHz inside the n^+ GaAs buffer layer are shown in Fig. 2.13. Results on the left side show how the current density extends into the buffer layer, whereas the results on the right side visualize the concentration of current density directly below the anode metal.

Ohmic Contacts The ohmic contact resistances R_{ohmic} from n^+ buffer layer to cathode metal are small compared to the other equivalent

[24] Ansys High Frequency Structure Simulator (HFSS™) offers standard impedance boundary (IB), layered impedance boundary (LIB) and finite conductivity boundary (FCB).

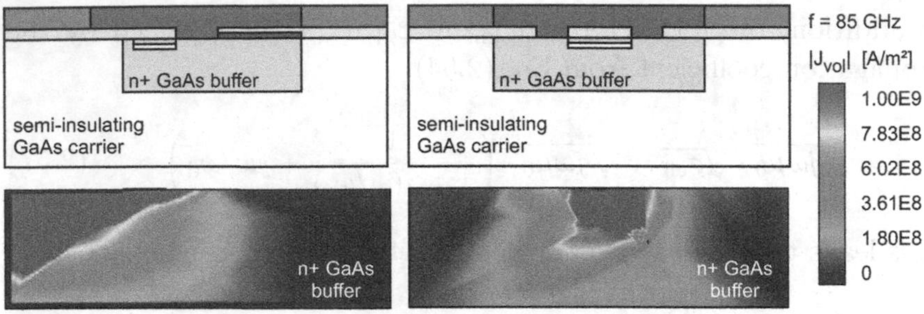

Figure 2.13: Simulated volume current density vector $\mathbf{J}_{\mathrm{vol}}$ at f = 85 GHz inside the n^+ GaAs buffer layer. Cross-sectional view in parallel to (left) and perpendicular to (right) to the direction of wave propagation.

circuit elements in Fig. 2.9. In general, ohmic contacts between metal and semiconductor are either realized by choosing an appropriate metal with work function $q_e\Phi_M$ lower than the semiconductor's work function $\Phi_s = \Phi_\chi - V_n$, resulting in negative built-in potential $\Phi_{bi} < 0$. Then electrons emit from metal to semiconductor and vice versa without a potential barrier (left side of Fig. 2.14).

$$q_e\Phi_{bi} < 0 \quad q_e\Phi_m < q_e\left(\Phi_\chi + V_n\right) \tag{2.59}$$

Another way to realize ohmic contacts is very high doping (n^{++}) of the semiconductor. According to Eq. (2.19), the thickness of the depletion width is indirect proportional to the square root of doping density.

$$t_d \sim \sqrt{1/N_D} \tag{2.60}$$

At very high doping, field emission (electron tunneling) becomes the dominant current mechanism through the thin barrier (right side of Fig. 2.14).

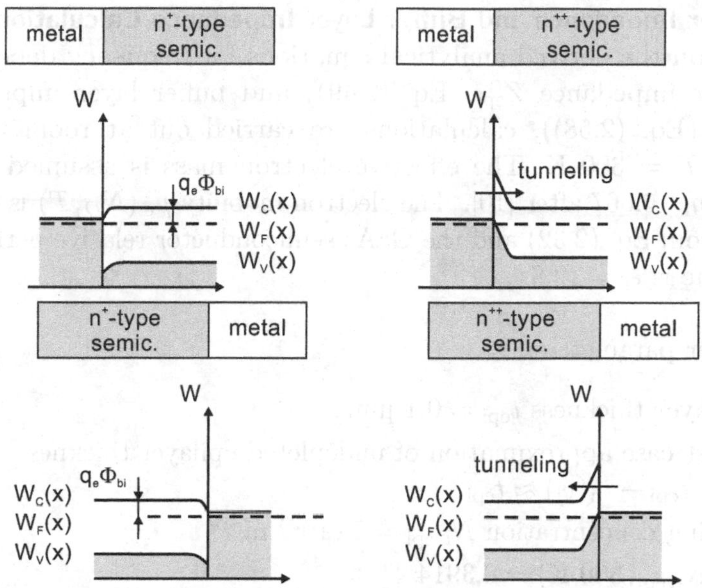

Figure 2.14: Energy band diagrams of ohmic contacts between metal and n-type semiconductor material. On the left side, the metal work function is lower than the semiconductor's work function, and on the right side, the ohmic contact is realized by field emission.

Details on ohmic contacts from a technological point of view can be found in [16, 31]. The ohmic contact resistance R_{ohmic} depends on the effective ohmic surface area S_{ohmic} and the specific ohmic contact resistance $(\ell_{\text{ohmic}}/\sigma_{\text{ohmic}})$, which is in the order of $10^{-10}\ \Omega\text{m}^2$ [32].

$$R_{\text{ohmic}} = \left(\frac{\ell_{\text{ohmic}}}{\sigma_{\text{ohmic}}}\right)\frac{1}{S_{\text{ohmic}}} \tag{2.61}$$

In case of $S_{\text{ohmic}} = 100/200/400/600\ \mu\text{m}^2$, the resulting ohmic resistance values are $R_{\text{ohmic}} = 10/2.5/0.63/0.3\ \text{m}\Omega$.
Hence, it is $R_{\text{ohmic}} \ll |Z_{\text{spr}}|, |Z_{\text{skin}}|, |Z_{\text{epi}}|$ and R_{ohmic} is negligible.

Epilayer Impedance and Buffer Layer Impedance Calculations
Based on the derived analytical equations for frequency dependent epilayer impedance Z_{epi} (Eq. (2.49)) and buffer layer impedance Z_{buffer} (Eq. (2.58)), calculations are carried out at room temperature $T = 300$ K. The effective electron mass is assumed to be $m^{\star} = m_{\mathrm{e}} \cdot 0.067$ after [10]. The electron mobility $\mu_{\mathrm{e}}(N_D, T)$ is calclulated from Eq. (2.32) and the GaAs semiconductor relative permittivity value is $\epsilon_{\mathrm{r}} = 12.9$.

Epilayer parameters:

- epilayer thickness $t_{\mathrm{epi}} = 0.1$ µm,
- worst case approximation of undepleted epilayer thickness $t_{\mathrm{u}} = t_{\mathrm{epi}} - t_{\mathrm{d}}(v_{\mathrm{j}}) \approx t_{\mathrm{epi}}$,
- doping concentration $N_{\mathrm{Depi}} = 2 \cdot 10^{23}\ \mathrm{m}^{-3}$,
- $\mu_{\mathrm{e}}(N_{\mathrm{Depi}}, 300\ \mathrm{K}) = 0.3914\ \frac{\mathrm{m}}{\mathrm{Vs}}$,
- maximum drift velocity $v_{\mathrm{dmax}} \approx 2 \cdot 10^5 \frac{\mathrm{m}}{\mathrm{s}}$,
- anode width AW = 5 µm, anode length AL = 1 µm
  → effective anode diameter $a \approx 2.5$ µm,

With the above epilayer parameters, the DC conductivity value becomes $\sigma_{\mathrm{DC}} = \sigma_{\mathrm{epi}} = 1.2542 \cdot 10^4$ S/m, and the effective scattering frequency, dielectric relaxation frequency and plasma frequency are $\omega_{\mathrm{seff}}/(2\,\pi) = 1.39$ THz, $\omega_{\mathrm{d}}/(2\,\pi) = 17.48$ THz and $\omega_{\mathrm{p}}/(2\,\pi) = 4.92$ THz.

Buffer layer parameters:

- buffer layer thickness $t_{\mathrm{b}} \gg b$,
- doping concentration $N_{\mathrm{Db}} = 5 \cdot 10^{24}\ \mathrm{m}^{-3}$,
- $\mu_{\mathrm{e}}(N_{\mathrm{Db}}, 300\ \mathrm{K}) = 0.1826\ \frac{\mathrm{m}}{\mathrm{Vs}}$,
- buffer width BW, buffer length BL,
- effective buffer radius $b \approx 250$ µm,

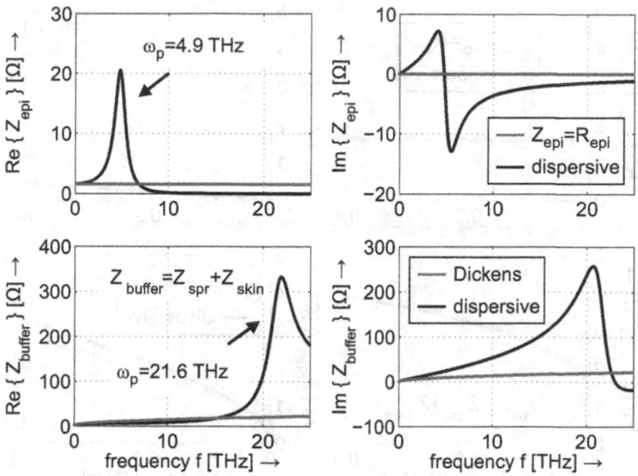

Figure 2.15: Calculated real and imaginary parts of epilayer impedance (top) Z_{epi} and buffer impedance (bottom) Z_{buffer} versus frequency.

With the above buffer layer parameters, the DC conductivity value becomes $\sigma_{\text{DC}} = \sigma_{\text{b}} = 1.4627 \cdot 10^5$ S/m, and the scattering frequency, dielectric relaxation frequency and plasma frequency are $\omega_{\text{s}}/(2\,\pi) =$ 2.28 THz, $\omega_{\text{d}}/(2\,\pi) = 203.82$ THz and $\omega_{\text{p}}/(2\,\pi) = 21.60$ THz.

Fig. 2.15 compares calculated real and imaginary parts of epilayer impedance (top) Z_{epi} and buffer impedance (bottom) Z_{buffer} versus frequency. Differences between the conventional solutions (red) and the high frequency ones (black) are significant only if frequencies are in the order of the plasma frequencies. The thin epitaxial layer has much lower plasma frequency than the buffer layer.

Fig. 2.16 contains the same data but illustrates the frequency range DC to 400 GHz only. Obviously, up to 400 GHz, there is no need to utilize the extended models. It further shows, the real part of Z_{buffer} varies between its static value and about 4 Ω at 400 GHz. According to Fig. 2.17, this frequency dependency is due to the skin effect, whereas the spreading impedance is almost a real value from DC to 400 GHz. The diode parameters roughly belong to the UMS

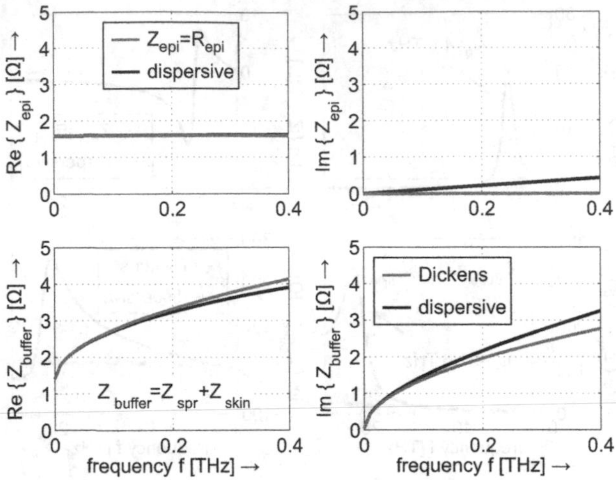

Figure 2.16: Calculated real and imaginary parts of epilayer impedance (top) Z_{epi} and buffer impedance (bottom) Z_{buffer} versus frequency. Maximum frequency $f = 400$ GHz.

DBES105a diode of section 2.6 and section 2.7, which is modelled with constant series resistance throughout this work.

The presented analysis allows for analytic calculation of the Schottky junction parameters (subsection 2.2.1) and the linear parts of the diode model (subsection 2.2.3). For the design of diode based frequency multipliers, mixers and detectors with commercial circuit simulators, the active part of the diode (Schottky junction) is best covered by the common Berkeley SPICE model DIODE [6, 33]. The linear diode parts are either described by the given equations or the effective material parameters are used in conjunction with 3D EM fullwave simulators, which is used to design the circuits of sections 2.5, 2.6, 2.7.

Detailed information about diode materials and geometry constitutes a precondition and are barely provided by the diode manufacturer. To offer accurate models without revealing business secrets, diode manufactures pay third party companies to build and sell black box models (e.g. Modelithics Inc. [34]).

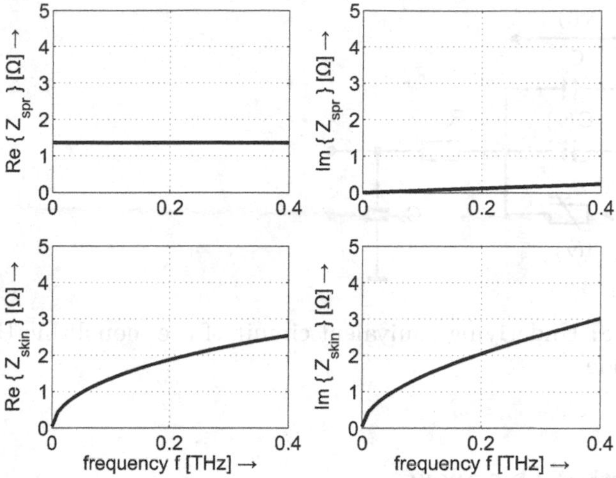

Figure 2.17: Calculated real and imaginary parts of spreading impedance (top) Z_{spr} and skin impedance (bottom) Z_{skin} versus frequency. Maximum frequency $f = 400$ GHz.

Semiconductor foundries offer design kits including diode models. These are often so called compact models. All parameters are given as a function of diode geometry and therefore the single compact model is valid for all the available diodes of the semiconductor process. Fig. 2.18 shows the equivalent circuit of the monolithic BES105 diode from the UMS buried epitaxial layer process (BES). Epilayer doping concentration and thickness (0.1 µm) are fixed by the BES process The diode has an anode length of AL = 1 µm, ideality factor $\eta = 1.204$ and built-in potential $\Phi_{\mathrm{bi}} = 0.97$ V. Anode widths AW are available in the range of 3 to 10 µm. The saturation current I_{S} and junction capacitance C_{j} are given as a function of the anode surface area $S_{\mathrm{j}} = \mathrm{AL} \cdot \mathrm{AW}$ in Eq. (2.62). Such compact models are also available for other semiconductor factories [35–37].

$$\begin{aligned} I_{\mathrm{S}} &= 7.04 \cdot 10^{-3} \frac{\mathrm{A}}{\mathrm{m}^2} \cdot S_{\mathrm{j}} = 7.04 \cdot 10^{-3} \frac{\mathrm{A}}{\mathrm{m}^2} \cdot 1\ \mu\mathrm{m} \cdot \mathrm{AW} \\ C_{\mathrm{j}}(\mathrm{fF}) &= C_{\mathrm{j0}} + C_{\mathrm{j1}} / (1 - v_{\mathrm{j}}/\Phi_{\mathrm{bi}})^M \qquad M = 1.82 \end{aligned} \tag{2.62}$$

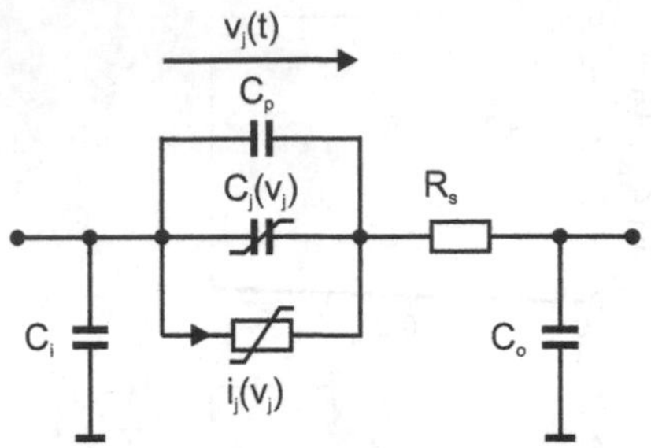

Figure 2.18: Underlying equivalent circuit of the monolithic UMS BES105 diode.

2.2.4 Market Overview

Table 2.1 summarizes characteristic parameters of several silicon and gallium arsenide Schottky diodes. It is average values, taken from the manufacturers datasheets. It includes the diode series resistance R_s, the junction capacitance at zero bias C_{j0}, saturation current I_s, ideality factor η and forward voltage at a certain DC current.

In addition, the cut-off frequency $f_c = (2\pi R_s C_T)^{-1}$ is calculated from R_s and the total capacitance $C_T = C_{j0} + C_{parasitic}$. It serves as a rough figure of merit for the maximum operating frequency (compare explanations of [8] page 333 et seqq.). In case of antiparallel, series tee or quad diode configuration, the data corresponds to a single junction. Table 2.1 illustrates that commercial Schottky diodes are barely characterized by the manufacturers. For circuit design, individual parameter extractions from DC (current-voltage characteristics) and low frequency (junction capacitance) measurements are mandatory.

The reader should keep in mind, the diodes of Table 2.1 are either used within the author's circuit designs or constitute alternatives. Beside these, there are a lot more diodes available for millimeter-wave and sub-THz applications, like

- Technical Research Center (VTT) of Finland, www.vtt.fi,

- Advanced Compound Semiconductor Technologies (ACST) GmbH, spin-off from Technical University Darmstadt, Germany, www.acst.de,
- Jet Propulsion Laboratory, Pasadena, United States, www.jpl.nasa.gov,
- Hughes Research Laboratories (HRL), Malibu, United States, www.hrl.com,
- Teratech Components Ltd, spin-off from Science & Technology Facilities Council (STFC) Rutherford Appleton Laboratory (RAL), United Kingdom, www.teratechcomponents.com,
- Terahertz and Millimeter-Wave Laboratory at Chalmers University of Technology, Göteborg, Sweden, www.chalmers.se.

Table 2.1: Market Overview of Commercial Si and GaAs Schottky Diodes

diode	R_s [Ω]	C_{j0} [fF]	f_c [THz]	I_S [µA]	η	v_j [mV] at i_j [mA]	mat.	diode arch.	package	dim. [µm]
DBES105a ⋆	4	9.5	2.60	35E-9	1.2	750 at 1	GaAs	series tee	flipchip	530, 230, 100
after dicing ⋆ [38]	4	9.5	2.60	35E-9	2.2	750 at 1	GaAs	single	flipchip	140, 60, 30
VDI ZBD ⊛	19	15	0.34	11.9	1.21	70 at 0.1	GaAs	single	flipchip	600, 250, 100
VDI SD ⊛	4	n/a	1.11	n/a	n/a	475 at 1E-6	GaAs	single	flipchip	600, 250, 100
VDI AP ⊛	4	n/a	0.67	n/a	n/a	495 at 1E-6	GaAs	antiparallel	flipchip	600, 290, 100
VDI ST ⊛	4	n/a	1.17	n/a	n/a	500 at 1E-6	GaAs	series tee	flipchip	740,250,100
MGS801 †	7	50	0.45	(1.6 to 16)E-6	1.4	700 at 1	GaAs	antiparallel	flipchip	685, 275, 125
MGS801A †	5	75	0.42	(1.6 to 16)E-6	1.4	700 at 1	GaAs	antiparallel	flipchip	685, 285, 200
MGS904 †	7	60	0.38	(1.6 to 16)E-6	1.4	700 at 1	GaAs	quad ring	beamlead	450, 450, 90
MZBD-9161 †	50	30	0.11	12	1.2	$\Phi_{bi} = 230$	GaAs	single	beamlead	300, 300, 90
MSS30 PCR46 B47 †	10	100	0.16	(9.1 to 43)E-3	1.01	280 at 1	Si	crossed quad ring	beamlead	254, 254, 90
DC1346 •	8	10	0.66	n/a	1.28	720 at 2.5	GaAs	single	beamlead	800, 290, 50
HSCH-9161 ◇	20	35	0.17	12	1.2	$\Phi_{bi} = 230$	GaAs	single	beamlead	250, 250, 70
MA4E1317 ◦	5.5	20	0.64	n/a	n/a	700 at 1	GaAs	single	flipchip	660, 330, 200
MA4E1310 ◦	5.5	10	0.72	n/a	n/a	700 at 1	GaAs	single	flipchip	660, 330, 200
MA4E1318 ◦	5.5	20	0.64	n/a	n/a	700 at 1	GaAs	antiparallel	flipchip	660, 330, 200
MA4E1319-1 ◦	5.5	20	0.64	n/a	n/a	700 at 1	GaAs	series tee	flipchip	700, 475, 200
MA4E2160 ◦	5.5	20	0.64	n/a	n/a	700 at 1	GaAs	antipar., uncon.	flipchip	370, 330 200
MS8150 ‡	3	45	0.82	200E-9	1.2	$\Phi_{bi} = 850$	GaAs	single	flipchip	660, 330, 140
MS8151 ‡	7	25	0.51	320E-9	1.01	$\Phi_{bi} = 850$	GaAs	single	flipchip	660, 330, 140
MS8250 ‡	3	45	0.82	320E-9	1.01	$\Phi_{bi} = 850$	GaAs	antiparallel	flipchip	660, 330, 140
MS8251 ‡	7	25	0.51	320E-9	1.01	$\Phi_{bi} = 850$	GaAs	antiparallel	flipchip	660, 330, 140
MS8350 ‡	3	45	0.82	320E-9	1.01	$\Phi_{bi} = 850$	GaAs	series tee	flipchip	710, 480, 140
MS8351 ‡	7	25	0.51	320E-9	1.01	$\Phi_{bi} = 850$	GaAs	series tee	flipchip	710, 480, 140
GC9901-QR1 ‡	20	100	0.08	n/a	n/a	340 at 1	Si	quad ring	beamlead	470, 470, 40
DMK2790 ∗	7	35	0.41	n/a	n/a	700 at 1	GaAs	single	flipchip	660, 330, 100
DMK2308 ∗	7	35	0.41	n/a	n/a	700 at 1	GaAs	antiparallel	flipchip	660, 330, 100

⋆ United Monolithic Semiconductors (UMS), www.ums-gaas.com, ⊛ Virginia Diodes Inc. (VDI), www.vadiodes.com, † Aeroflex Metelics, www.aeroflex.com,
• Linwave Technology Ltd., www.linwave.co.uk, ◇ Agilent Technologies, www.home.agilent.com, ◦ M/A-COM Technology Solutions, www.macom.com
‡ Microsemi Corp., www.microsemi.com, ∗ Skyworks Solutions Inc., www.skyworksinc.com

2.3 Single Tone Large Signal (LS) Analysis

Single tone large signal analysis is required to analyze frequency multiplier circuits and is the basis of large signal / small signal (LSSS) mixer analysis in subsection 4.2.1. The latter assumes small signal RF and IF stimulus.

Following the design procedure of section 2.4, all linear circuit parts (embedding impedances) are covered by linear multiport scattering parameters from 3D EM simulation. Preferably, these do also include the linear parts of the diode models and are shown on the left side of Fig. 2.19.

Nonlinear models of the N Schottky junctions exist as SPICE models (right side of Fig. 2.19). This configuration is used to derive the

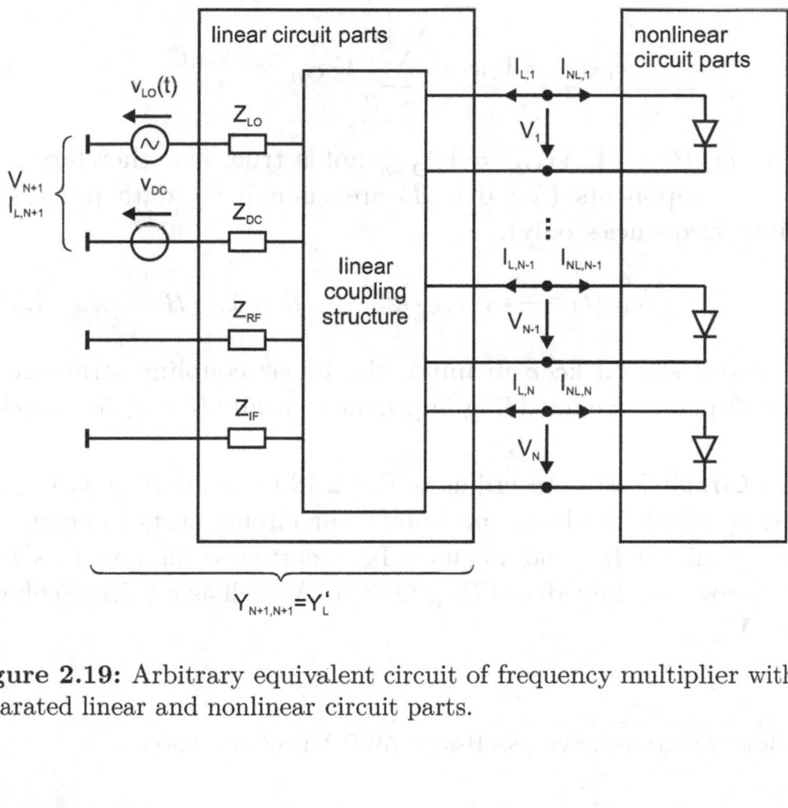

Figure 2.19: Arbitrary equivalent circuit of frequency multiplier with separated linear and nonlinear circuit parts.

harmonic balance equation in the following. Commercially available circuit simulators like Agilent ADS, Ansys Designer and many others[25] provide single and multi tone harmonic balance simulation, which is in addition easily combinable with many predefined equivalent models of commonly occurring devices.

A single large signal local oscillator voltage component $v_{\mathrm{LO}}(t) = \hat{v}_{\mathrm{LO}} \cos(\omega_{\mathrm{LO}} t + \varphi_{\mathrm{LO}})$ and DC voltage component V_{DC} are applied and lead to a countably infinite number of voltage and current harmonics at each of the N nonlinear ports in Fig. 2.19. Up to H harmonics are considered in the following. A complex double sided Fourier series representation of such a large signal $v_{\mathrm{LS}}(t)$ including DC component is given by Eq. (2.63). Hence, two complex exponential functions $V_{\mathrm{LO}\,h} \cdot e^{hj\omega_{\mathrm{LO}} t}$ and $V_{\mathrm{LO}\,-h} \cdot e^{-hj\omega_{\mathrm{LO}} t}$ at each nonlinear port and harmonic number are required.

$$v_{\mathrm{LS}}(t) = V_{\mathrm{DC}} + \sum_{h=-H}^{h=+H} V_{\mathrm{LO}\,h} \cdot e^{hj\omega_{\mathrm{LO}} t} \tag{2.63}$$

Due to $v_{\mathrm{LS}}(t) \in \mathbb{R}$, $V_{\mathrm{LO}\,h} = V^{\star}_{\mathrm{LO}\,-h}$ holds true, and therefore only $(H+1)$ components $(h = 0 \dots H)$ are enough for data processing (positive frequencies only).

$$v_{\mathrm{LS}}(t) \circ\!\!-\!\!\bullet\; V_{\mathrm{LO}\,h} \cdot e^{h\omega_{\mathrm{LO}} t} \quad h = 0 \dots H \tag{2.64}$$

The reader should keep in mind, the linear coupling structure in Fig. 2.19 provides embedding impedances at all $(H+1)$ frequencies.

Linear Circuit Part According to Fig. 2.19 there are N $(n = 1 \dots N)$ ports, at which the linear and nonlinear circuit parts interact. At each port, linear $\mathbf{I}_{\mathrm{L}\,n}$ and nonlinear $\mathbf{I}_{\mathrm{NL}\,n}$ current column vectors with $(H+1)$ rows are introduced (Eq. (2.65)). As well as a voltage column vector $\mathbf{V}_n$.

[25]Cadence SpectreRF, Synopsis Hspice, AWR Microwave Office.

$$\begin{aligned} \mathbf{I}_{\mathrm{NL}\,n} &= [I_{\mathrm{NL}\,n,0}, \dots I_{\mathrm{NL}\,n,H}]^{\mathrm{T}} \quad \mathbf{I}_{\mathrm{L}\,n} = [I_{\mathrm{L}\,n,0}, \dots I_{\mathrm{L}\,n,H}]^{\mathrm{T}} \\ \mathbf{V}_n &= [V_{n,0}, \dots V_{n,H}]^{\mathrm{T}} \end{aligned} \tag{2.65}$$

An additional pair of linear voltage and current column vectors $\mathbf{V}_{\mathrm{L}N+1}, \mathbf{I}_{\mathrm{L}N+1}$ account for the DC and LO stimulus (Eq. (2.66)).

$$\begin{aligned} \mathbf{I}_{\mathrm{L}\,N+1} &= [I_{\mathrm{L}\,N+1,0}, \dots I_{\mathrm{L}\,N+1,H}]^{\mathrm{T}} \\ \mathbf{V}_{N+1} &= [V_{N+1,0}, \dots V_{N+1,H}]^{\mathrm{T}} \end{aligned} \tag{2.66}$$

These $(N+1)$ linear voltage and current vectors are components of the vectors $\mathbf{V}', \mathbf{I}'_{\mathrm{L}}$, as shown in Eq. (2.67).

$$\mathbf{V}' = \begin{bmatrix} \mathbf{V}_1 \\ \dots \\ \mathbf{V}_N \\ \mathbf{V}_{N+1} \end{bmatrix} \quad \mathbf{I}'_{\mathrm{L}} = \begin{bmatrix} \mathbf{I}_{\mathrm{L}\,1} \\ \dots \\ \mathbf{I}_{\mathrm{L}\,N} \\ \mathbf{I}_{\mathrm{L}\,N+1} \end{bmatrix} \tag{2.67}$$

Due to Ohm'slaw, there is a $(N+1) \times (N+1)$ admittance matrix $\mathbf{Y}'_{\mathrm{L}}$ transforming $\mathbf{V}'$ into $\mathbf{I}'_{\mathrm{L}}$.

$$\mathbf{I}'_{\mathrm{L}} = \mathbf{Y}'_{\mathrm{L}} \mathbf{V}' \tag{2.68}$$

The submatrices $\mathbf{Y}_{\ell k}$ of $\mathbf{Y}'_{\mathrm{L}}$ in Eq. (2.69) are diagonal matrices of size $(H+1) \times (H+1)$, that contain DC and harmonic transfer admittances between port ℓ and k. $\quad \ell, k \in \{1, \dots, N+1\}$.

$$\begin{bmatrix} \mathbf{I}_{L,1} \\ \dots \\ \mathbf{I}_{L,N} \\ \mathbf{I}_{L,N+1} \end{bmatrix} = \underbrace{\begin{bmatrix} \mathbf{Y}_{1,1} & \dots & \mathbf{Y}_{1,N+1} \\ \dots & \dots & \dots \\ \mathbf{Y}_{N,1} & \dots & \mathbf{Y}_{N,N+1} \\ \mathbf{Y}_{N+1,1} & \dots & \mathbf{Y}_{N+1,N+1} \end{bmatrix}}_{\mathbf{Y}'_{\mathrm{L}}} \begin{bmatrix} \mathbf{V}_{L,1} \\ \dots \\ \mathbf{V}_{L,N} \\ \mathbf{V}_{L,N+1} \end{bmatrix} \tag{2.69}$$

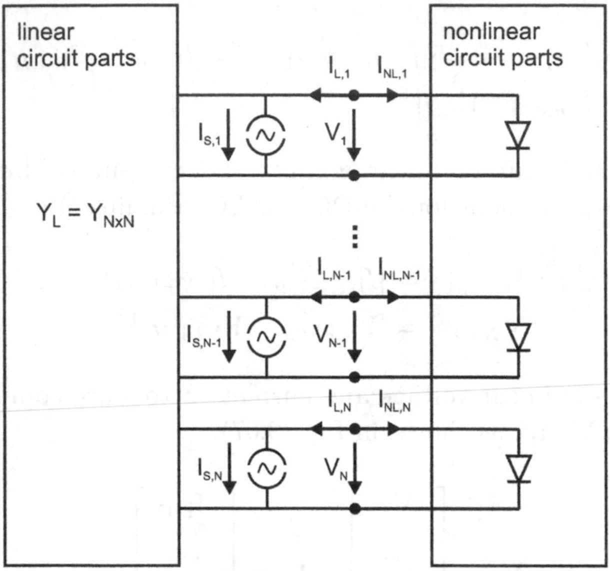

Figure 2.20: Arbitrary equivalent circuit of frequency multiplier with separated linear and nonlinear circuit parts, after transforming all sources into equivalent current sources.

Eq. (2.69) completely describes the linear circuit part of Fig. 2.19. In order to get a more compact description of the problem, the equivalent circuit is modified. The original sources of Fig. 2.19 are transformed into N current sources $\mathbf{I}_\mathrm{s}$ at the junctions between linear and nonlinear circuit parts, as shown in Fig. 2.20. The last column of $\mathbf{Y}_\mathrm{L}^{'}$ transforms the voltage stimulus $\mathbf{V}_{N+1}$ of Eq. (2.66) into $\mathbf{I}_\mathrm{s}$.

$$\mathbf{I}_\mathrm{s} = \begin{bmatrix} \mathbf{I}_{s,1} \\ \dots \\ \mathbf{I}_{s,N} \end{bmatrix} \quad \mathbf{I}_\mathrm{s} = \begin{bmatrix} \mathbf{Y}_{1,N+1} \\ \dots \\ \mathbf{Y}_{N,N+1} \end{bmatrix} \begin{bmatrix} \mathbf{V}_{N+1} \end{bmatrix} \tag{2.70}$$

The $(N \times N)$ admittance matrix $\mathbf{Y}_\mathrm{L}$ consists of the first N rows and N columns of $\mathbf{Y}_\mathrm{L}^{'}$. The column vectors $\mathbf{I}_\mathrm{L}, \mathbf{V}$ equal $\mathbf{I}_\mathrm{L}^{'}, \mathbf{V}^{'}$ without the last row. The latter is considered by the current sources $\mathbf{I}_\mathrm{s}$. Then the linear circuit part of Fig. 2.20 is entirely covered by Eq. (2.71), which is the description of the Norton[26] equivalent circuit in a multiport, multi frequency scenario.

$$\mathbf{I}_\mathrm{L} = \mathbf{I}_\mathrm{s} + \mathbf{Y}_\mathrm{L}\mathbf{V} \tag{2.71}$$

Nonlinear Circuit Part Whereas the linear circuit part is discussed in the frequency domain, time domain analysis is preferred for the nonlinear circuit part. The charge per unit surface area as a function of Schottky junction voltage $Q_j/S_j\,(v_j)$ and the corresponding nonlinear displacement current $i_C(v_j, t)$ are derived in section 2.2, compare Eq. (2.20) and Eq. (2.24), respectively. As well as the nonlinear resistive Schottky diode junction current $i_j(v_j) = i_\mathrm{G}$ (Eq. (2.26)).

In the notation of this section, the nonlinear current vector $\mathbf{I}_\mathrm{NL}$ in Fig. 2.20, with its N components $\mathbf{I}_{\mathrm{NL}\,n}$, $n \in \{1, \ldots, N\}$, is composed of a resistive $\mathbf{I}_\mathrm{G}$ and displacement current $\mathbf{I}_\mathrm{C}$.

$$\begin{aligned} \mathbf{I}_\mathrm{NL} = \mathbf{I}_\mathrm{G} + \mathbf{I}_\mathrm{C} &= [\mathbf{I}_{\mathrm{NL}\,1}, \ldots, \mathbf{I}_{\mathrm{NL}\,N}]^\mathrm{T} \\ \mathbf{I}_{\mathrm{NL}\,n} &= [I_{\mathrm{NL}\,n,0}, \ldots I_{\mathrm{NL}\,n,H}]^\mathrm{T} \end{aligned} \tag{2.72}$$

There are N resistive current time waveforms $i_{\mathrm{G}\,n}(t)$ and N charge time waveforms $Q_n(t)$, both dependent on all N components of the voltage vector $\mathbf{V}$. These consist of $(H+1)$ spectral components in the frequency domain or a single voltage time waveform $\mathbf{V}_n = [V_{n,0}, \ldots V_{n,H}]^\mathrm{T} \bullet\!\!-\!\!\circ\, v_n(t)$.

$$\begin{aligned} i_{\mathrm{G}\,n}(t) &= i_{\mathrm{G}\,n}\left(t, v_1(t) \ldots v_N(t)\right) \\ Q_n(t) &= Q_n\left(t, v_1(t) \ldots v_N(t)\right) \end{aligned} \tag{2.73}$$

[26] Edward Lawry Norton (1898–1983), American engineer.

Fourier transform[27] maps the current waveform of the diode's conductance and charge time waveform of the diode's capacitor to the frequency domain. When choosing the number of harmonics as an integer power of $H = 2^p, p \in \mathbb{N}^+$, the number of time samples should be greater than $2H$ by an appropriate amount of time oversampling.

$$\begin{aligned} \{i_{\mathrm{G}\,1}(t) \ldots i_{\mathrm{G}\,N}(t)\} &\circ\!\!-\!\!\bullet\ \mathbf{I}_{\mathrm{G}} = [\mathbf{I}_{\mathrm{G}\,1}, \ldots, \mathbf{I}_{\mathrm{G}\,N}]^{\mathrm{T}} \\ \{Q_1(t) \ldots Q_N(t)\} &\circ\!\!-\!\!\bullet\ \mathbf{Q} = [\mathbf{Q}_1, \ldots, \mathbf{Q}_N]^{\mathrm{T}} \end{aligned} \tag{2.74}$$

The resistive current vector $\mathbf{I}_{\mathrm{G}}$ is known from Eq. (2.74). Displacement currents are the time derivatives of charge time waveforms $\mathrm{d}Q_n(t)/\mathrm{d}t$, which become multiplications with $j\omega$ in the frequency domain.

$$i_{\mathrm{C}\,n}(t) = \frac{\mathrm{d}Q_n(t)}{\mathrm{d}t} \circ\!\!-\!\!\bullet\ \mathbf{I}_{\mathrm{C}\,n} = [0, j\omega_{LO}Q_{n,1}, \ldots, jH\omega_{LO}Q_{n,H}]^{\mathrm{T}} \tag{2.75}$$

Hence, the displacement current vector $\mathbf{I}_{\mathrm{C}}$ is given by Eq. (2.76), where $\boldsymbol{\Omega}$ is a diagonal matrix containing all considered frequencies.

$$\begin{aligned} \mathbf{I}_{\mathrm{C}} &= [\mathbf{I}_{\mathrm{C}\,1}, \ldots, \mathbf{I}_{\mathrm{C}\,N}]^{\mathrm{T}} = j\boldsymbol{\Omega}\mathbf{Q} \\ \boldsymbol{\Omega} &= \mathrm{diag}\,[0, \omega_{LO}, \ldots, H \cdot \omega_{LO}] \end{aligned} \tag{2.76}$$

Harmonic Balance Equation Applying Kirchhoff's[28] law (Eq. (2.77)) at each of the N ports in Fig. 2.20 and inserting Eq. (2.71), Eq. (2.72), Eq. (2.76) leads to the harmonic balance equation in Eq. (2.78).

$$\mathbf{I}_{\mathrm{L}} + \mathbf{I}_{\mathrm{NL}} = \mathbf{0} \tag{2.77}$$

[27] In case of multi tone harmonic balance, mappings different from classical Fourier transform are required (FFT-like).

[28] Gustav Robert Kirchhoff (1824–1887), German physicist.

$$\mathbf{F}(\mathbf{V}) = \mathbf{I}_s + \mathbf{Y}_L\mathbf{V} + \mathbf{I}_G + \underbrace{j\mathbf{\Omega}\mathbf{Q}}_{\mathbf{I}_C} \stackrel{!}{=} \mathbf{0} \tag{2.78}$$

Every set of voltages $\mathbf{V}$ with the current error vector $\mathbf{F}(\mathbf{V}) = 0$ is a solution of the nonlinear large signal problem. A similar formulation can be found on the basis of scattering parameters instead of admittance parameters.

Starting with an estimated initial solution $v_n(t)^{\#0} \circ\!\!-\!\!\bullet\ \mathbf{V}^{\#0}$ at iteration step $\#s = \#0$, the harmonic balance equation is solved by numerical iteration techniques. Although implementation and user interface are different, commercial circuit simulators predominantly implement Newton[29]-Raphson[30] method.

$$\mathbf{J} = \frac{d\mathbf{F}(\mathbf{V})}{d\mathbf{V}} = \mathbf{Y}_L + \frac{\partial \mathbf{I}_G}{\partial \mathbf{V}} + j\mathbf{\Omega}\frac{\partial \mathbf{Q}}{\mathbf{V}} \tag{2.79}$$

The latter utilizes the Jacobian[31] matrix $\mathbf{J}$ of the current-error vector $\mathbf{F}(\mathbf{V})$, shown in Eq. (2.79), to solve for $\Delta\mathbf{V}$ of Eq. (2.80).

$$\mathbf{F}(\mathbf{V}^{\#s}) - \underbrace{\left.\frac{d\mathbf{F}(\mathbf{V})}{d\mathbf{V}}\right|_{\mathbf{V}=\mathbf{V}^{\#s}}}_{\mathbf{J}} \Delta\mathbf{V} = 0 \tag{2.80}$$

$$\Delta\mathbf{V} = \left(\mathbf{V}^{\#s} - \mathbf{V}^{\#s+1}\right) \tag{2.81}$$

This allows for calculating the estimate of the next iteration step $\mathbf{V}^{\#s+1}$ entirely from known data of step $\#s$ (Eq. (2.82)).

$$\mathbf{V}^{\#s+1} = \mathbf{V}^{\#s} - \mathbf{J}^{-1}\mathbf{F}(\mathbf{V}^{\#s}) \tag{2.82}$$

The iterative process of solution finding stops if $|\mathbf{F}(\mathbf{V})|$ reaches a user defined minimum.

[29] Sir Isaac Newton (1642–1727), English physicist and mathematician.
[30] Joseph Raphson (1648–1715), English mathematician.
[31] Carl Gustav Jacob Jacobi (1804–1851), German mathematician.

The commercial circuit solvers used throughout this work offer two different methods for solving the system of linear equations in Eq. (2.80).

- Direct solver: LU decomposition
- Krylov solver: GMRES

The first implements Lower Upper (LU) decomposition and requires more computation time and memory than the Krylov solver but is more robust. The Krylov[32] subspace method known as generalized minimal residual method (GMRES) [39] is more effective with respect to computational time and memory consumption but suffers from convergence problems. The author uses Krylov method as a default and direct solver in case of convergence or accuracy problems only.

2.4 Frequency Multiplier Design

2.4.1 Multiplier Architectures

The series and shunt mounted diodes with individual signal branches for every spectral component of Fig. 2.1 and Fig. 2.2 in section 2.1 constitute the most versatile equivalent circuit models, capable of modelling every mixer / frequency multiplier scenario.

Utilizing frequency dependent impedances $Z_{\mathrm{e}}(\omega)$ and admittances $Y_{\mathrm{e}}(\omega)$ allows for building single diode equivalent circuits, as shown in Fig. 2.21. Summarizing multiple diodes is possible even if they are different by simultaneously modifying the embedding network $Z_{\mathrm{e}}, Y_{\mathrm{e}}$.

Within the scope of this section it is assumed, the single tone large signal (LS) analysis from section 2.3 has been applied to find a steady state solution of the nonlinear problem. Hence, all large signals are known as a function of time and at interesting frequencies $2\pi f = \omega = 0, \ldots, h\omega, \ldots, H\omega$ and power levels. The solution depends on all embedding impedances, including the diode mounting structure, and the diode's characteristics. The time to frequency domain mapping of

[32] Aleksey Nikolaevich Krylov (1863–1945), Russian mathematician.

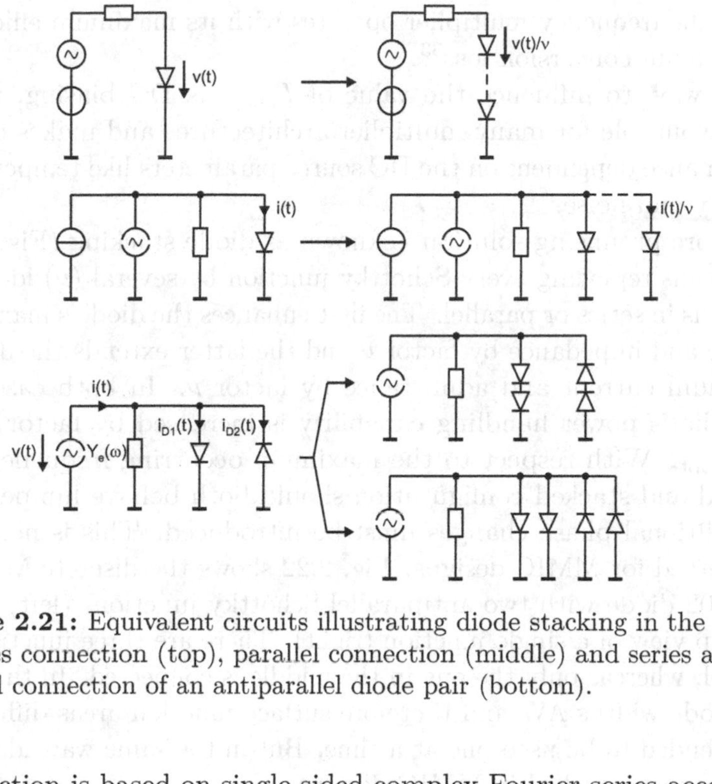

Figure 2.21: Equivalent circuits illustrating diode stacking in the form of series connection (top), parallel connection (middle) and series and parallel connection of an antiparallel diode pair (bottom).

this section is based on single sided complex Fourier series according to Eq. (2.83).

$$v(t) \circ\!\!-\!\!\bullet\ V_h \quad h = 0 \dots H$$
$$v(t) = \mathrm{Re}\left\{ \sum_{h=0}^{\infty} V_h \cdot e^{jh\omega t} \right\} \tag{2.83}$$

Diode Stacking To Influence Power Handling Capability In general, the diode's large signal impedance does not fit to the surrounding circuit at very low input power levels P_{in}. With increasing input power level the diode's impedance lowers until the best fit to the input and output embedding impedances is reached. At this input power level

$P_{\text{in,opt}}$ the frequency multiplier operates with its maximum efficiency or minimum conversion loss[33].

One way, to influence the value of $P_{\text{in,opt}}$ is DC biasing, which is unfavourable for many multiplier architectures and makes device performance dependent on the DC source parameters like temperature stability and noise.

A more promising solution is known as diode stacking (Fig. 2.21) and means replacing every Schottky junction by several (ν) identical junctions in series or parallel. The first enhances the diode's maximum voltage and impedance by factor ν and the latter extends the diode's maximum current and admittance by factor ν. In both cases the multiplier's power handling capability is increased by factor $\nu \rightarrow \nu \cdot P_{\text{in,opt}}$. With respect to the maximum occurring frequency, the original and stacked configuration should both behave lumped and no additional phase changes must be introduced. This is normally guaranteed for MMIC designs. Fig. 2.22 shows the discrete Metelics MGS802 diode with two antiparallel Schottky junctions (left) and a close up view of a single junction (right). There are three junctions in parallel, whereas only the one in the middle is connected. In this case the anode widths AW and therefore surface junction areas differ and are intended to be used one at a time. But in the same way, identical junctions are stacked in MMIC designs.

In the way, Fig. 2.22 illustrates, diode manufacturers offer different diode versions in the same package, to allow for diode stacking in hybrid designs. From section 2.2, the approximative[34] relations in Eq. (2.84) are known.

$$\begin{aligned} &I_{\text{S}}, C_{\text{j0}} && \sim S_{\text{j}} && R_{\text{s}} \approx R_{\text{epi}} && \sim N_{\text{Depi}} \\ &R_{\text{s}} \approx R_{\text{epi}} && \sim S_{\text{j}}^{-1} && C_{\text{j0}} && \sim N_{\text{Depi}}^{-1} \end{aligned} \tag{2.84}$$

[33]This holds true for every resistive frequency multiplier, although the occurring optimum condition does not necessarily equal the minimum achievable conversion loss.

[34]Saturation current depends on doping concentration and the actual metal-semiconductor combination as well, which is ignored in Eq. (2.84).

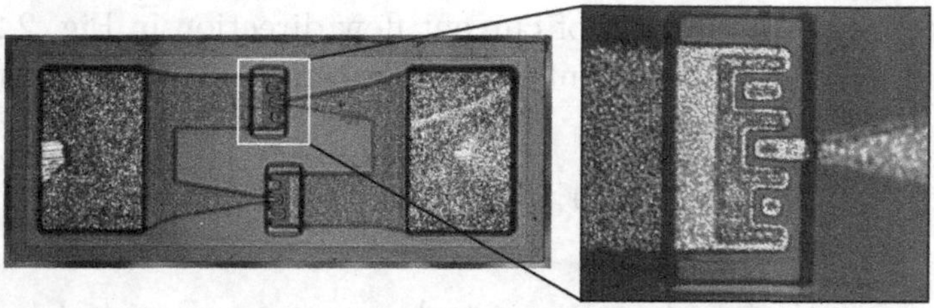

Figure 2.22: Photograph of discrete Aeroflex Metelics MGS802 diode with antiparallel configuration (left) and close up view of a single junction (right).

Hence, different anode surface areas (Fig. 2.22) lead to different but constant series resistance times maximum junction capacitance products $R_s C_{j0}$ = constant. In addition, for a given geometry S_j, increasing the doping concentration[35] allows for lowering R_s while increasing $C_{\mathrm{j}0}$ and vice versa. This is basically the same as it is with parallel and series connection of identical Schottky junctions.

Stacking several discrete diodes (Fig. 2.5) in hybrid designs on the same carrier material does allow for changing the power handling capability as well, but leads to a different electromagnetic configuration and usually requires a complete redesign of the entire frequency multiplier.

Antiparallel Diodes The frequency multipliers in sections 2.5, 2.6, 2.7 make use of the antiparallel diode configuration to allow for balanced output circuitry in case of frequency doublers and to exhibit inherent filter effects in case of frequency triplers. The latter are discussed in the following, based on the simple equivalent circuit of Fig. 2.21. The time to frequency domain mapping of currents, voltages and admittances is shown in Eq. (2.85).

$$i(t), v(t), i_{\mathrm{D}1}(t), i_{\mathrm{D}2}(t), y_{\mathrm{e}}(t) \circ\!\!-\!\!\bullet\; I_h, V_h, I_{\mathrm{D}1\,h}, I_{\mathrm{D}2\,h}, Y_{\mathrm{e}\,h} \quad h \in \{0, \dots, H\} \tag{2.85}$$

[35]Higher doping concentration does also increase reverse breakdown voltage.

Considering the indication of current flow direction in Fig. 2.21, an expression of the total current time waveform is given by Eq. (2.86).

$$i(t) = \mathrm{Re}\left\{\sum_{h=0}^{H} \left(Y_{\mathrm{e}\,h} V_h + I_{\mathrm{D1}\,h} + I_{\mathrm{D2}\,h}\right) e^{jh\omega t}\right\} \tag{2.86}$$

Assuming identical diodes, the spectral current components $I_{\mathrm{D1}\,h}, I_{\mathrm{D2}\,h}$ of both diodes are equal in magnitude but π-phase shifted.

$$I_{\mathrm{D2}\,h} + I_{\mathrm{D1}\,h} e^{-jh\pi} = 0 \tag{2.87}$$

Hence, at fundamental frequency and odd order harmonics $(2h+1)$, both components are equal in magnitude and phase (Eq. (2.88)).

$$\begin{aligned} I_{\mathrm{D2}\,2h+1} &= I_{\mathrm{D1}\,2h+1} \underbrace{(-1)e^{-j(2h+1)\pi}}_{e^{-j\pi-j2h\pi-j\pi}=1} = I_{\mathrm{D1}\,2h+1} \\ i_{\mathrm{odd}}(t) &= \mathrm{Re}\left\{\sum_{h=0}^{H} \left(Y_{\mathrm{e}\,2h+1} V_{2h+1} + 2I_{\mathrm{D1}\,2h+1}\right) e^{j(2h+1)\omega t}\right\} \end{aligned} \tag{2.88}$$

Whereas, at DC and even order harmonics $(2h)$ the diode current components cancel each other out, building a virtual short circuit at the common node (Eq. (2.89)). The even order currents circulate within the loop of the diodes and incorporate power loss, if there are resistive parts of the diode equivalent model.

$$\begin{aligned} I_{\mathrm{D2}\,2h} &= -I_{\mathrm{D1}\,2h} \quad V_{2h} = 0 \\ i_{\mathrm{even}}(t) &= \mathrm{Re}\left\{\sum_{h=0}^{H} \left(\underbrace{Y_{\mathrm{e}\,2h} V_{2h}}_{=0} + \underbrace{I_{\mathrm{D1}\,2h} - I_{\mathrm{D1}\,2h}}_{=0}\right) e^{j(2h)\omega t}\right\} = 0 \end{aligned} \tag{2.89}$$

Some textbooks (e.g. [7]) reduce multi diode frequency multipliers and mixers to single diode Thévenin or Norton equivalent circuits, as shown in Fig. 2.21, for comparison reasons. In the author's opinion, this is not a good idea. As explained before, the large signal quantities have to be determined by an iterative numerical method (section 2.3) and the solution depends on all embedding impedances and input drive level. In this sense, an antiparallel diode configuration with its inherent filter effects constitutes a totally different case compared to series, shunt or bridge mounted diode configurations. Anyway, the statements of the preceding paragraph about diode stacking hold true for more complex multi diode circuits, as well, if stacking preserves the original diode configuration. The bottom of Fig. 2.21 shows series and parallel connection of an antiparallel diode pair.

Antiparallel Diodes With Single Ended Stimulus And Balanced Load The frequency doublers of section 2.6 and section 2.7 are based on the simple equivalent circuit of Fig. 2.23. The diodes are antiparallel with respect to the source and therefore the explanations of the preceding paragraph hold true. Fundamental and odd mode $(2h+1)$ grounding is established by the transformer circuit[36]. The currents i_{odd} of Eq. (2.88) flow through $Y_{\mathrm{e\,in}\,2h+1}$ but are isolated from $Z_{\mathrm{e\,out}}$. $Y_{\mathrm{e}\,2h+1}$ is chosen to equal zero for frequency doubling. The DC and even mode currents circulate within the diode loop and build a virtual short circuit at the common node but also produce load currents through $Z_{\mathrm{e\,out}}$ at the secondary coil of the transformer. Hence the circuit in Fig. 2.23 is suitable to build even order frequency multipliers with inherent isolation between source and load.

[36]The fundamental and odd mode currents constitute an even mode excitation to the primary coil of the transformer and are therefore short circuited in the transformer's center tap. The reader should keep in mind, several planar balun realizations have open circuit behaviour at even mode excitation instead. In such cases, circuit modifications are necessary to establish grounding.

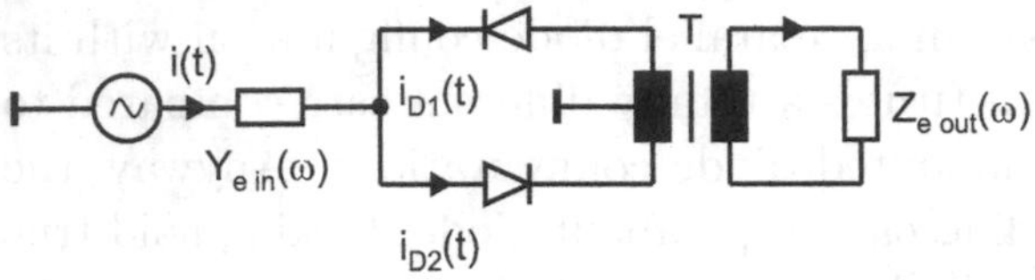

Figure 2.23: Simplified frequency doubler equivalent circuit with antiparallel diode configuration, single ended stimulus and balanced load.

2.4.2 Optimum Embedding Impedances

It is known from chapter 2.3 that all large signal quantities of any frequency multiplier circuit depend on all embedding impedances, the diode's characteristics and the input power level.

The manufactured frequency multipliers of sections 2.5, 2.6, 2.7 are planar realizations exclusively, which brings further restrictions into effect. The design flow is discussed in the next subsection.

In the following, general limitations of frequency doubler's and tripler's optimum performance with respect to the chosen embedding impedances are outlined. According to Fig. 2.24, a single diode Norton equivalent circuit of the frequency multiplier is discussed. A current source with an available power P_{in} (compare chapter 4) and shunt mounted, frequency dependent resistive $R(\omega)$ and reactive $X(\omega)$ elements forming the embedding network. In this context, the input, output and idle impedances are given by Eq. (2.90).

$$\begin{aligned} R_{\text{in}}, \quad X_{\text{in}} &= R(1 \cdot \omega), X(1 \cdot \omega) \\ R_{\text{out}}, \quad X_{\text{out}} &= R(d \cdot \omega), X(d \cdot \omega) \\ R_{\text{idle}}, \quad X_{\text{idle}} &= R(h \cdot \omega), X(h \cdot \omega) \quad h \in \mathbb{N}_0 \setminus \{1, d\} \end{aligned} \tag{2.90}$$

The results in Fig. 2.24 are not restricted to an exemplary scenario, but are of general significance. The available source power is the same in all cases. The minimum conversion loss $(P_{out}/P_{in})_{dB\,min}$[37] or maximum real power across R_{out}, turns out, if the input and output impedances are tuned to their optimum values. These optimum values are different for every chosen series resistance and junction capacitance. All other impedances are either infinite (open idlers) or zero (short idlers). These procedures allow for determination of general limitations, more realistic than the limits of chapter 2.1, but still difficult to realize, especially in case of ultrawideband frequency multipliers.

The top of Fig. 2.24 shows an almost linear dependency between the minimum conversion loss and the diode's series resistance. Short idlers lead to higher conversion loss because of current flow through the diode's series resistance at the idle frequencies. The difference, compared to open idlers is about 2 dB in case of frequency doublers and about 5 dB in case of triplers. Both, frequency doubler's and tripler's minimum conversion loss, show little dependency of the diode's maximum junction capacitance if all idle impedances are short circuits. A slight increase is observed with open idlers. The bottom of Fig. 2.24 depicts the input and output impedances, necessary to achieve minimum conversion loss. The results of Fig. 2.24 correspond to a single frequency and are different at each frequency. There are planar design techniques, which allow for providing real source and load impedances to the diode, including mounting structure, over large bandwidth (compare Bode[38]-Fano[39] criterion). Whereas realization of defined reactive impedances X_{in}, X_{out} is restricted to narrow operating bandwidth. R and X are shunt mounted components, therefore the higher the reactive components are, the more likely they can be ignored. Hence, ultrawideband realizations of frequency multipliers are more easily achieved if the idle impedances are set to zero. Although the minimum conversion loss values are higher than it is with open idlers.

[37]Output power is given by $P_{out} = 1/2\,\mathrm{Re}\left\{R_{out} \cdot I^2(d \cdot \omega)\right\}$ if complex phasors are related to peak amplitudes.

[38]Hendrik Wade Bode (1905–1982), American engineer.

[39]Robert Mario Fano (1917), Italian engineer.

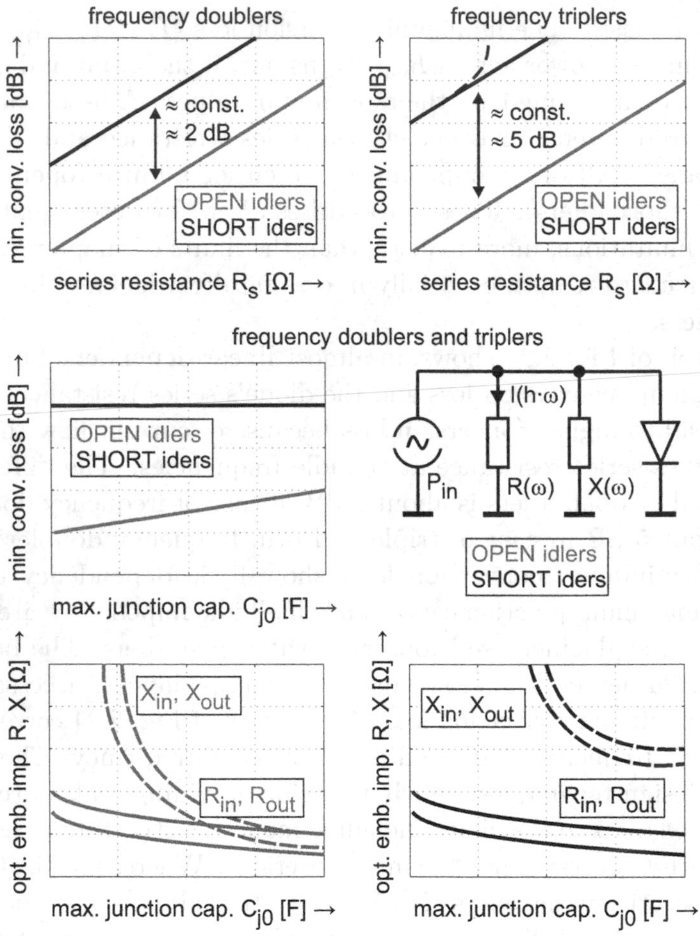

Figure 2.24: Performance dependency of frequency doublers and frequency triplers from open circuit (red) and short circuit embedding / idle impedances (black). Minimum conversion loss versus series resistance R_s (top), minimum conversion loss versus maximum junction capacitance C_j0 (middle) and optimum input $R_\mathrm{in}, X_\mathrm{in}$ and output embedding impedances $R_\mathrm{out}, X_\mathrm{out}$ versus C_j0 (bottom).

2.4.3 Design Flow

Fig. 2.25 illustrates the design flow of frequency multiplier synthesis and the correlation of several optimization objectives, which is mostly self-explaining. The following enumeration contains the most important requirements on the input and output circuitry in descending order.

1. Input: passband behaviour at 1st harmonic and open / short circuit behaviour at desired output harmonic.
2. Output: passband at desired output harmonic and open / short circuit behaviour at 1st harmonic.
3. Reflection phases from diodes to input / output circuitry below 180 ° to avoid destructive interference of 1st and desired harmonic.
4. Find compromise between 1st harmonic rejection and additional loss[40] of output filter.
5. Find compromise between input and output circuit symmetry (linear circuit part, not diode symmetry) and rejection of undesired harmonics.
6. Manipulate 4th harmonic idle impedance in case of doublers and 5th harmonic in case of triplers for minimum conversion loss (not possible for ultrawideband designs due to overlapping frequency ranges).
7. Tune other idle impedances.

Minimum conversion loss strongly depends on the operating frequency range and instantaneous bandwidth. For ultrawideband frequency doublers / triplers 10 dB / 15 dB conversion loss constitute excellent results. To achieve such values, the design must not violate

[40]Prediction of loss is best derived from thruline measurements and experiences, rather than simulations. The latter always underestimate real loss mechanisms.

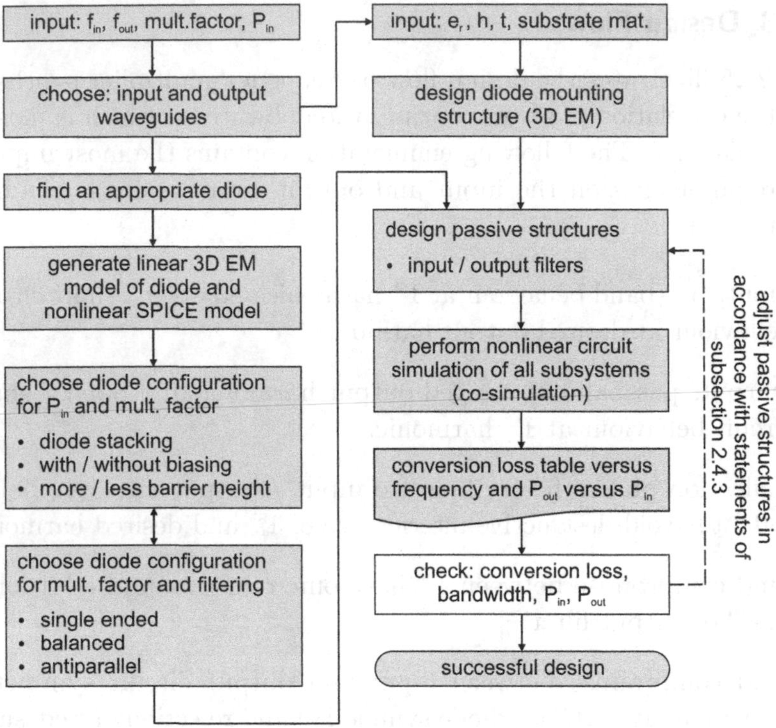

Figure 2.25: Flow chart illustrating the design flow of frequency multiplier synthesis.

one of the first three points in the above list. These requirements underly mutual influences and do also influence the idle impedances at undesired harmonics. These explanations together with facts of the preceding subsection make clear, optimizing conversion loss behaviour by defined tuning of single idle impedances is of importance for narrowband frequency multipliers only.

2.5 Octave Bandwidth Tripler for 20 to 40 GHz

Superheterodyne transceivers for synthetic automatic test systems require high frequency phase locked stimulus at the local oscillator mixer ports. Especially systems with high instantaneous signal bandwidth as shown in Fig. 4.34 of section 4.4 or Fig. 1.4 of chapter 1 make use of many different mixing cases. Fundamental, subharmonic, upper and lower sideband mixing is often utilized in a single module to ensure best spurious tone suppression at different input / output frequency ranges. This section presents a frequency tripler for 20 to 40 GHz, as it is required (LO) to downconvert signals below 20 GHz to several intermediate frequencies in lower sideband mixing configuration $\langle \mathrm{RF}, \mathrm{LO} \rangle = \langle -1, 1 \rangle$.

The presented tripler is an improved version of the author's earlier work [G2]. An equivalent circuit and photographs of the utilized antiparallel Schottky diodes from VDI are shown in Fig. 2.26. The basic circuit idea is from Albin, who reported a tripler design with unilateral finlines on fused silica substrate for Hewlett Packard source modules in 1988 [40]. The assembled diodes offer negligible deviations of both Schottky junctions. The parameters of a single junction are $I_\mathrm{S} = 157.5$ fA, $R_\mathrm{s} = 2.6\ \Omega$, $\eta = 1.175$ and $C_{\mathrm{j0}} = 0.07$ pF.

The tripler is manufactured on woven fiberglass reinforced, ceramic filled, PTFE based composite material AD600 $\epsilon_\mathrm{r} = 6.15$ from Arlon Micowave Materials [41]. Standard PCB processing is applied. Fig. 2.27 depicts a 3D model of the tripler for EM simulation. A transition from 1.85 mm coaxial connector to grounded coplanar waveguide (gCPW) is used at the circuit input and output. The fundamental signal in gCPW mode enters an 11th order stepped impedance gCPW lowpass filter (LPF). Simulated scattering parameters of the LPF are given in Fig. 2.28. The first and last filter elements are shunt capacitances to ensure short circuit behaviour at stopband frequencies, as it is required for this circuit topology (Fig. 2.26). Fig. 2.29 illustrates the cross-sectional views of the low (C) and high (L) impedance gCPW, magnitude of the electric field strengths at 18 GHz and real parts of the characteristic impedances $\mathrm{Re}\,(Z_{\mathrm{gCPW}})$ of 50 Ω gCPW,

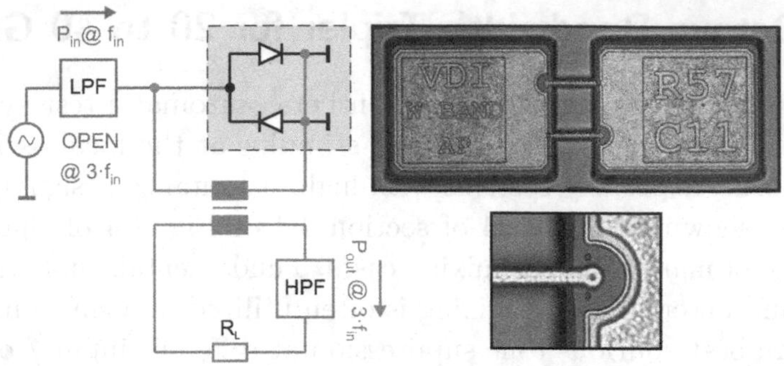

Figure 2.26: Equivalent circuit of the proposed 20 to 40 GHz frequency tripler and a photograph of Virginia Diodes Inc. (VDI) discrete anti-parallel Schottky diode with close up view of the anode finger. Diode dimensions are (250×600×100) µm^3.

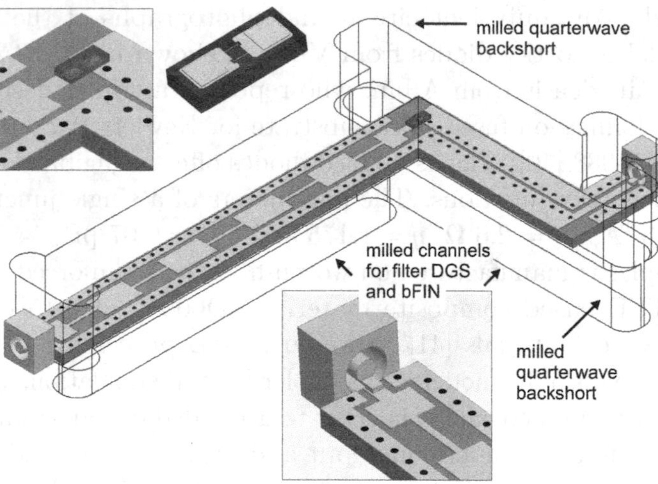

Figure 2.27: 3D model of the tripler for EM simulation and close up views of diode and transition to coaxial waveguide.

high and low gCPW versus frequency. Table 2.2 summarizes the filter parameters.

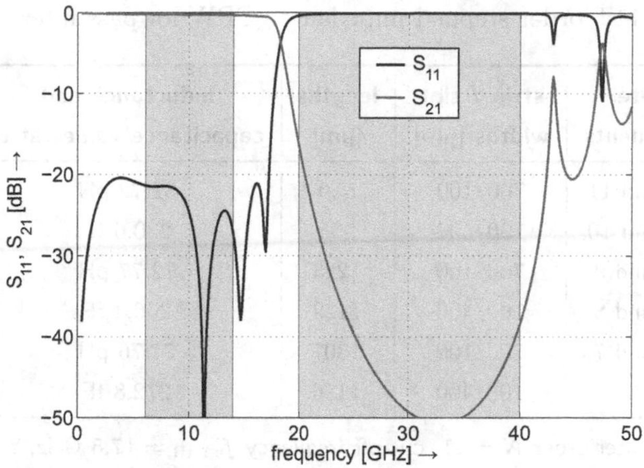

Figure 2.28: Simulated transmission (black) and reflection coefficients (red) of 11th order stepped impedance gCPW lowpass filter.

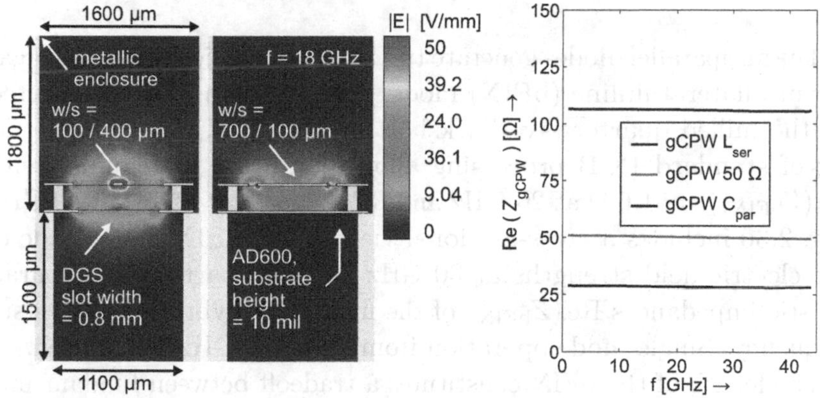

Figure 2.29: Cross-sectional views of the low (C) and high (L) impedance gCPW, magnitude of the electric field strengths at 18 GHz and real parts of the characteristic impedances $\mathrm{Re}(Z_{\mathrm{gCPW}})$ of 50 Ω gCPW, high and low gCPW versus frequency. $\mathrm{Re}(Z_{\mathrm{gCPW-L}}) = 106.2\ \Omega, \mathrm{Re}(Z_{\mathrm{gCPW-C}}) = 28.5\ \Omega$ and the corresponding effective permittivities are $\epsilon_{\mathrm{reff-L}} = 3.00, \epsilon_{\mathrm{reff-C}} = 4.07$ at $f = 18$ GHz.

Table 2.2: 11th order stepped impedance gCPW lowpass filter

filter elements	strip / slot widths [μm]	lengths [μm]	inductance and capacitance values at f_c
1 and 11	700/100	550	32.72 pH
2 and 10	100/400	935	220.6 fF
3 and 9	700/100	1223	72.77 pH
4 and 8	100/400	1129	266.4 fF
5 and 7	700/100	1307	77.76 pH
6	100/400	1156	272.8 fF

filter order $N = 11$, cut-off frequency $f_{c\,1\,\mathrm{dB}} = 17.5$ GHz, $f_{c\,3\,\mathrm{dB}} = 18.1$ GHz, passband ripple $r_{\mathrm{dB}} = 0.01$ dB, stopband rejection of 25 dB at $1.2 \times f_{c\,3\,\mathrm{dB}}$.

The antiparallel diodes generate two third harmonic signals propagating in bilateral finline (bFIN) mode. One of them is back reflected by the milled quarterwave backshort of Fig. 2.27. The design rule set of standard PCB processing allows for bFIN waveguides with $\mathrm{Re}\,(Z_{\mathrm{bFIN1}}) = 51.6\ \Omega$ at 20 GHz and $\mathrm{Re}\,(Z_{\mathrm{bFIN1}}) = 42.8\ \Omega$ at 40 GHz. Fig. 2.30 includes a cross-sectional view of the bFIN, magnitude of the electric field strengths at 30 GHz and real parts of the characteristic impedances $\mathrm{Re}\,(Z_{\mathrm{bFIN}})$ of the first three hybrid modes versus frequency. Single mode operation from 20 to 40 GHz is achieved.

The length of the bFIN constitutes a tradeoff between fundamental signal rejection and third harmonic insertion loss. After the bFIN highpass filter, another transition to gCPW (Fig. 2.31) and 1.85 mm coaxial connector is utilized. Simulated scattering parameters of the transition from bFIN to gCPW are shown in Fig. 2.32. The fully assembled frequency tripler with its Au plated brass housing is depicted in Fig. 2.33.

The measured 3rd harmonic output power levels P_{out} at the output frequencies $f_{\mathrm{out}} = 21.4,\ 25,\ 32,\ 40,\ 43$ GHz versus the fundamental

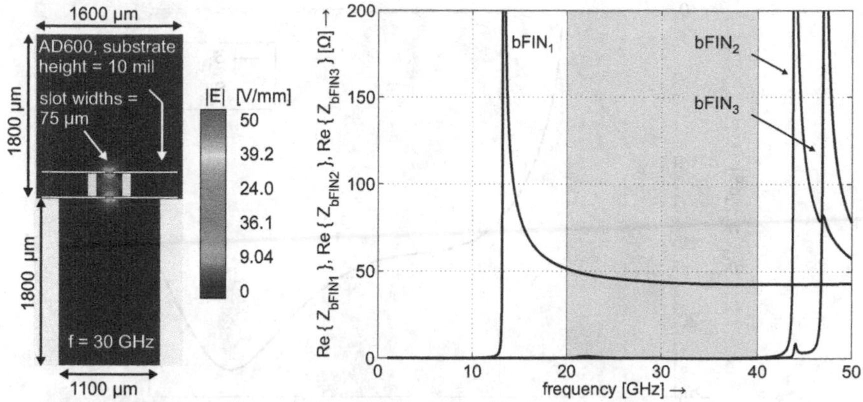

Figure 2.30: Cross-sectional view of the bilateral finline, magnitude of the electric field strengths at 30 GHz and real parts of the characteristic impedances Re (Z_{bFIN}) of the first three hybrid modes versus frequency.

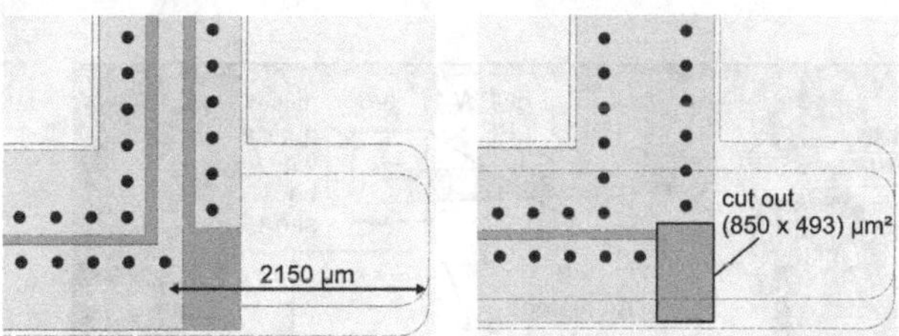

Figure 2.31: Top (left) and bottom (right) view of transition from gCPW to bilateral finline.

input power level P_{in} are given by Fig. 2.34. All measurement data from 21.4 to 40 GHz are within the gray shaded area. At center frequency a saturated output power level of -5 dBm is achieved.

Fig. 2.35 compares simulated and measured conversion loss values versus output frequency f_{out} at three input power levels P_{in} = 5.6, 7.6, 9.6 dBm. Results from co-simulation (section 2.4) are in reasonable agreement with measurement results. The measured conversion loss values are between 16 and 20 dB from 21 to 40 GHz. The frequency tripler is designed to substitute the MMIC tripler TGC1430G

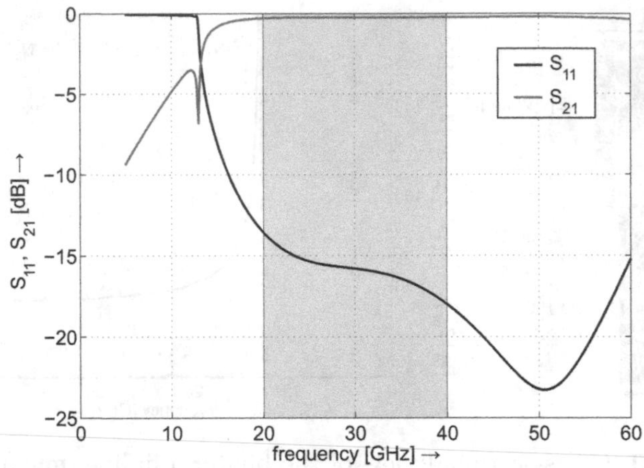

Figure 2.32: Simulated transmission (red) and reflection coefficients (black) of the transition from gCPW to bFIN, which is used for highpass filtering.

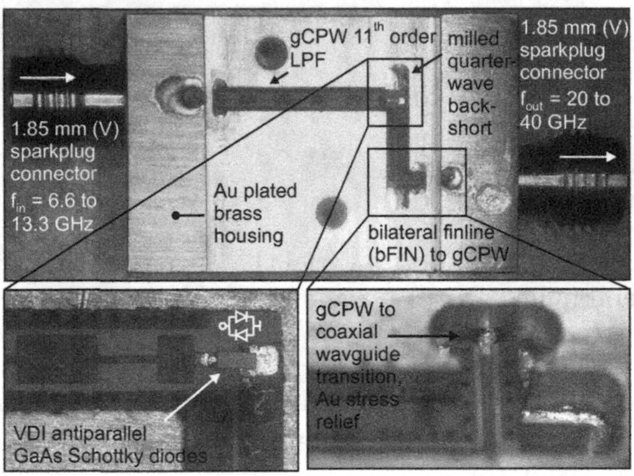

Figure 2.33: Photographs of the assembled frequency tripler with close up view of the diode mounting structure and transition from gCPW to coaxial connector.

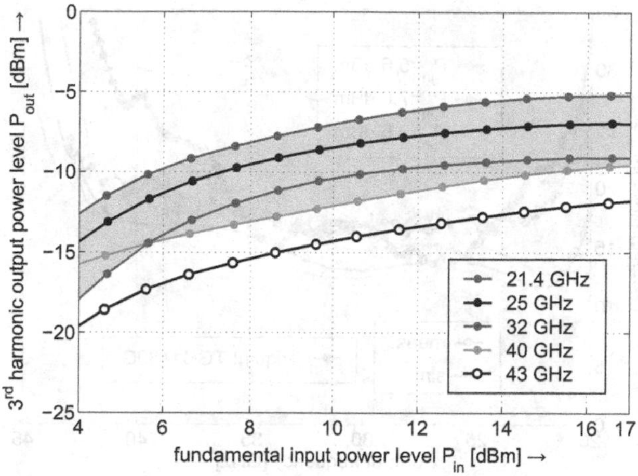

Figure 2.34: Measured 3rd harmonic output power levels P_{out} at the output frequencies $f_{\text{out}} = 21.4,\ 25,\ 32,\ 40,\ 43$ GHz versus the fundamental input power level P_{in}. All measurement data from 21.4 to 40 GHz are within the gray shaded area.

from TriQuint Semiconductor [42]. Compared to TGC1430G (blue curve), the proposed tripler offers larger bandwidth with similar conversion efficiency. The reader should keep in mind, measurement results of the proposed tripler include insertion loss of the coaxial connectors and transitions to planar waveguide. Whereas the blue curve in Fig. 2.35 corresponds to bare die measurements.

Measured spectral output power levels up to the 5th harmonic versus output frequency f_{out} at input power level $P_{\text{in}} = 5.6$ dBm are shown in Fig. 2.36. Rejection of the even order harmonics is higher than 30 dB from 21 to 40 GHz. As a consequence of the antiparallel diode configuration, suppression of the 5th harmonic is quite poor. A filter bank of only two BPFs in cascade can fix this problem. The fundamental rejection is directly connected to the length of the bFIN waveguide. At the expense of conversion loss higher fundamental rejection is possible.

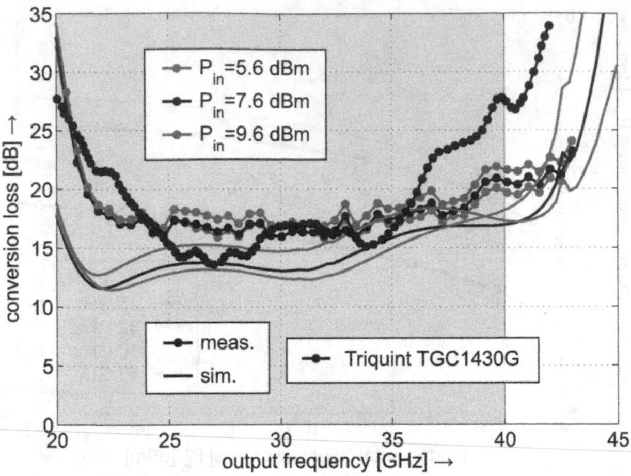

Figure 2.35: Simulated and measured conversion loss versus output frequency f_{out} at three input power levels $P_{\text{in}} = 5.6$, 7.6, 9.6 dBm. The blue curve belongs to the commercial MMIC tripler TGC1430G from TriQuint Semiconductor.

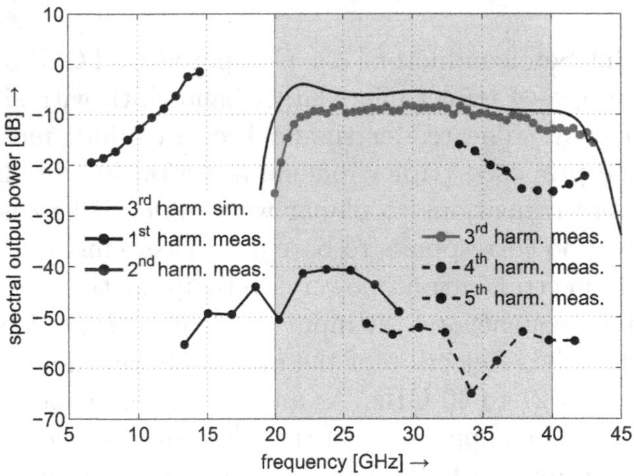

Figure 2.36: Measured spectral output power levels up to the 5th harmonic versus output frequency f_{out} at input power level $P_{\text{in}} = 5.6$ dBm. Compared to simulated 3rd harmonic output power levels (black, without markers).

2.6 Frequency Doublers and Triplers for 50/60 to 110 GHz

Although there is a persistent tendency to higher operating frequencies of direct signal generation concepts, frequency multipliers are still used to provide phase locked signal stimuli at millimeter-wave frequencies and beyond. Operation over preferably large bandwidths is especially required in system design of frequency extension modules for vector network analyzers (VNA), microwave signal generators and front end modules of semiconductor automatic testsystems (ATS). Varactor based multipliers have much better power efficiency, but they can hardly serve the bandwidth requirements (2.6). There is a constant output power level times output frequency range product ($P_{\text{out}} \cdot \Delta f_{\text{out}} = \text{const.}$) with varactor multipliers, which restricts the broadband varactor multipliers to rather low output power levels. Monolithic integrated circuit (MIC) technology is ideally suited to design active and passive varactor and varistor multipliers for the aforementioned fields of application [43–45], not least because of the opportunity to integrate the diode or transistor architecture into the optimization procedure of the multiplier design [46–49]. Another advantage is the integration of amplifier stages in cascade on the same die [50–52]. Nevertheless, hybrid designs offer advantages with respect to the achievable fundamental signal isolation, filter performance and manufacturing cost in case of low to medium quantities. The modules used in the envisaged fields of application have high complexity and great functionality. Therefore, multi-chip module realizations rather than fully MIC constructions are preferred. Ease of maintenance, module cost and thermal engineering reasons determine which component is best integrated as MIC, hollow waveguide component or hybrid planar device.

In the following, we focus on three planar resistive frequency triplers (#1, #2, #3, $f_{\text{out}} = [60, 110]$ GHz) and two frequency doublers (#4, #5, $f_{\text{out}} = [50, 110]$ GHz) based on commercially available gallium arsenide (GaAs) Schottky diodes and conventional thin-film processed alumina (Al_2O_3) substrate with 5 µm Ni/Au metallization [I6, J2]. Synthesis is based on a co-simulation procedure between 3D electromagnetic field and nonlinear harmonic balance circuit simulations as explained in subsection 2.4.3.

If common substrate widths are used, restricted by dicing technology, 110 GHz constitutes almost the maximum operating frequency for planar designs on 5 mil Al_2O_3 substrates due to propagation of higher order modes. The reader should further keep in mind, plated-through vias establish diode ground connection. Hence, the mechanical length of this ground path equals the sum of substrate thickness and half of the via pad diameter. At higher frequencies, low permittivity substrates (e.g. fragile quartz SiO_2, BCB or PFTE) with thicknesses $h \leq 5$ mil have to be used (section 2.7). Innovative planar circuit designs are presented to overcome these restrictions. Different architectures are presented to account for the individual demands associated with multi-chip module designs.

Scalar and spectral measurement data over the focused output frequency ranges are presented. Conversion loss values around 18 dB from 60 to 95 GHz and below 22 dB from 60 to 110 GHz are achieved with frequency tripler #1. Frequency doubler #5 achieves conversion loss values below 15 dB from 50 to 89 GHz and below 22 dB from 50 to 110 GHz. A comprehensive comparison of the presented work with reported frequency multipliers is included (Table 2.6).

Schottky Junction Modelling and Co-Simulation Procedure The proposed multipliers are optimized for use with the United Monolithic Semiconductors (UMS) DBES105a diode, which is a discrete component with two Schottky junctions in series tee configuration from the UMS buried epitaxial layer (BES) process with anode finger dimensions of (5 µm × 1 µm). Each solder pad is equipped with two gold bumps ($\varnothing = 20$ µm). Fig. 5.14 depicts a 3D model for electromagnetic field (EM) simulation of the utilized diode. Within the proposed multiplier design, the junction geometry is small compared to the minimum wavelength $\lambda_{\text{min}} = c_0/(\sqrt{\epsilon_{\text{reff}}} \cdot f_{\text{max}})$, which allows for lumped port modelling of the junctions in 3D EM simulation.

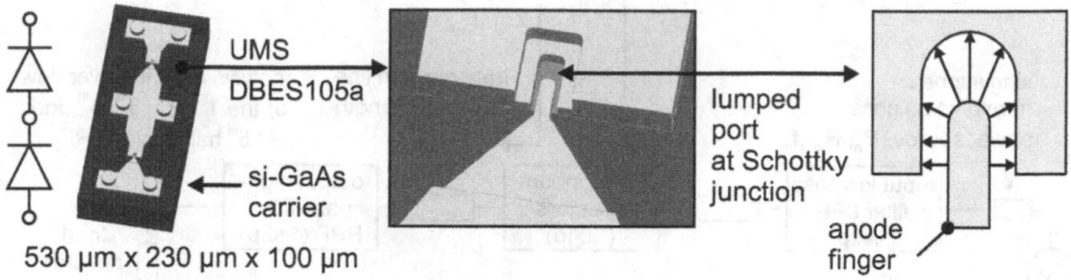

Figure 2.37: 3D EM model of the utilized commercial GaAs Schottky diode UMS DBES105a with close up view of lumped port Schottky junction modelling.

The multiplier synthesis is based on co-simulation between finite element 3D EM simulation in the frequency domain[41] and nonlinear harmonic balance circuit simulation[42]. As long as the chosen architectures allow for partitioning of the multiplier into a preferably high number of subsystems, which can be synthesized and optimized individually, the simulation time is significantly reduced. These subsystems are various required waveguide transitions, the input lowpass filter (LPF), the diode mounting structure including a 3D EM model of the diode and the output highpass filter (HPF). Accurate 3D modelling of the diode, including the semi-insulating GaAs carrier and the first metallization layer, allows for sufficient prediction of parasitic effects, such as pad to pad capacitance and harmonic generation due to asymmetries. Within the harmonic balance circuit simulator, the linear parts are connected with a rather simple circuit model of the Schottky junctions, Berkeley SPICE: DIODE, according to Fig. 2.38. The junction parameters are series resistance R_{S}, junction capacitance at zero bias C_{j0}, ideality factor η and saturation current I_{S}.

A more sophisticated junction modelling approach, e.g. [53], would be promising, but it requires disproportionate measurement effort or dimensions and material parameters of the Schottky junction that are difficult to obtain in case of proprietary diodes. With all

[41] Ansys, High Frequency Structure Simulator (HFSS™)
[42] Agilent, Advanced Design System (ADS™)

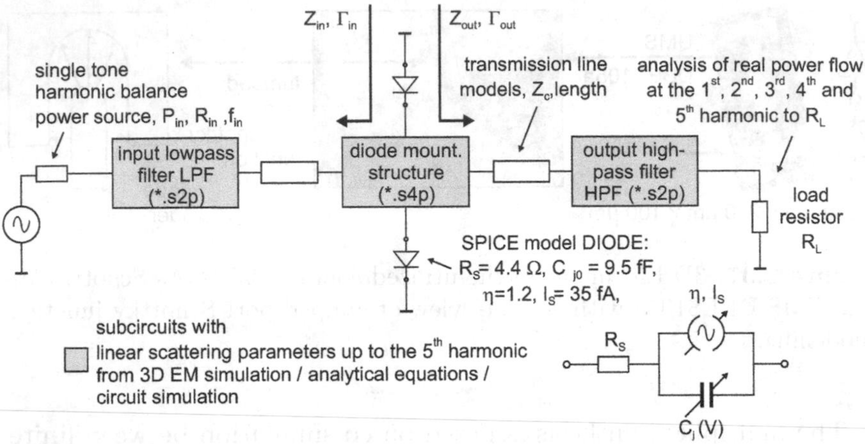

Figure 2.38: Block diagram of the harmonic balance schematic for frequency multiplier synthesis.

subsystems connected in a parameterized co-simulation, the multipliers conversion efficiency is optimized by tuning the embedding impedances $Z_{\text{in}}(f)$, $Z_{\text{out}}(f)$ at all involved harmonics [54]. The input lowpass filter in Fig. 2.38 provides resistive input impedance, which allows for ultrawideband operation, contrary to reactive matching with severe bandwidth limitations. To account for the strong near-field interaction of all subcircuits, it is necessary to perform a final simulation with all subsystems combined within the 3D EM simulation. Without the use of parallelization algorithms, such a 3D EM simulation takes less than two hours using an ordinary personal computer. The broadband frequency multipliers of this chapter [G2, I6, I7] are based on the outlined co-simulation approach. The same holds true for the triple balanced mixer of chapter 4 [I4] and the power detectors of chapter 5 [I8].

Waveguide Transitions The planar multipliers are intended to be used within highly integrated front end modules and are normally not directly connected to the module's output. Hence the multipliers

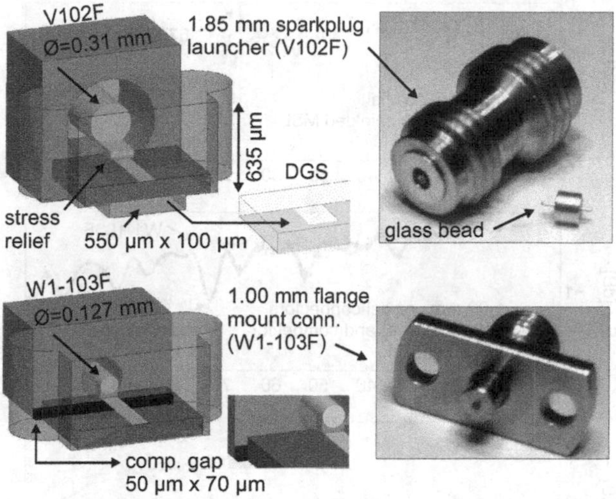

Figure 2.39: 3D EM models of transitions from 1.85 mm / 1.00 mm coaxial connectors to shielded microstrip line.

have to be connected to MICs or hollow waveguides or other planar substrates. Transitions to 1.85 mm and 1.00 mm coaxial waveguide are designed to ease measurement characterization. The proprietary coaxial connectors V102F and W1-103F from Anritsu GmbH and the corresponding transitions are shown in Fig. 2.39. The parasitics from both transitions can be modelled as shunt capacitances in equivalent circuits. Compensation in case of the 1.85 mm connector with stress relief is realized with a rectangular shaped defected ground structure (DGS). The equivalent shunt capacitance of the 1.00 mm connector is much smaller. A compensation gap of 50 µm width and 70 µm height is sufficient. The 3D EM models include the transitions from shielded microstrip line (MSL) to the first air-filled coaxial waveguide section, not the entire 1.85 mm and 1.00 mm connector. Fig. 2.39 further shows the dimensions of the milled channels that are chosen to avoid parasitic higher order modes like TE_{10} up to the maximum operating frequency.

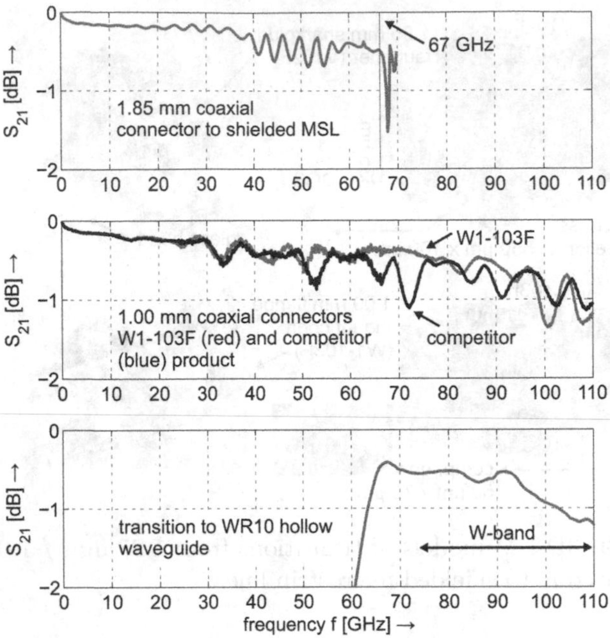

Figure 2.40: Measured magnitude of transmission coefficient S_{21} of single 1.85 mm connector, two different single 1.00 mm connectors and a single transition to WR-10 waveguide versus frequency.

There is no complete 3D model of the proprietary coaxial connectors available, therefore back to back measurements have been performed. Together with on-wafer measurements after enhanced-line-reflect-reflect-match (eLRRM) calibration [55] of the utilized Al_2O_3 thrulines, scattering parameters of the single connectors have been extracted. According to Fig. 2.40, insertion loss of the 1.85 mm connector is below 1 dB up to 67 GHz. Results from two different 1.00 mm connectors show insertion loss values below 1.5 dB from DC to 110 GHz (Fig. 2.40).

The 1.00 mm coaxial connector constitutes a costly but ultrawideband interconnect solution, covering the frequency range from DC to 110 GHz. If smaller output frequency ranges are acceptable, cost

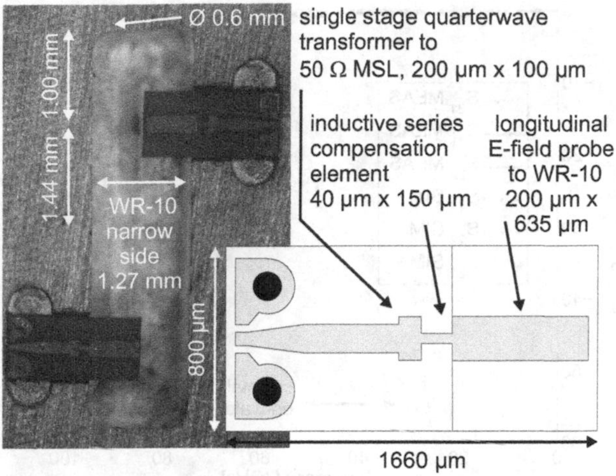

Figure 2.41: Photograph of MSL to WR-10 transition for back to back on-wafer measurement (second half of split-block not shown) and close up view of the planar transition.

effective hollow waveguides are preferred. Therefore longitudinal E-field probes to WR-10 waveguide have been developed. The transition in Fig. 2.41 allows for bondwire interconnection to the proposed frequency multipliers, other MICs or planar substrates. It is a split block construction, whereas Fig. 2.41 only shows the lower half of the mechanical housing. The E-field probe has half the length of the WR-10 hollow waveguide's narrow wall side. A large width of the probe is preferred to ensure high coupling, but it is limited by higher order modes. The probe's input impedance has a shunt capacitive behaviour. In conjunction with an inductive series compensation element, the remaining resistive input impedance is slightly below 50 Ω. A single stage quarterwave transformer allows for matching to 50 Ω. Dimensions are given in Fig. 2.41. The distance from E-field probe to the milled hollow waveguide backshort and the diameter of the milling tool determine the center frequency and influence the capacitance value of the probe. In back to back configuration with 2.88 mm WR-10 waveguide, a total insertion loss below 2.5 dB and

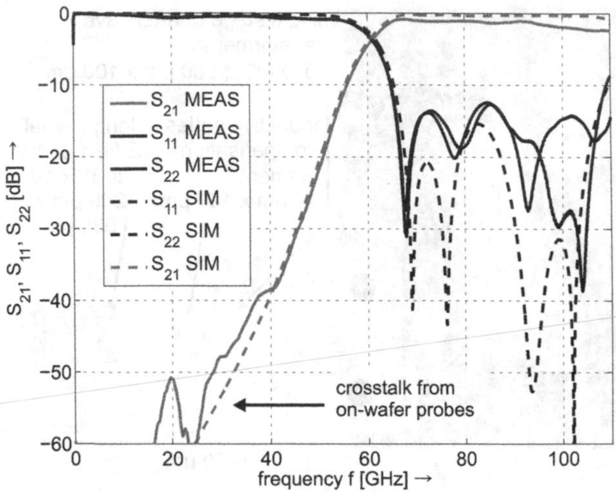

Figure 2.42: Scattering parameters of MSL to WR-10 transition in back to back configuration from 3D EM simulation and on-wafer measurement after eLRRM calibration versus frequency.

return loss greater than 13 dB from 63 to 110 GHz could be achieved (Fig. 2.42), which is an excellent result. Many similar transitions have been reported, e.g. [56], achieving less bandwidth and less return loss while incorporating comparable insertion loss values. With an Al_2O_3 substrate width of 0.8 mm, the same transition is easily redesigned for WR-15 and WR-12 waveguides. Lower permittivity and less substrate width is necessary to build similar transitions to WR-6 and WR-3 (section 2.7). Scattering parameters of a single transition have been extracted (Fig. 2.40). Insertion loss values are below 1.2 dB from 63 to 110 GHz.

Input Lowpass Filter Design Reflection of the generated harmonics at the circuit input of the proposed frequency multipliers is achieved by an 11th order MSL stepped impedance LPF (Fig. 2.43) with 1 dB cut-off frequency at $f_{c\,1\,\text{dB}} = 52.5$ GHz and $f_{c\,3\,\text{dB}} = 54$ GHz. The design is based on the analytical synthesis procedure from Matthaei

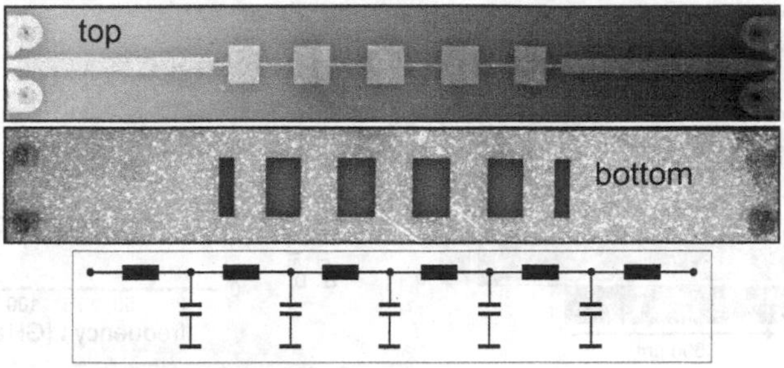

Figure 2.43: Photograph and circuit schematic of the 11th order stepped impedance MSL LPF filter. Top housing is not shown.

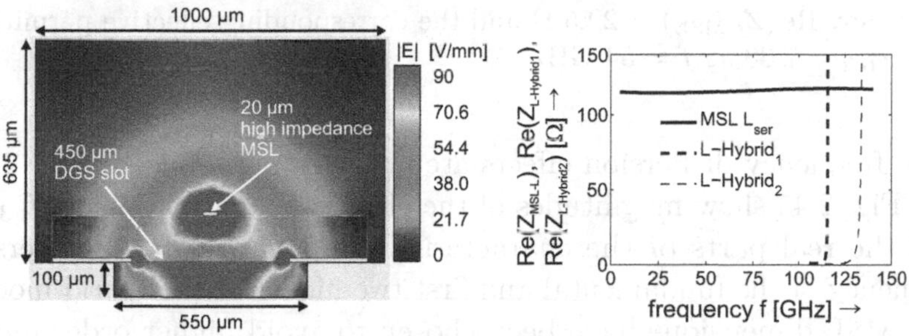

Figure 2.44: Cross-sectional view of the high impedance MSL with DGS, magnitude of the electric field strengths at 54 GHz and real parts of the characteristic impedances $\mathrm{Re}\,(Z_c)$ of MSL and the first two hybrid modes versus frequency. $\mathrm{Re}\,(Z_{c\,\mathrm{MSL}}) = 118.0\ \Omega$ and the corresponding effective permittivity is $\epsilon_{\mathrm{reff}} = 4.56$ at $f = 54$ GHz.

[57] with passband ripple of 0.01 dB, corresponding to 26 dB passband return loss, and stopband rejection of 25 dB at $1.2 \times f_{c\,3\,\mathrm{dB}} = 64.8$ GHz. The first and last filter elements are inductive to achieve open circuit stopband behaviour. The necessary inductance and capacitance values, extracted from the lowpass prototype filter at $f_{c\,3\,\mathrm{dB}}$, are summarized in column four of Table 2.3. 2D EM simulations in the frequency domain of the high (Fig. 2.44) and low (Fig. 2.45) impedance MSL cross-sections allow for extraction of effective permittivities ϵ_{eff} and real parts of the characteristic impedances $\mathrm{Re}\,(Z_c)$ at $f_{c\,3\,\mathrm{dB}}$. The

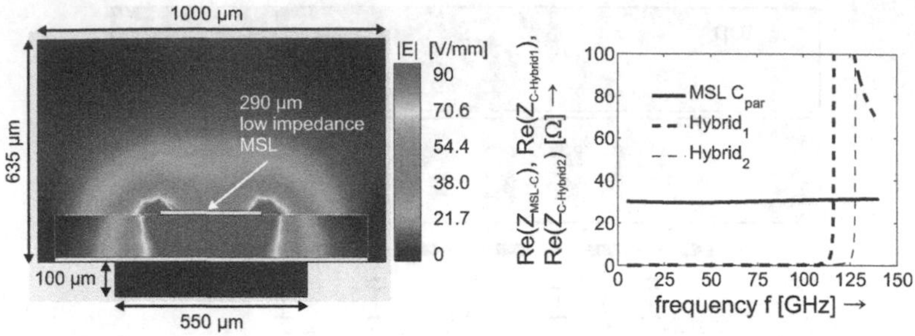

Figure 2.45: Cross-sectional view of the low impedance MSL, magnitude of the electric field strengths at 54 GHz and real parts of the characteristic impedances $\mathrm{Re}\,(Z_c)$ of MSL and the first two hybrid modes versus frequency. $\mathrm{Re}\,(Z_{c\,\mathrm{MSL}}) = 29.5\ \Omega$ and the corresponding effective permittivity is $\epsilon_{\mathrm{reff}} = 7.38$ at $f = 54$ GHz.

MSL frequency dispersion effects are not very pronounced. Fig. 2.44 and Fig. 2.45 show magnitudes of the electric field strengths at $f_{c\,3\,\mathrm{dB}}$ and the real parts of the characteristic impedances $\mathrm{Re}\,(Z_c)$ versus frequency of the fundamental and first two higher order hybrid modes. The MSL dimensions have been chosen to avoid higher order mode propagation up to 110 GHz. Hence, there is no need for ferrite absorber material within the milled channel. Due to the distributive nature of MSL, there is a second parasitic passband. The peak isolation and the achievable bandwidht of stopband increase with an increasing ratio of the high and low characteristic impedances $\mathrm{Re}\,(Z_{c\,\mathrm{L}})\,/\mathrm{Re}\,(Z_{c\,\mathrm{C}})$. To achieve a ratio of 4 and therefore greater than 20 dB stopband rejection from 60 to 110 GHz, the high impedance sections have been designed with 20 µm strip width and 450 µm DGS slot. To overcome insufficient peak isolation and bandwidth of stopband of preselector filters in ATS front end modules, several LPFs are arranged in cascade (compare Fig. 4.38). In case of ultrawideband frequency multipliers, we have to further design for preferably low reflection phase min $(\arg S_{11})$ to avoid destructive interference of the incident desired harmonic with the backscattered one. Therefore, only filters with inherently good stopband behaviour are of interest. The analytical synthesis results in

Table 2.3: 11th order stepped impedance MSL lowpass filter

filter elements	strip widths [µm]	lengths [µm]	inductance and capacitance values at f_c
1 and 11	20	126	105.8 pH
2 and 10	290	236	72.4 fF
3 and 9	20	281	236.0 pH
4 and 8	290	285	87.5 fF
5 and 7	20	301	252.8 pH
6	290	292	89.6 fF

filter order $N = 11$, cut-off frequency $f_{c\,1\,\mathrm{dB}} = 52.5$ GHz, $f_{c\,3\,\mathrm{dB}} = 54$ GHz, passband ripple $r_{\mathrm{dB}} = 0.01$ dB, stopband rejection of 25 dB at $1.2 \times f_{c\,3\,\mathrm{dB}}$.

LPFs with slightly different cut-off frequency. A single scaling factor $0.5 < \ell_{\mathrm{scale}} < 1.5$ for the lengths of the inductive and capacitive filter elements within parameterized 3D EM model is sufficient to map the 1 dB cut-off frequency to 52.5 GHz. The final dimensions are summarized in Table 2.3.

3D EM simulated scattering parameters including the transition to 1.85 mm connector from Fig. 2.39 are compared to on-wafer measurements after eLRRM calibration in Fig. 2.46. Passband return loss values are greater than 20 dB. The parasitic passband and higher order mode effects only occur at frequencies greater than 120 GHz. The isolation discrepancies at around 95 GHz are due to crosstalk of the on-wafer probes. These results have been checked with thru-reflect-line (TRL) calibration [58], [B1]. To underpin suitability of the designed LPF, the gray shaded areas in Fig. 2.46 indicate the input $f_{\mathrm{in}} = [20, 36.67]$ GHz and output frequency ranges $f_{\mathrm{out}} = [60, 110]$ GHz of frequency tripler #1 to #3. To build frequency doublers with the same output frequency range, a filter order of at least $N = 15$ is required to establish passband behaviour at the highest input fre-

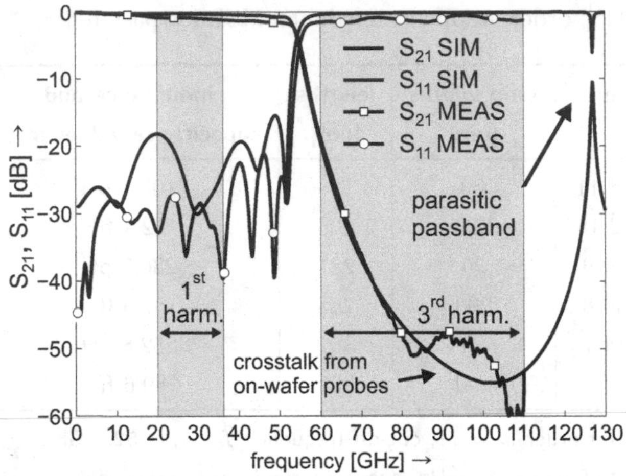

Figure 2.46: Scattering parameters of the 11th order LPF from 3D EM simulation (including transition to 1.85 mm) and on-wafer measurement after eLRRM calibration versus frequency.

quency along with sufficient stopband rejection at the lowest output frequency. However, within the applied dicing process the maximum substrate length of 6 mm does not allow for higher filter orders. Utilizing the presented LPF for frequency doubling (subsections 2.6.4 and 2.6.5), the achievable input and output frequency ranges are $f_{\text{in}} = [27.5, 53]$ GHz, $f_{\text{out}} = [53/55, 106]$ GHz. Fig. 2.47 compares simulated and measured insertion loss of the LPF. The input frequency ranges of both, triplers and doublers are indicated.

Multiplier Assembly As already outlined, coaxial connectors are used to ease experimental characterization. Fig. 5.15 shows the mechanical housing with 1.85 mm / 1.00 mm connectors. Within an integrated front end module, either split block constructions or lid (Fig. 5.15) can be used. The glass bead of the 1.85 mm connector from Fig. 2.39 is soldered to the Au-plated brass housing utilizing solder with high melting temperature ($T_{\text{M}} \approx 219$ °). The further assembly steps are illustrated in Fig. 2.49. Although the DBES105a diode pads are

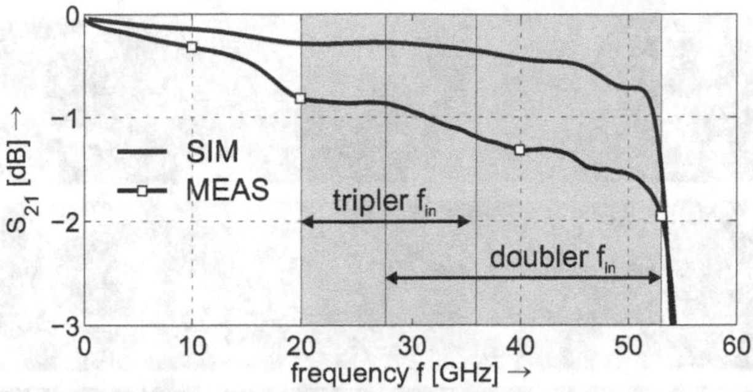

Figure 2.47: Magnitude of transmission coefficient S_{21} of the 11th order LPF from 3D EM simulation (including transition to 1.85 mm) and on-wafer measurement after eLRRM calibration versus frequency.

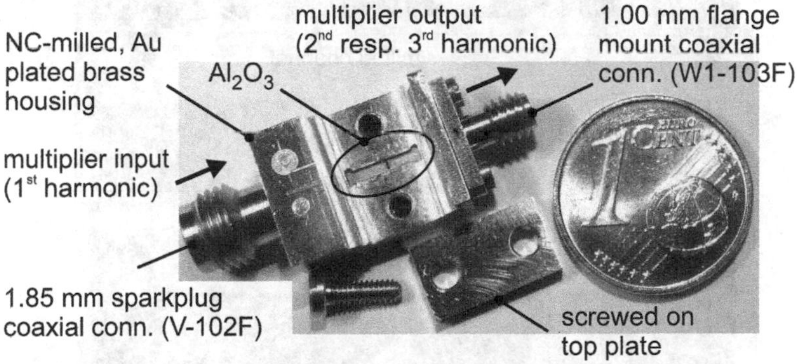

Figure 2.48: Mechanical housing of the proposed frequency multipliers with 1.85 mm / 1.00 mm coaxial connectors (top view).

equipped with Au bumps to allow for thermo-sonic bonding assembly, conventional solder paste (62Sn-36Pb-2Ag, $T_M = 179$ °) with particle size of 20 to 45 µm is used to mount the diode on the Al_2O_3 substrates (0.9 mm × 6.0 mm). As it is shown in Fig. 2.49, while soldering the substrate is fixed with high temperature resistant polyimide tape with silicon adhesive on a hotplate. As a second solder step with the same solder paste, substrate and diode are assembled to the Au-plated brass housing. Finally the stress relief of the 1.85 mm connector

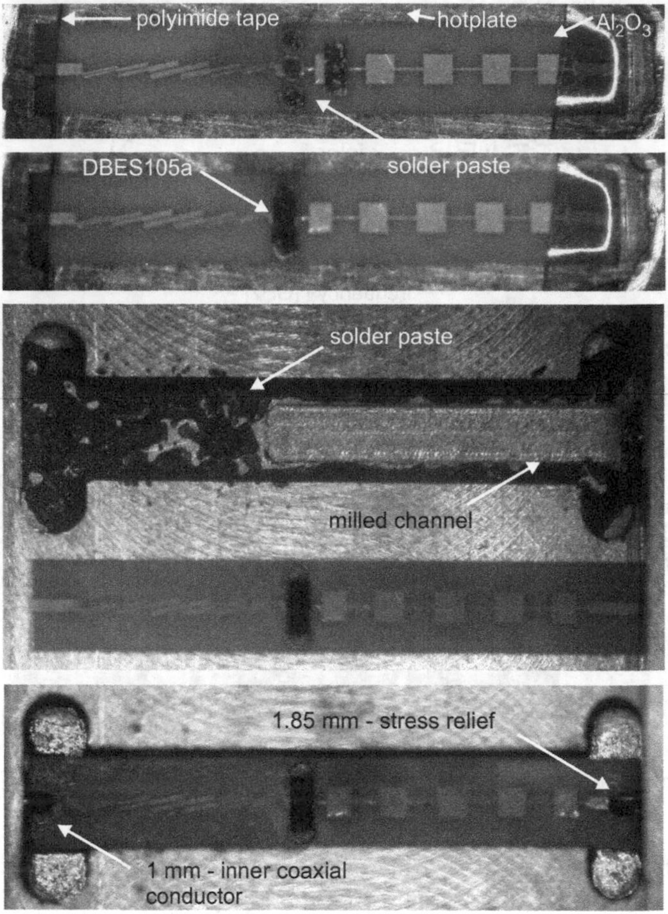

Figure 2.49: Frequency multiplier assembly steps.

(Fig. 2.39) and the inner coaxial conductor of the 1.00 mm connector are soldered to MSL with an indium- (In) ($T_\mathrm{M} = 154$ °) or bismuth-based (Bi) ($T_\mathrm{M} = 140$ °) solder. Alternatively conductive adhesive is used. Table 2.4 summarizes electrical conductivity properties and melting points of several solders and conductive adhesives.

Table 2.4: Comparison of solder alloys and conductive adhesives for multiplier assembly

solder or conductive adhesive	IACS [%]	el. cond. σ [MS/m]	approx. melt. T_M point [°]
96.5Sn-3Ag-0.5Cu	16	9.3	219
SC126 (62Sn-36Pb-2Ag)	14	8.1	179
Indalloy 2 (80In-15Pb-5Ag)	13	7.5	154
Indalloy 1E (52In-48Sn)	11.7	6.8	118
Indalloy 282 (57Bi-42Sn-1Ag)	4.5	2.6	140
EPO-TEK® H20E	0.69	0.4	-
Panacol Elecolit® 325	0.34	0.2	-

IACS compares conductivity to copper material (Cu), $\kappa_{Cu} = 58$ MS/m and $IACS_{Cu} = 100$ %.

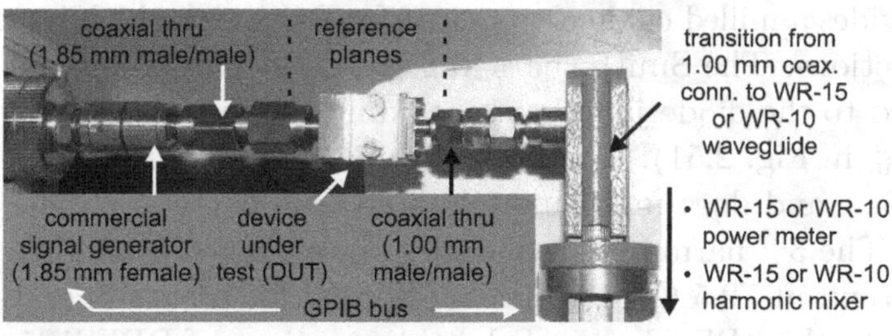

Figure 2.50: Measurement setup of the proposed multipliers with 1.85 mm / 1.00 mm coaxial connectors.

Measurement Setup The fully assembled multipliers with 1.85 mm / 1.00 mm coaxial connectors are driven by a commercial signal generator, 5 dBm $\leq$ $P_{in} \leq$ 18 dBm, within the input frequency range. Fig. 2.50 illustrates the measurement setup. The scalar output power level is measured with commercial V- and W-band power meters. All measurement results are corrected for the additional insertion loss of the measurement setup, a 1.00 mm thru connector and transitions to

WR-15 and WR-10 waveguide, but include the influence of the multiplier's coaxial connectors (reference planes in Fig. 2.50). Power meter measurements constitute the most precise method to capture scalar power levels, but do not allow for distinguishing spectral components of the harmonic content. Spectral measurements with decreased accuracy are performed with calibrated V- and W-band harmonic mixers in front of a spectrum analyzer to verify the power meter measurements consider the envisaged spectral components and to measure the multiplier's rejection of undesired harmonics (subsection 2.7.2).

2.6.1 Frequency Tripler x3_HE$_1$ (#1)

Top and bottom view of the manufactured x3_HE$_1$ (#1) frequency tripler [I6] Al_2O_3 substrates are shown in Fig. 2.51.

After the 1.85 mm connector or bondwire transition within integrated modules, the fundamental signal passes the input LPF, which ensures reflection of the multiplied harmonics. The mechanical housing provides a milled channel, required for DGS of the high impedance LPF sections. The Smith chart in Fig. 2.52 depicts the impedances provided to the diode junctions, including the diode mounting structure (Γ_{in} in Fig. 2.51). The 1st harmonic should be matched, which is the case and does unfortunately hold true partly for the 2nd harmonic. The 3rd harmonic frequencies are arranged around the ideal open circuit at 79.5 GHz, which is essential for proper operation. In cascade to the LPF, the two Schottky junctions of DBES105a diode are excited in antiparallel configuration.

Assuming equal Schottky junction behaviour, the antiparallel diode configuration ensures an effective suppression of the 2nd and 4th harmonic (short idlers, [54]). Within this balanced configuration no measures can be adopted to reject the 5th harmonic.

The same waveguide (MSL) is used at the input and output of the diode mounting structure. This is critical, as the electrical length from the diode junctions to the output HPF has to stay below 180 degree at f_{in}. Highpass filtering is established by exciting a HE$_1$ like mode with 53 GHz cut-off frequency in the area of its electric field

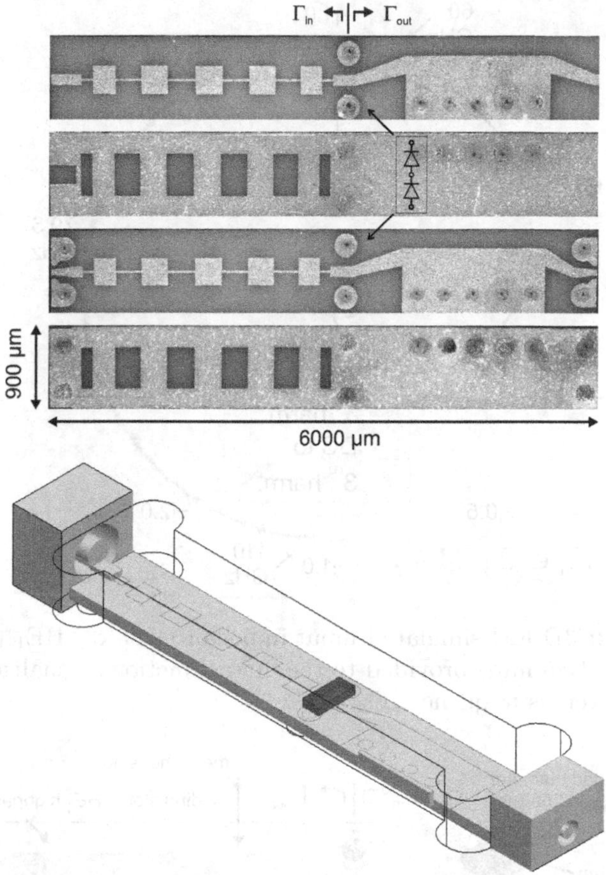

Figure 2.51: Top and bottom view of the manufactured x3_HE_1 frequency tripler Al_2O_3 substrates (0.9 mm × 6.0 mm) and 3D model for coaxial interconnection.

maximum (Fig. 2.53). The HE_1 mode is guided by a slot between the top metallization and the mechanical housing. This class of HPFs has first been suggested by Chramiec in 1981 [59–61] and has also been discussed as a degenerated case of transmission line comb HPFs recently in [62].

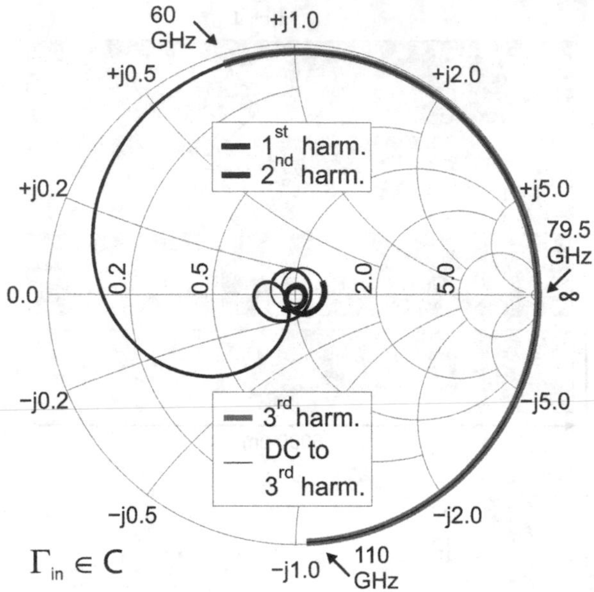

Figure 2.52: 3D EM simulated input impedances of x3_HE$_1$ (including transition to 1.85 mm) provided to the diode junctions visualized in the Smith chart versus frequency. $\Gamma_{\text{in}} \in \mathbb{C}$.

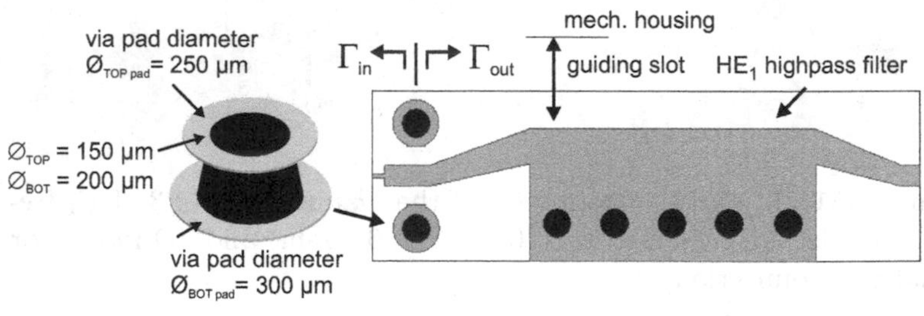

Figure 2.53: Detail view of the transition from x3_HE$_1$ diode mounting structure to HE$_1$ waveguide.

Fig. 2.54 shows a cross-sectional view of the HE$_1$ waveguide with 250 μm gap to the metallic enclosure, the magnitude of the HE$_1$ mode electric field strength at 80 GHz and the real parts of the characteristic

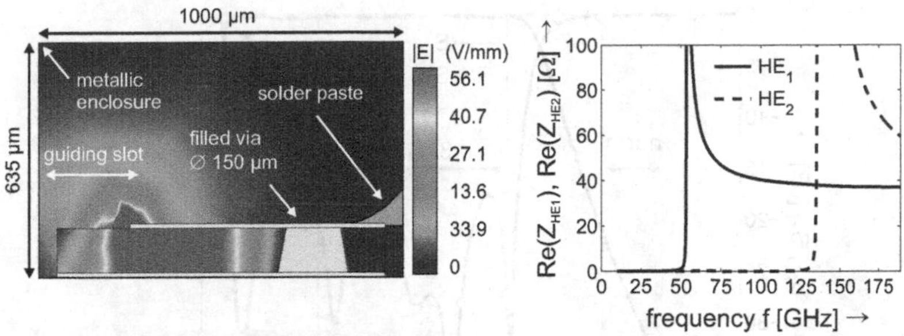

Figure 2.54: Cross-sectional view of the HE_1 waveguide, magnitude of the electric field strengths at 80 GHz and real parts of the characteristic impedances of HE_1 and HE_2 versus frequency.

impedances of HE_1 and HE_2 versus frequency. The HE_1 waveguide is in single mode operation up to ≈ 130 GHz and exhibits much lower impedance (power-current definition) values than conventional unilateral finlines (uFIN). Therefore, it can be matched more easily to 50 Ω MSL. Fig. 2.55 shows scattering parameters of the output HPF (including transition to 1.00 mm) from 3D EM simulation with return loss values greater than 15 dB at f_{out}. Selectivity is nearly proportional to the HE_1 waveguide length. The chosen length of 1.55 mm is a compromise between additional loss and 1st harmonic rejection.

The Smith chart in Fig. 2.56 shows the impedances provided to the diodes at the output port, including the diode mounting structure (Γ_{out} in Fig. 2.51).

The HPF is an ideal short circuit at DC. The 1st harmonic is rejected with nearly open circuit idle impedances (ideal open circuit at $f =$ 34 GHz). The 2nd harmonic is idled capacitively. The 3rd harmonic is matched. The feeding MSL length behind the diodes of this HPF could be designed much shorter to stay below 180 degree at f_{in}, but to reduce the coupling between HE_1 and the diode mounting structure a minimum length is required. Otherwise the circuit symmetry is affected and the even order harmonic suppression of the antiparallel diode configuration deteriorates drastically.

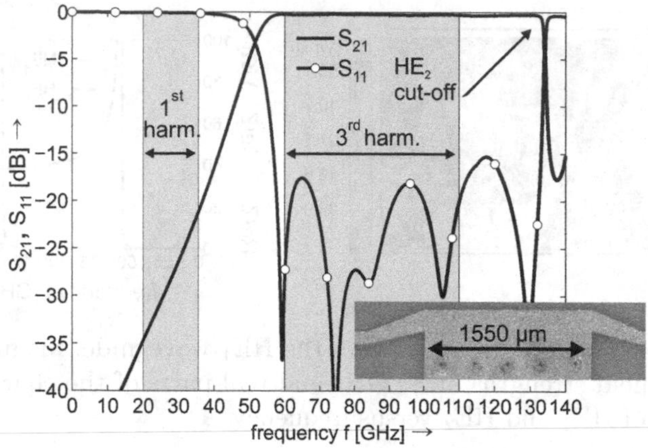

Figure 2.55: 3D EM simulated scattering parameters of the output HPF (including transition to 1.00 mm) versus frequency.

Fig. 2.57 shows the 3rd harmonic output power levels P_{out} at the output frequencies $f_{\text{out}} = 60,\ 66,\ 75,\ 82,\ 95,\ 100,\ 110$ GHz versus the fundamental input power level P_{in}. Measurement data in the frequency range of 60 to 110 GHz are within the gray shaded area.

A minimum input power level of $P_{\text{in}} = 10$ dBm to 14 dBm (at the high end of f_{out}) should be provided. At an input drive level of 18 dBm, the achieved output power levels are within the range of −3 to 1 dBm. A comparison of the measured conversion loss (solid lines) at $P_{\text{in}} = 12,\ 14,\ 16,\ 18$ dBm with results from co-simulation (dashed lines) is given in Fig. 2.58. The conversion loss averages out at 18 dB from 60 to 95 GHz and is below 22 dB up to 110 GHz, which is 2 to 6 dB less output power than simulation results predict. Spectral measurements up to the 5th harmonic have been performed. Fundamental rejection behaves according to Fig. 2.55, which means 20 dB at 40 GHz. At $P_{\text{in}} = 18$ dBm suppression of the 2nd and 4th harmonic is better than 15 dBc. In the frequency range of 100 to 110 GHz rejection of the 5th harmonic is at least 8 dBc (Fig. 2.59).

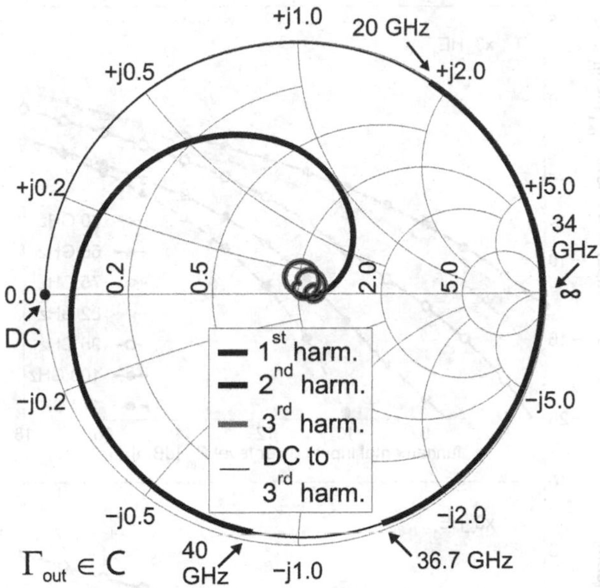

Figure 2.56: 3D EM simulated output impedances of x3_HE$_1$ (including transition to 1.00 mm) provided to the diode junctions visualized in the Smith chart versus frequency. $\Gamma_{\text{out}} \in \mathbb{C}$.

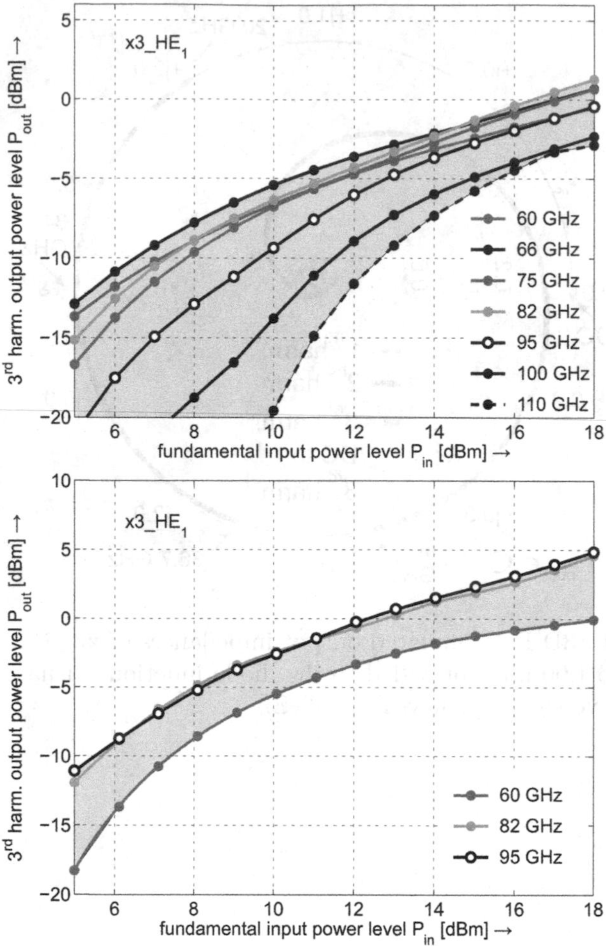

Figure 2.57: 3rd harmonic output power levels P_{out} at the output frequencies $f_{\mathrm{out}} = 60,\ 66,\ 75,\ 82,\ 95,\ 100,\ 110$ GHz versus the fundamental input power level P_{in}. Measurement data (top) and simulation data (bottom) in the frequency range of 60 to 110 GHz are within the gray shaded area.

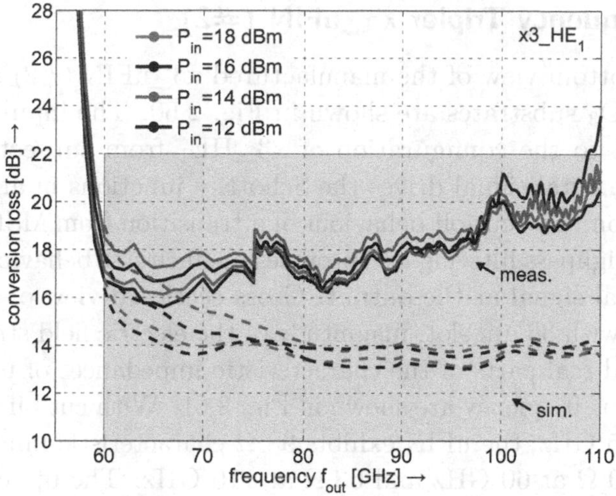

Figure 2.58: Comparison of measured (solid) and simulated (dashed) conversion loss versus output frequency f_{out}. Input power levels P_{in} = 12, 14, 16, 18 dBm.

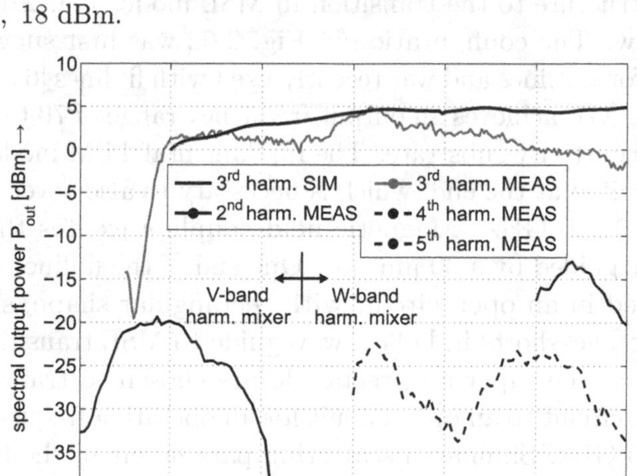

Figure 2.59: Measured spectral output power levels up to the 5th harmonic versus output frequency f_{out} at input power level P_{in} = 18 dBm. Compared to simulated 3rd harmonic output power levels with P_{in} = 18 dBm (black, solid).

2.6.2 Frequency Tripler x3_uFIN (#2)

Top and bottom view of the manufactured x3_uFIN (#2) frequency tripler Al_2O_3 substrates are shown in Fig. 2.60. The input circuitry is identical to the configuration of x3_HE$_1$ from subsection 2.6.1. The fundamental signal drives the Schottky junctions in antiparallel configuration. The cut-off behaviour of a transition from MSL to uFIN allows for highpass filtering and provides open circuit behaviour for the fundamental signal at the output. Cross-sectional view of the uFIN waveguide with 30 µm slot, magnitude of the electric field strengths at 80 GHz and real parts of the characteristic impedances of uFIN$_1$ and uFIN$_2$ versus frequency are shown in Fig. 2.61. With cut-off frequency of $f_c = 39.5$ GHz, the uFIN exhibts 98 Ω characteristic impedance at 50 GHz, 80 Ω at 60 GHz and 63 Ω at 110 GHz. The upper limit of operating frequency range is 129 GHz due to the next higher order mode uFIN$_2$. The generated harmonics propagate from the diode mounting structure to the transition in MSL mode. Fig. 2.62 depicts a detail view. The configuration in Fig. 2.62 was first suggested by Gupta [63] for slotlines and was recently used with finlines [64, 65]. The transition in [G1] achieves an output frequency range of 70 to 110 GHz on a low permittivity substrate. The fundamental TEM mode at MSL is short circuited at the end, which is necessary to also cover the lower frequencies $\geq$ 60 GHz. Magnetic field coupling excites the hybrid finline mode guided by a 30 µm slot. One end of the finline waveguide is terminated in an open circuit with rectangular shape, similar to quarterwave backshorts in hollow waveguide to MSL transitions. The dimensions of the taper and rectangle are chosen to transform the finline short circuit to an open circuit for an operating frequency range of 60 to 110 GHz. Simulated scattering parameters of the transition in back to back configuration with a total length of $\ell = 2500$ µm are shown in Fig. 2.63 and predict return loss values in the order of 20 dB. Due to the high permittivity substrate, the necessary quarterwave backshort dimensions (450 µm × 50 µm) are quite small compared to similar realizations on SiO_2 or teflon (PTFE). This is advantageous to keep the substrate width small enough for single mode operation

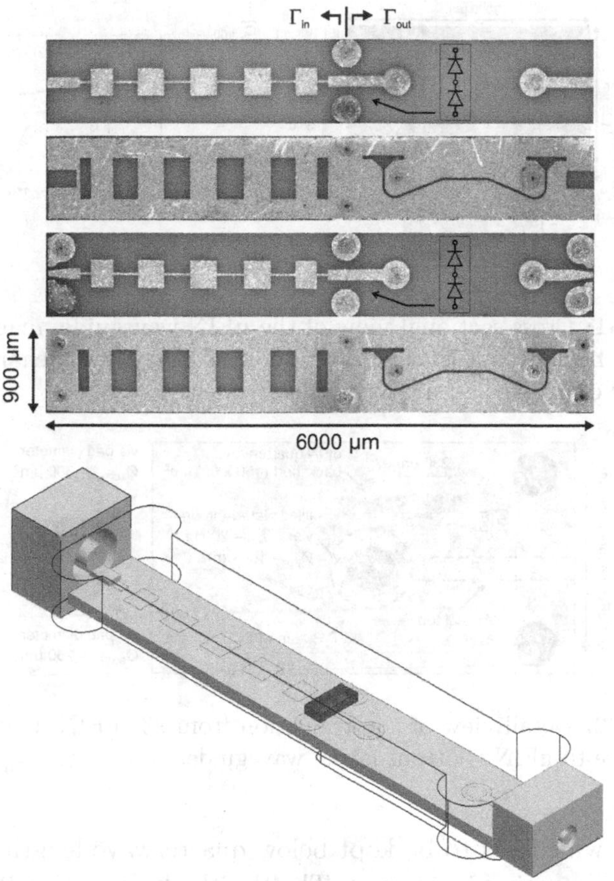

Figure 2.60: Top and bottom view of the manufactured x3_uFIN frequency tripler Al_2O_3 substrates (0.9 mm × 6.0 mm) and 3D model for coaxial interconnection.

up to 110 GHz but leads to more sensitive device performance in presence of tolerances. Contrary to x3_HE_1 from subsection 2.6.1, the filled plated-through vias are drilled from top to bottom layer. Consequently the greater via diameter of $\varnothing_{TOP} = 200$ µm appears at the top layer. This is necessary to allow for a short distance between the position of maximum magnetic field coupling and the

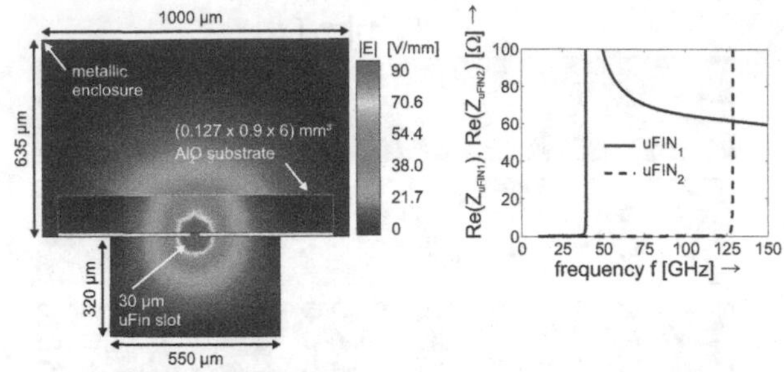

Figure 2.61: Cross-sectional view of the uFIN waveguide, magnitude of the electric field strengths at 80 GHz and real parts of the characteristic impedances of $uFIN_1$ and $uFIN_2$ versus frequency.

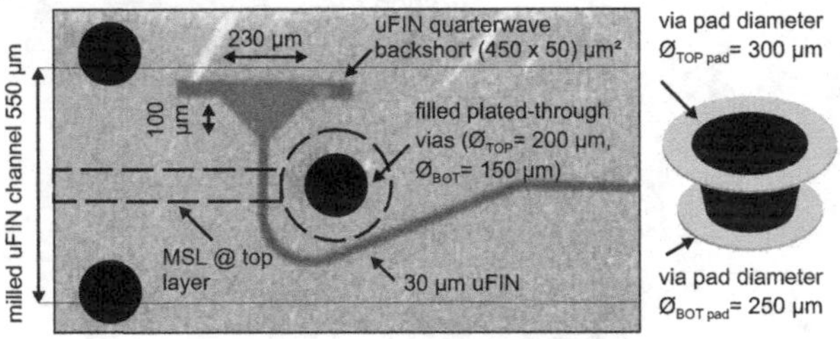

Figure 2.62: Detail view of the transition from x3_uFIN diode mounting structure to uFIN (bottom layer) waveguide.

via barrel, which has to be kept below quarterwave length up to the maximum operating frequency. The Smith chart in Fig. 2.64 shows the impedances provided to the diodes at the output port including the diode mounting structure (Γ_{out} in Fig. 2.60). The HPF is an ideal short circuit at DC. The 1st harmonic frequency range is rejected with inductive behaviour at the lower band limit and nearly open circuit behaviour at the upper band limit. The 2nd harmonic is idled capacitively and partly matched at the frequencies that overlap with the desired output frequency range f_{out}. The 3rd harmonic is matched.

Fig. 2.65 shows the 3rd harmonic output power levels P_{out} at the output frequencies f_{out} = 60, 65, 72, 80, 95, 100, 110 GHz versus the fundamental input power level P_{in}. Measurement data in

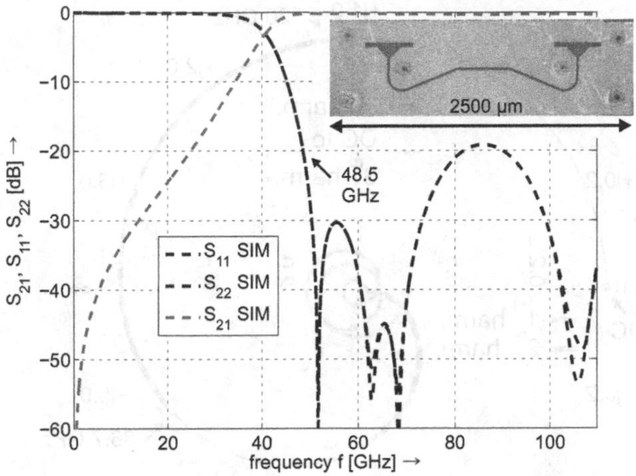

Figure 2.63: Photograph and 3D EM simulated scattering parameters of MSL to uFIN transition in back to back configuration versus frequency (total length $\ell = 2500$ µm).

the frequency range of 60 to 110 GHz are within the gray shaded area. A comparison of the measured conversion loss (solid lines) at $P_{\text{in}} = 12$, 14, 16, 18 dBm with results from co-simulation (dashed lines) is given in Fig. 2.66. With $P_{\text{in}} = 18$ dBm the conversion loss is below 19 dB from 60 to 100 GHz and is below 22 dB up to 103 GHz. There is a sharp increase of conversion loss above 103 GHz. In Fig. 2.62 the milled channel with width of 550 µm is marked. Optical measurements of the assembled multiplier have shown that due to a slightly off-center position of the substrate within the mechanical housing the backshort geometry is partly overlaid by the housing. Therefore, the assembled structure is not capable of operating up to 110 GHz. The problem can be solved by using marginally greater channel width, at least in the area of the backshort geometry. Reliable soldering is still possible with less contact area. Spectral measurements up to the 5th harmonic have been performed. At $P_{\text{in}} = 18$ dBm suppression of the 2nd, 4th, and 5th harmonic is better than 10 dBc from 60 to 110 GHz (Fig. 2.67).

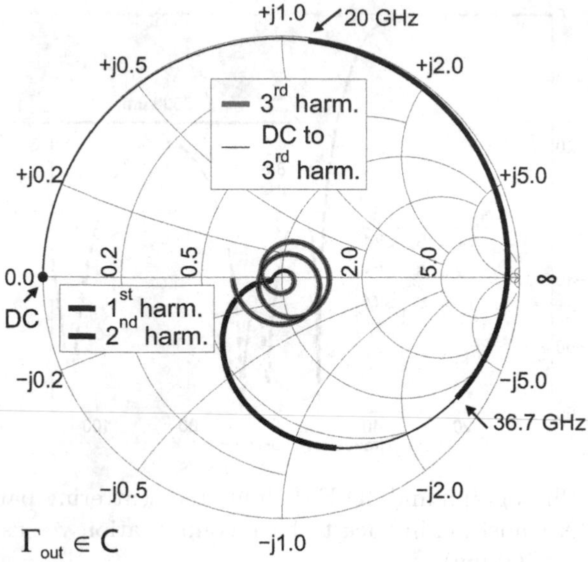

Figure 2.64: 3D EM simulated output impedances of x3_uFIN (including transition to 1.00 mm) provided to the diode junctions visualized in the Smith chart versus frequency. $\Gamma_{\text{out}} \in \mathbb{C}$.

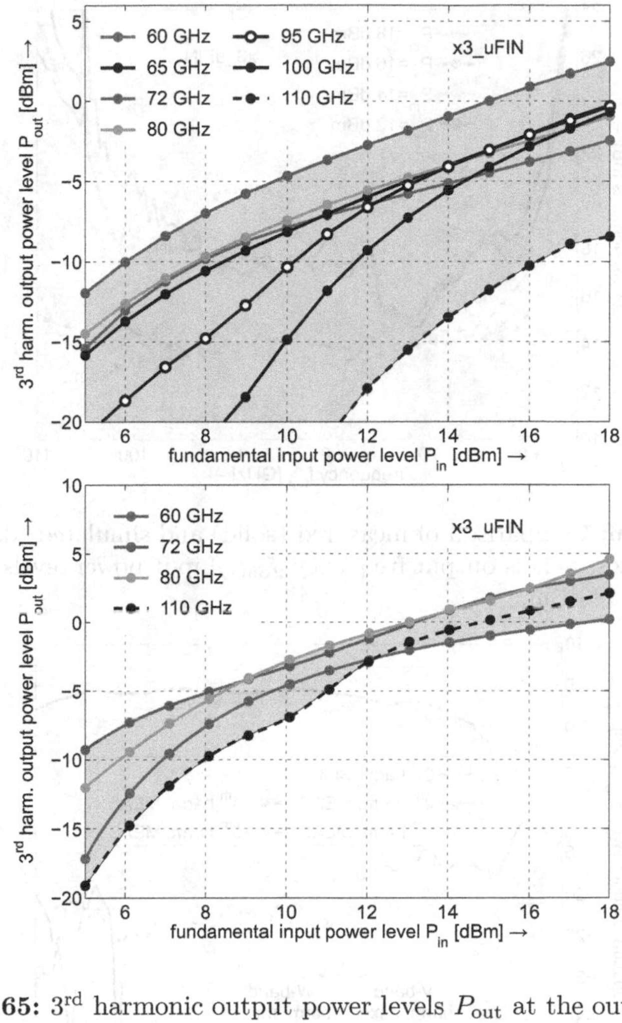

Figure 2.65: 3rd harmonic output power levels P_{out} at the output frequencies f_{out} = 60, 65, 72, 80, 95, 100, 110 GHz versus the fundamental input power level P_{in}. Measurement data (top) and simulation data (bottom) in the frequency range of 60 to 110 GHz are within the gray shaded area.

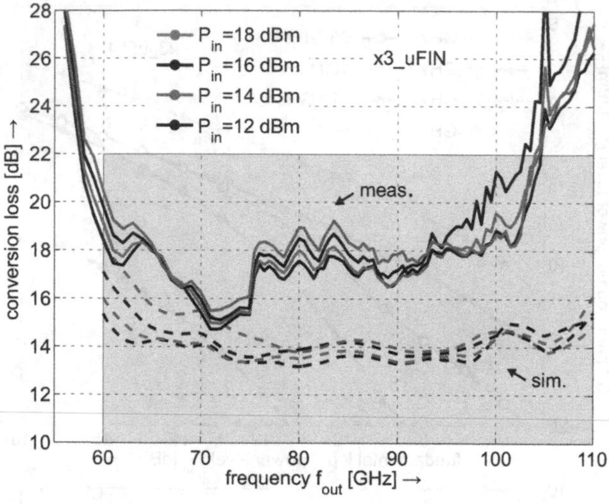

Figure 2.66: Comparison of measured (solid) and simulated (dashed) conversion loss versus output frequency f_{out}. Input power levels P_{in} = 12, 14, 16, 18 dBm.

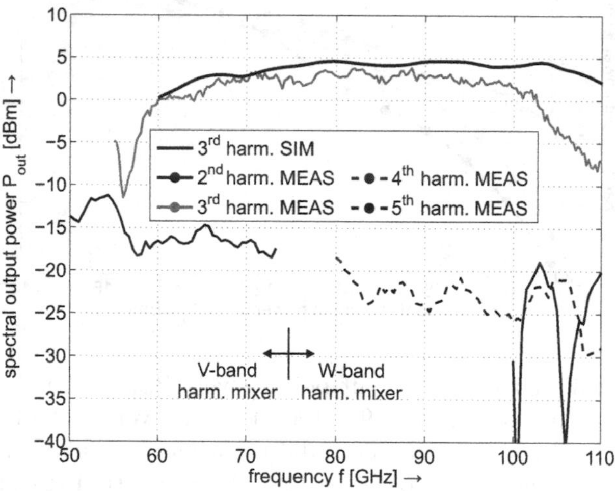

Figure 2.67: Measured spectral output power levels up to the 5th harmonic versus output frequency f_{out} at input power level P_{in} = 18 dBm. Compared to simulated 3rd harmonic output power levels with P_{in} = 18 dBm (black, solid).

2.6.3 Frequency Tripler x3_SIDE (#3)

Top and bottom view of the manufactured x3_SIDE (#3) frequency tripler Al_2O_3 substrates are shown in Fig. 2.68. The input circuitry of x3_SIDE is identical to multiplier #1 and #2. The two Schottky junctions are driven in antiparallel configuration by the fundamental signal. Beside the fundamental signal, all harmonics with different power levels exist across the Schottky junctions. An input LPF and output HPF allows for controlling spectral harmonic content at the multiplier's input and output. With ultrawideband frequency multiplier design, it is a key function to avoid destructive interference of the fundamental and desired harmonic $\left(3^{\text{rd}}\right)$ at the Schottky junctions over the input and output frequency ranges. This is achieved with filter structures that show minimum reflection phase $\min(\arg S_{11})$. As already mentioned, this normally prohibits the use of many LPFs in cascade. The frequency triplers #1 and #2, as well as #4 and #5 all make use of different modes of propagation at the input and output. Utilizing the cut-off behaviour as HPF is an excellent solution with respect to minimum reflection phase. Literal HPF resp. BPF based on lowpass to highpass transformation with quarterwave or halfwave resonators according to Matthaei [57] are difficult to implement in ultrawideband frequency multipliers. Realizing octave bandwidth BPF is challenging and requires high filter order, which is associated with greater insertion loss and reflection phase. The presented multipliers in this section utilize HE_1 waveguide and several uFIN solutions. In section 2.5 [G2] bilateral finlines are successfully applied to build up a frequency tripler with output frequency range of 20 to 40 GHz, which is another promising concept for E- and W-band frequency multipliers. Nevertheless, utilization of hybrid modes at the output complicates the mechanical housing and to some extent makes device performance conditional to the housing. There are fields of application that benefit from the use of resonator-based output filters in multiplier designs, although the maximum achievable bandwidth is reduced. Therefore, multiplier #3 with halfwave resonator BPF at the output (Fig. 2.69) has been developed. The 5th order BPF is

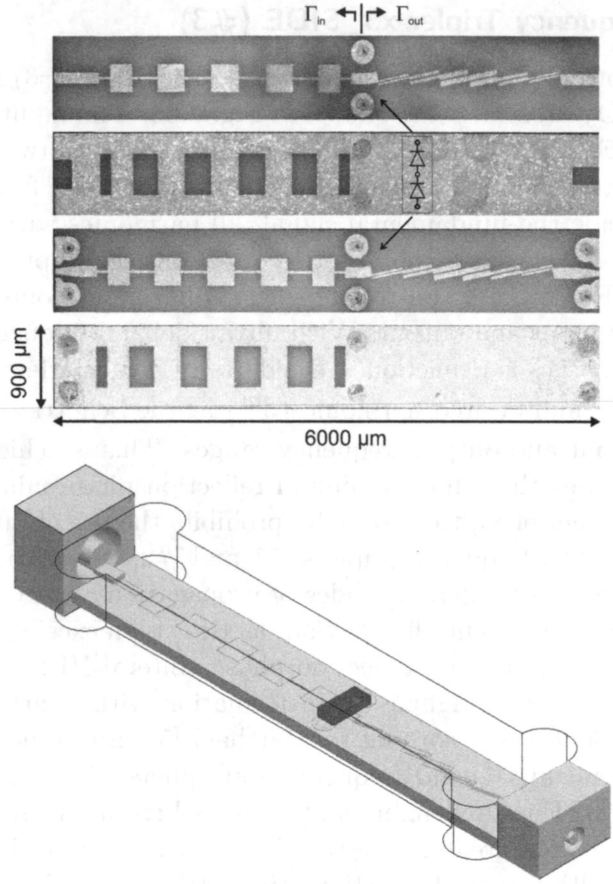

Figure 2.68: Top and bottom view of the manufactured x3_SIDE frequency tripler Al_2O_3 substrates (0.9 mm × 6.0 mm) and 3D model for coaxial interconnection.

based on side-coupled striplines (cMSL), each of them a quarterwave length long, to act as impedance inverters $J_k, k \in \mathbb{N}$ for the MSL halfwave parallel resonators. The center frequency of the design is $f_0 = \sqrt{70 \cdot 110}$ GHz $= 87.7$ GHz, which means fractional bandwidth of FBW $= 46$ %. Fig. 2.70 illustrates the cross-sectional view of the first J_1 and last J_6 inverter's cMSL and magnitudes of the electric

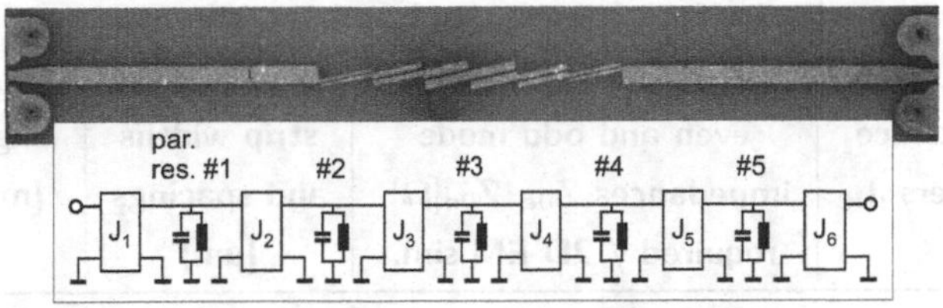

Figure 2.69: Photograph of the 5th order coupled microstrip halfwave resonator BPF with corresponding block diagram.

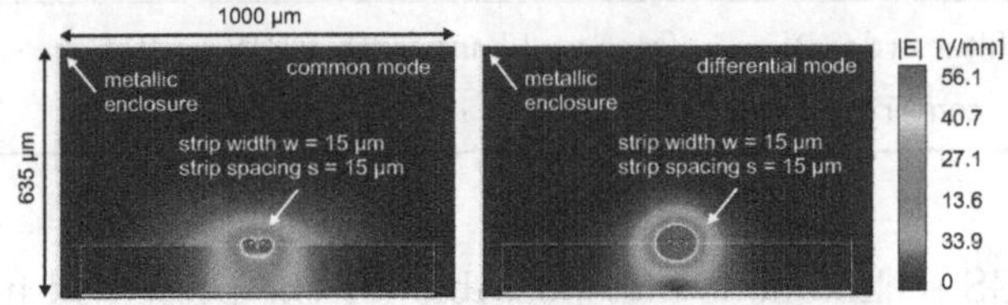

Figure 2.70: Cross-sectional views of the first side-coupled microstrip line (cMSL, first and last J inverter) and magnitudes of the electric field strengths of common and differential modes at 87.7 GHz.

field strengths of common and differential modes at $f_c = 87.7$ GHz. Stepped impedance LPF designs benefit from a large ratio of high and low characteristic impedances of the incorporated filter elements to achieve great bandwidth of stopband. Similarly the BPF requires a high ratio between the even mode characteristic impedance of the outer J inverters compared to the J inverters in the filter center to achieve great passband frequency range. The concepts described in [66] based on DGS technique allow for realization of greater even mode impedances compared to fully backside metallized cMSL. Further on the center frequency of the second parasitic passband is increased if the even and odd mode of each J inverter experience the same electrical length when propagating along the quarterwave cMSL waveguide. This is critical with dispersive cMSL realizations. DGS technique allows for increasing the phase velocity of the even mode to match the odd mode phase velocity. Superior suppression of the second passband could be achieved in [66] for BPF operating from 10 to

Table 2.5: 5th order halfwave resonator bandpass filter

impedance inverters J_k	even and odd mode impedances $Z_{0e}, Z_{0o}[\Omega]$, required / 2D EM sim.	strip widths and spacings [μm]	lengths [mm]
1 and 6	135, 43 / 137.9, 44.6	15, 15	343
2 and 5	111, 40 / 113.3, 41.2	30, 20	324
3 and 4	94, 35 / 95.7, 36.0	50, 20	307

filter order $N = 5$, fractional bandwidth FBW = 46 %,
center frequency $f_0 = \sqrt{70 \cdot 110}$ GHz = 87.7 GHz.

20 GHz. DGS technique is indispensable for the presented input LPF at f_{in}, but complicates the analytical based design flow of the BPF. It is further difficult to keep cut-off frequency of higher order modes greater than the maximum operating frequency if DGS technique is used for cMSL at W-band frequencies and beyond. The designed BPF for multiplier #3 without DGS technique utilizes minimum strip widths and spacings of 15 μm, which is sufficient to realize passband frequency range of 70 GHz to 110 GHz. Filter performance is achieved without via interconnections. The characteristic impedances of each inverter's even and odd mode from analytical synthesis [57], together with corresponding values of the actual realization (2D EM simulation) are summarized in Table 2.5. Contrary to the presented LPF design, the analytical requirements cannot be realized as good with dispersive cMSL. Therefore, deviations from analytically calculated filter performance to 3D EM simulation and measurement have to be tolerated. Scattering parameters from 3D EM simulation and on-wafer measurement after eLRRM calibration are shown in Fig. 2.71. Poor isolation from 20 to 40 GHz is caused by crosstalk between the on-wafer probes and is not a problem within integrated front end modules. Simulation and measurement are in very good agreement, except for underestimation of real losses in 3D EM simulation. The achieved bandwidth and performance constitute almost

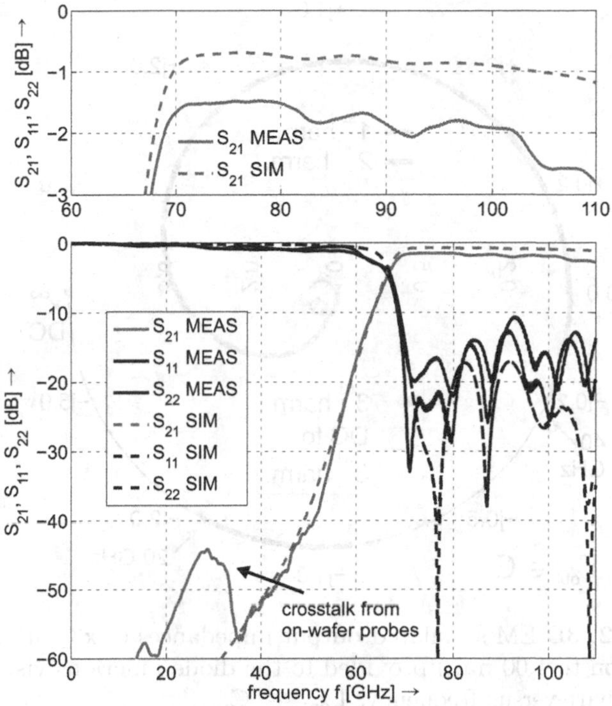

Figure 2.71: Scattering parameters of 5th order coupled microstrip halfwave resonator bandpass filter from 3D EM simulation and on-wafer measurement after eLRRM calibration versus frequency.

the limit for this filter architecture. The analytical synthesis results do not show the roll-off at the lowest and especially at the highest passband frequency, visible in Fig. 2.71, but could be observed by the author in many side-coupled BPF realizations with different center frequencies f_0. Other architectures are required to build filters with sharper edges and greater bandwidth, e.g. [67] based on broad-side coupled quarterwave resonators. The Smith chart in Fig. 2.72 depicts the impedances provided to the diodes at the output port including the diode mounting structure and the transition to 1.00 mm coaxial connector (Γ_{out} in Fig. 2.68). The filter is an ideal open circuit at

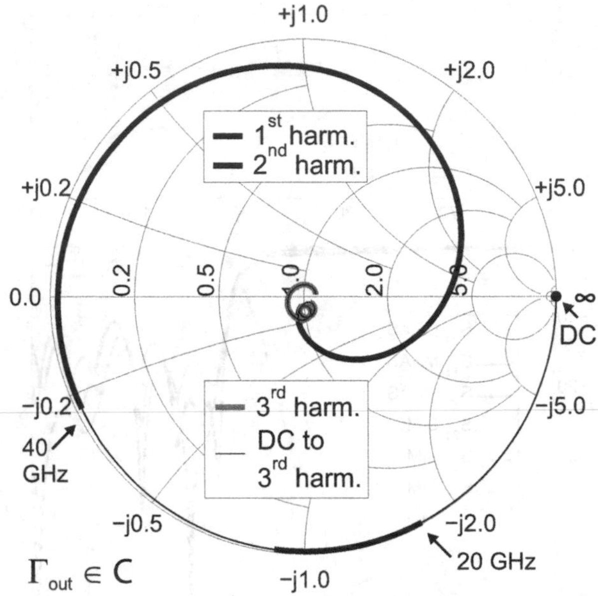

Figure 2.72: 3D EM simulated output impedances of x3_SIDE (including transition to 1.00 mm) provided to the diode junctions visualized in the Smith chart versus frequency. $\Gamma_{\text{out}} \in \mathbb{C}$.

DC. The 1st harmonic is rejected capacitively. The 2nd harmonic is arranged around the ideal short circuit and matched at frequencies > 70 GHz. The 3rd harmonic and part of the 4th harmonic's frequency range are matched. Suppression of the even order harmonics $\left(2^{\text{nd}} \text{ and } 4^{\text{th}}\right)$ is realized by the antiparallel diode configuration (short idlers). Fig. 2.73 shows the 3rd harmonic output power levels P_{out} at the output frequencies $f_{\text{out}} = 67,\ 75,\ 80,\ 85,\ 90,\ 100,\ 110$ GHz versus the fundamental input power level P_{in}. Measurement data in the frequency range of 67 to 110 GHz are within the gray shaded area. At an input power level of $P_{\text{in}} = 18$ dBm, the achieved output power levels are within the range of -5 to 3 dBm. A comparison of the measured conversion loss (solid lines) at $P_{\text{in}} = 14,\ 16,\ 18$ dBm with results from co-simulation (dashed lines) is given in Fig. 2.74.

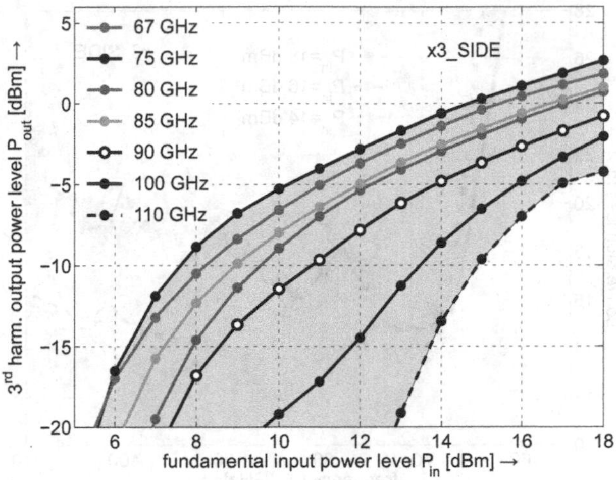

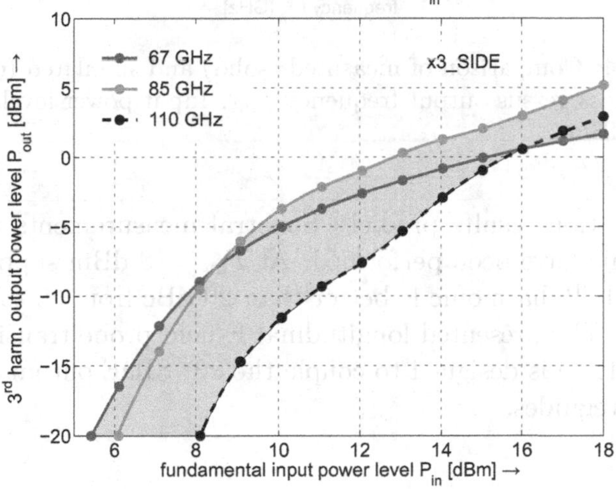

Figure 2.73: 3rd harmonic output power levels P_{out} at the output frequencies f_{out} = 67, 75, 80, 85, 90, 100, 110 GHz versus the fundamental input power level P_{in}. Measurement data (top) and simulation data (bottom) in the frequency range of 67 to 110 GHz are within the gray shaded area.

The conversion loss values are below 20 dB from 67 to 100 GHz and below 24 dB up to 110 GHz, which is 2 to 8 dB less output power

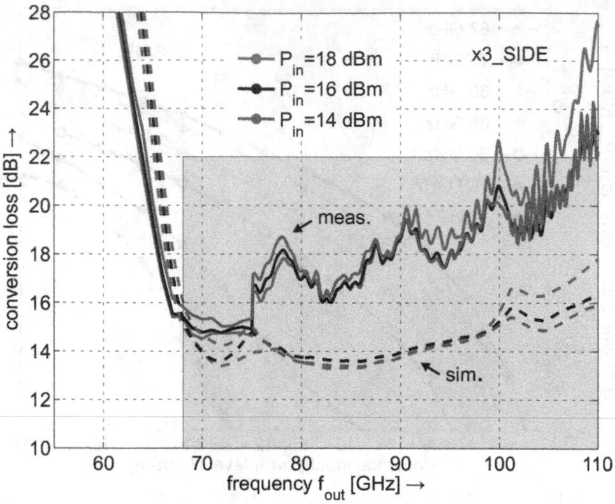

Figure 2.74: Comparison of measured (solid) and simulated (dashed) conversion loss versus output frequency f_{out}. Input power levels $P_{\text{in}} =$ 14, 16, 18 dBm.

than simulation results predict. Spectral measurements up to the 4th harmonic have been performed. At $P_{\text{in}} = 18$ dBm suppression of the 2nd and 4th harmonic is better than 20 dBc from 60 to 110 GHz (Fig. 2.75). The presented longitudinal E-field probe transition from MSL to WR-10 is designed to couple the x3_SIDE output signals to WR-10 waveguides.

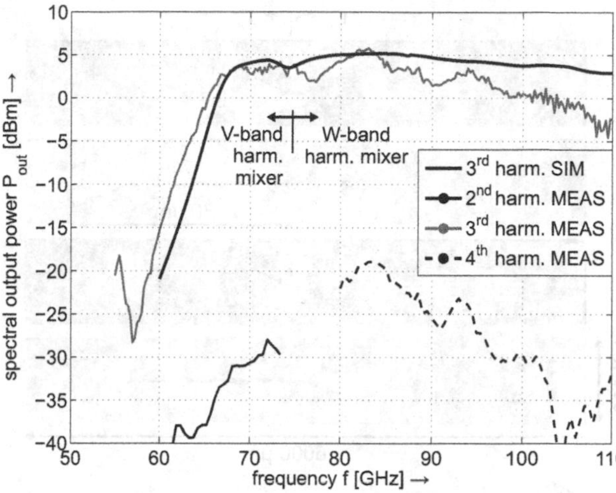

Figure 2.75: Measured spectral output power levels up to the 5th harmonic versus output frequency f_{out} at input power level P_{in} = 18 dBm. Compared to simulated 3rd harmonic output power levels with P_{in} = 18 dBm (black, solid).

2.6.4 Frequency Doubler x2_uFIN (#4)

Top and bottom view of the manufactured x2_uFIN (#4) frequency doubler Al_2O_3 substrates are shown in Fig. 2.76.

As already mentioned a minimum filter order of $N = 15$ is required for frequency doublers that cover an output frequency range of 60 to 110 GHz, establishing passband behaviour at the highest input frequency along with sufficient stopband rejection at the lowest output frequency. Within the applied dicing process the maximum substrate length of 6 mm does not allow for higher filter orders than N = 11. The achievable input and output frequency ranges are f_{in} = [27.5, 53] GHz, f_{out} = [53/55, 106] GHz, when using the presented LPF. The designed LPF will add up to 4 dB loss to a fundamental signal at 55 GHz, therefore efficiency at frequencies greater than 106 GHz is poor. Things are different at the lower end of the output frequency range. Fundamental signals at frequencies below 27.5 GHz pass the

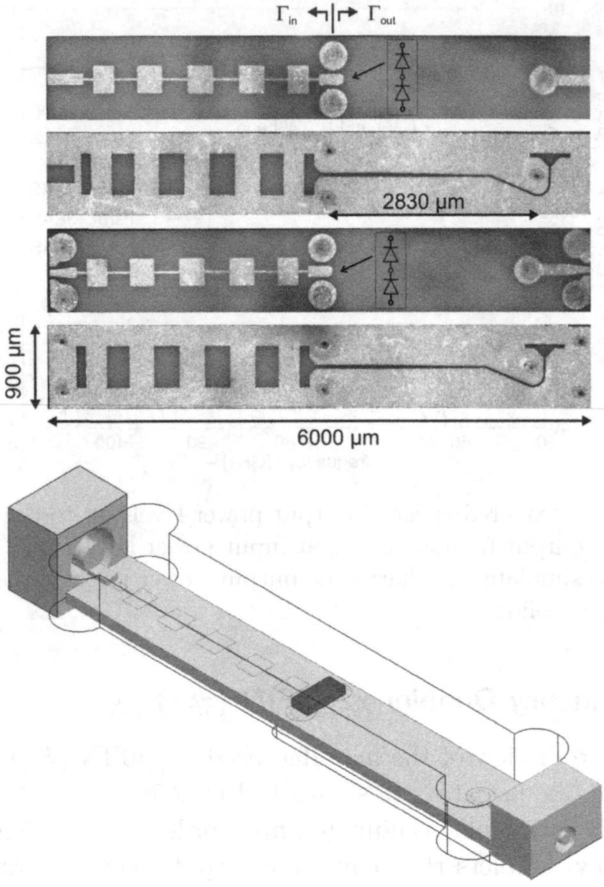

Figure 2.76: Top and bottom view of the manufactured x2_uFIN frequency doubler Al_2O_3 substrates (0.9 mm × 6.0 mm) and 3D model for coaxial interconnection.

LPF and generate harmonics at the Schottky junctions. Although, the amount of 2nd harmonic power that propagates to the matched input port is lost, there is 2nd harmonic power propagating in forward direction to the output load. This power splitting is rarely equal. Actually, it strongly depends on the frequency multiplier architecture

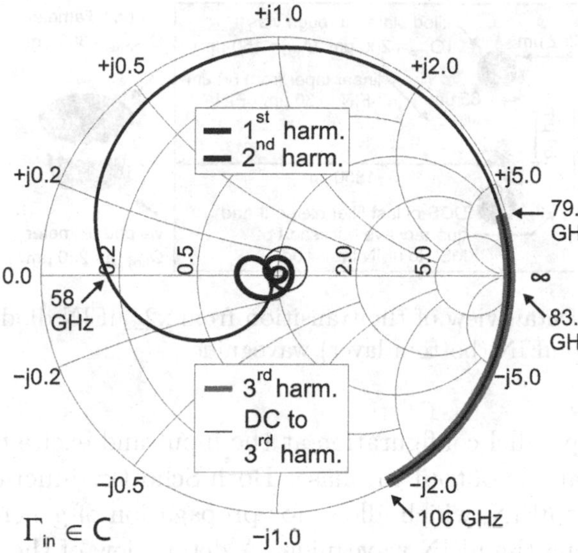

Figure 2.77: 3D EM simulated input impedances of x2 uFIN (including transition to 1.85 mm) provided to the diode junctions visualized in the Smith chart versus frequency. $\Gamma_{\text{in}} \in \mathbb{C}$.

and especially on the output circuitry. Therefore measurement results of x2_uFIN (this subsection) and x2_uFINring (subsection 2.6.5) cover the output frequency range of 50 to 110 GHz. The Smith chart in Fig. 2.77 shows the impedances provided by the LPF to the diode junctions including the diode mounting structure at the input, in case of frequency doubling (Γ_{in} in Fig. 2.76). The 1st harmonic is matched. Due to insufficient filter order the 2nd harmonic is not entirely reflected at the lower end of the output frequency range, but it is arranged around the ideal open circuit at 83.8 GHz at the higher output frequencies. The 3rd harmonic frequencies overlap with f_{out} and are rejected with open / inductive behaviour. The proposed frequency doubler x2_uFIN utilizes the cut-off behaviour of an uFIN with 30 µm slot as HPF. The same uFIN is used for frequency tripler x3_uFIN from subsection 2.6.2. It shows 98 Ω characteristic impedance at 50 GHz (Fig. 2.61). The two Schottky junctions are

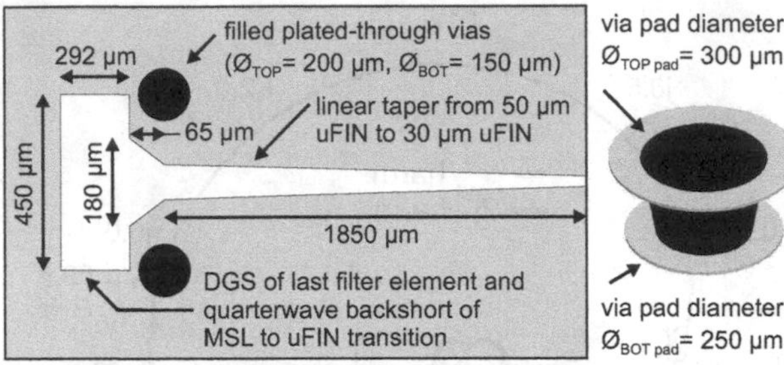

Figure 2.78: Detail view of the transition from x2_uFIN diode mounting structure to uFIN (bottom layer) waveguide.

driven in antiparallel configuration at the input and excite the hybrid finline mode at the output in phase. Both Schottky junctions are in series with the uFIN, which allows for propagation of generated even harmonics along the uFIN waveguide. A detail view of the transition from the diode mounting structure to uFIN is shown in Fig. 2.78. The DGS structure of the last LPF element is used as quarterwave backshort of the uFIN transition. The dimensions of the DGS are fixed, two linear tapers allow for adjusting the operating frequency of the backshort. The author successfully applied a similar approach to build D-band frequency doublers (section 2.7, [I7]) on PTFE with 50 µm thickness and WR-12 input and WR-6 output waveguide. The length of the finline (ℓ = 1850 µm) is chosen to achieve sufficient fundamental rejection (43 dB at 35 GHz) and to reduce reflections of the finline taper from 180 to 30 µm. At the output the uFIN to MSL transition from x3_uFIN (subsection 2.6.2) is used. As it is shown in Fig. 2.63 the transition has an operating frequency range from 48.5 to 110 GHz. Therefore the output circuitry is capable of covering the frequency range from 50 to 110 GHz, but the input LPF adds 3 dB to 4 dB additional loss at the maximum input frequency of 55 GHz (compare Fig. 2.47). With respect to the explanations about the applied co-simulation procedure in this section and subsection 2.6, multiplier x2_uFIN (#4) constitutes a multiplier architecture, which does not allow for extensive partitioning within the design process.

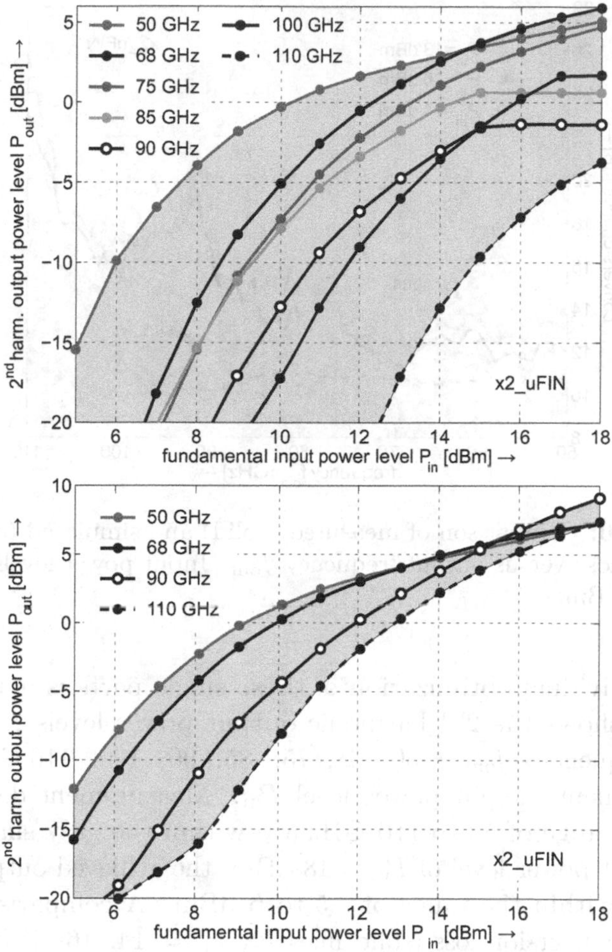

Figure 2.79: 2nd harmonic output power levels P_{out} at the output frequencies f_{out} = 50, 68, 75, 85, 90, 100, 110 GHz versus the fundamental input power level P_{in}. Measurement data (top) and simulation data (bottom) in the frequency range of 50 to 110 GHz are within the gray shaded area.

The diode mounting structure in Fig. 2.78 strongly interacts with the last LPF elements and the tapered uFIN parts at the output.

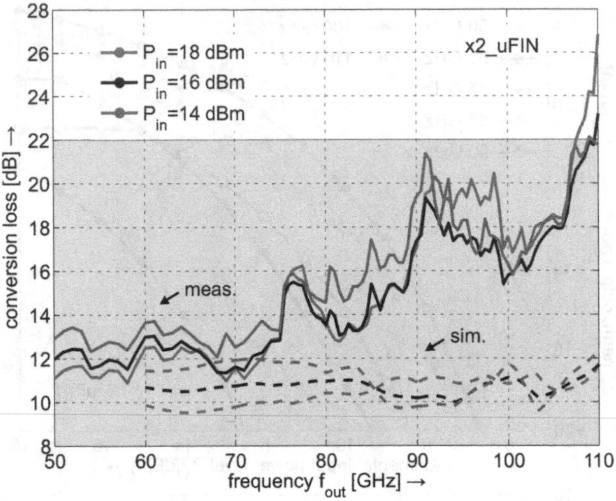

Figure 2.80: Comparison of measured (solid) and simulated (dashed) conversion loss versus output frequency f_{out}. Input power levels $P_{\text{in}} =$ 14, 16, 18 dBm.

Hence, individual optimization of these single parts is not possible. Fig. 2.79 shows the 2nd harmonic output power levels P_{out} at the output frequencies $f_{\text{out}} =$ 50, 68, 75, 85, 90, 100, 110 GHz versus the fundamental input power level P_{in}. Measurement data in the frequency range of 50 to 110 GHz are within the gray shaded area. At an input power level of $P_{\text{in}} =$ 18 dBm, the achieved output power levels are within the range of −5 to 6 dBm. A comparison of the measured conversion loss (solid lines) at $P_{\text{in}} =$ 14, 16, 18 dBm with results from co-simulation (dashed lines) is given in Fig. 2.80. The conversion loss values are below 14 dB from 50 to 75 GHz (entire V-band) and below 23 dB up to 110 GHz, which is 2 to 10 dB less output power than simulation results predict. Spectral measurements up to the 3rd harmonic have been performed. In case of frequency doublers for the output frequency range of 50 to 110 GHz, the 3rd harmonic is the only harmonic overlapping with the desired output

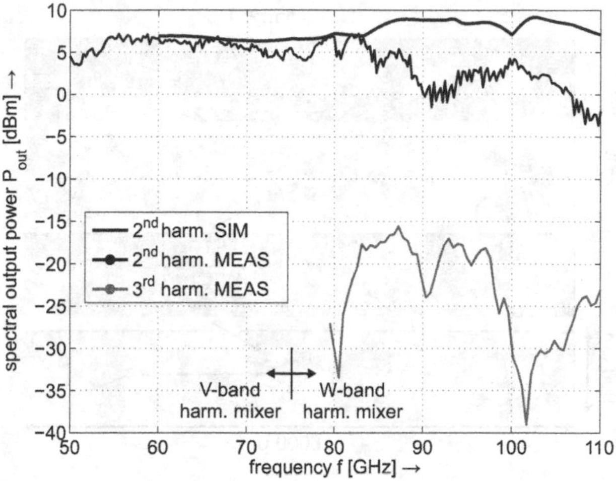

Figure 2.81: Measured spectral output power levels up to the 4th harmonic versus output frequency f_{out} at input power level P_{in} = 18 dBm. Compared to simulated 3rd harmonic output power levels with P_{in} = 18 dBm (black, solid).

signal. At P_{in} = 18 dBm suppression of the 3rd harmonic is better than 15 dBc (Fig. 2.81).

2.6.5 Frequency Doubler x2_uFINring (#5)

Top and bottom view of the manufactured x2_uFINring (#5) frequency doubler Al_2O_3 substrates are shown in Fig. 2.82. The explanations regarding the utilized LPF for frequency doublers from subsection 2.6.4 hold true for x2_uFINring. The fundamental signal exits the LPF in MSL mode. Through a filled plated-through via ($\varnothing_{TOP}$ = 200 µm), the signal enters a coplanar waveguide (CPW) at the bottom layer. In CPW mode the two Schottky junctions are driven in antiparallel configuration. A detail view of the transition is shown in Fig. 2.83. The 2nd harmonic at each Schottky junction excites uFIN modes (slot width 30 µm) in forward and backward direction. In backward direction, the two uFIN waves with opposite

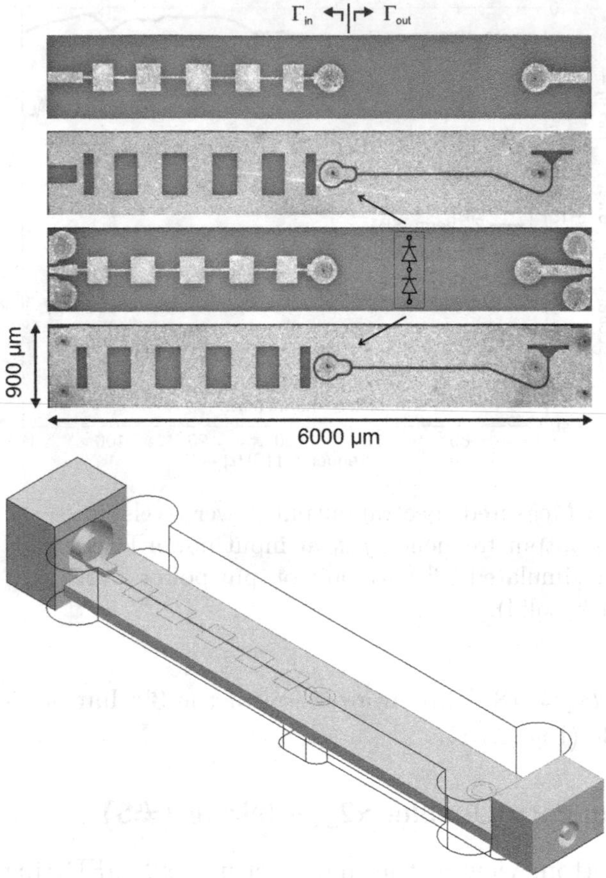

Figure 2.82: Top and bottom view of the manufactured x2_uFINring frequency doubler Al_2O_3 substrates (0.9 mm × 6.0 mm) and 3D model for coaxial interconnection.

phases propagate along the ring structure of Fig. 2.83, building a virtual ground behind the signal via. The distance from the diode junctions to the virtual ground is approximately a quarterwave length long to establish open circuit behaviour at the diode junctions. This ring structure was first introduced in [68] and modified in [69] to build single balanced mixers with RF frequency range of 40 to 70 GHz.

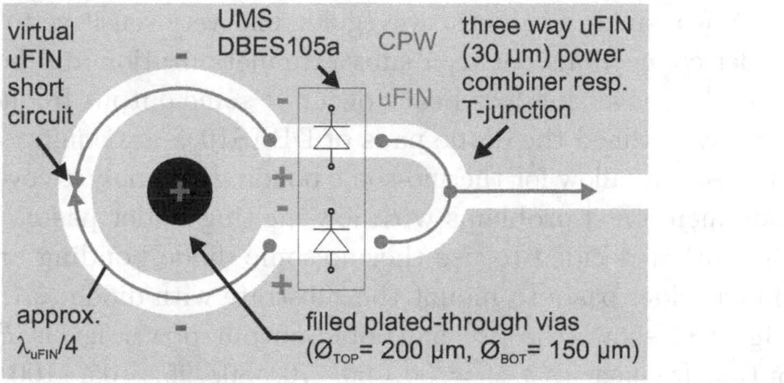

Figure 2.83: Detail view of the transition from x2_uFINring diode mounting structure to uFIN (bottom layer) waveguide.

The virtual ground already provides isolation between the input and output port, but the LPF is still needed to establish well-defined back reflection of the parasitic 2nd and 4th harmonic propagating in CPW mode. An uFIN three way power combiner allows for combination of both uFIN waves in forward direction. At the output the transition from uFIN to MSL of multiplier #2 from subsection 2.6.2 is used. The multiplier assembly procedure described in the beginning of this section has been used to assemble multiplier #1 to #4. In case of x2_uFINring the DBES105a diode is located at the bottom layer. Hence, if conventional solder paste (62Sn-36Pb-2Ag) is used for diode assembly to the substrate, conductive adhesive or low-melting solder based on In or Bi has to be used for assembly of substrate with diode to the Au-plated brass housing. Otherwise the diode solder connection would become affected. The low-melting solder pastes come with improved wettability, but lower electrical conductivity (compare Table 2.4). The author experienced problems with several assemblies, if low-melting solder paste interacts with conventional tin-based solder or soldering flux residues. Optical inspections of the soldered connections show discontinuous points and the joints are prone to cracking. As the contact area between Al_2O_3 substrate and mechanical housing is quite small, sufficient cleaning can hardly be done. The utilized uFIN is a single conductor waveguide and very sensitive to failures of solder connections between substrate metallization and housing. As a result, conversion loss peaks occur at some output frequencies. As already outlined the diode pads of DBES105a are equipped with Au bumps, that allow for thermo-sonic bonding assembly, to overcome the aforementioned problems with low-melting solder paste. Therefore, the author decided to use thermo-sonic diode bonding and conventional solder paste to mount the substrate with diode to the housing. Fig. 2.84 shows the 2nd harmonic output power levels P_{out} at the output frequencies f_{out} = 50, 60, 70, 80, 95, 105, 100, 110 GHz versus the fundamental input power level P_{in}. Measurement data in the frequency range of 50 to 110 GHz are within the gray shaded area. At an input power level of P_{in} = 18 dBm, the achieved output power levels are within the range of −3.4 to 6.8 dBm. A comparison of the measured conversion loss (solid lines) at P_{in} = 14, 16, 18 dBm with results from co-simulation (dashed lines) is given in Fig. 2.85. The conversion loss values are below 15 dB from 50 to 89 GHz and below 22 dB from 50 to 110 GHz. Spectral measurements up to the 3rd harmonic have been performed. At P_{in} = 18 dBm suppression of the 3rd harmonic is better than 15 dBc (Fig. 2.86).

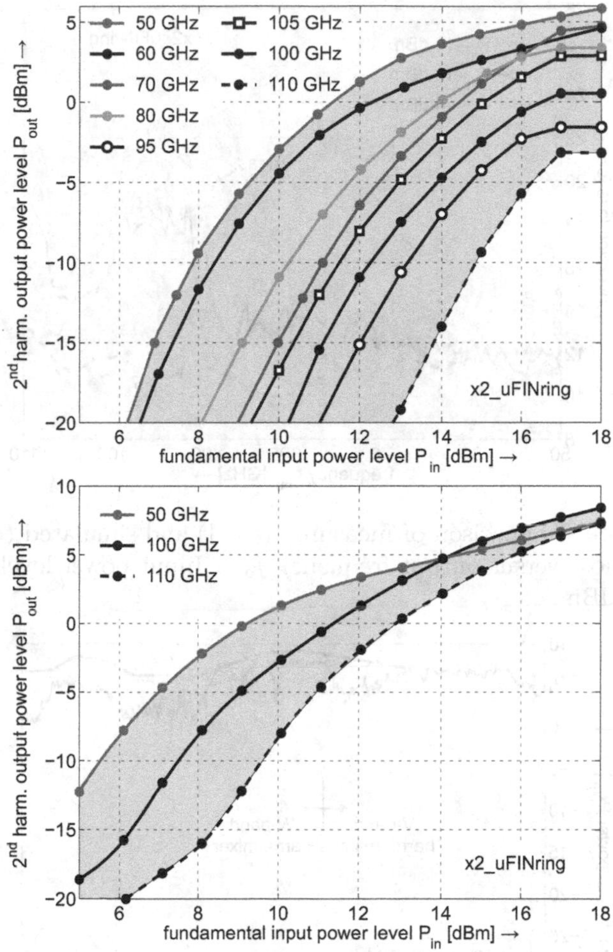

Figure 2.84: 2nd harmonic output power levels P_{out} at the output frequencies f_{out} = 50, 60, 70, 80, 95, 105, 100, 110 GHz versus the fundamental input power level P_{in}. Measurement data (top) and simulation data (bottom) in the frequency range of 50 to 110 GHz are within the gray shaded area.

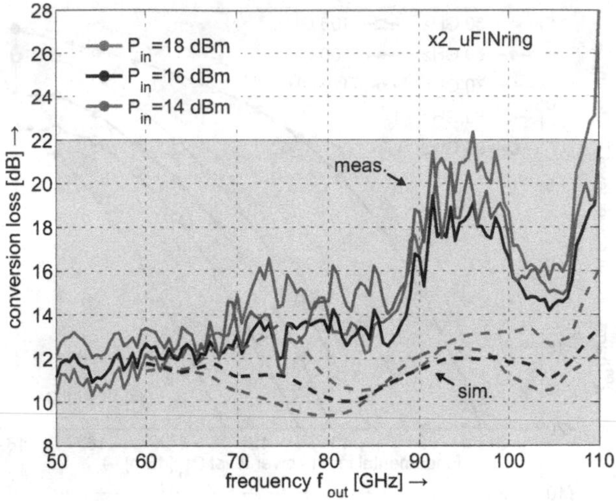

Figure 2.85: Comparison of measured (solid) and simulated (dashed) conversion loss versus output frequency f_{out}. Input power levels P_{in} = 14, 16, 18 dBm.

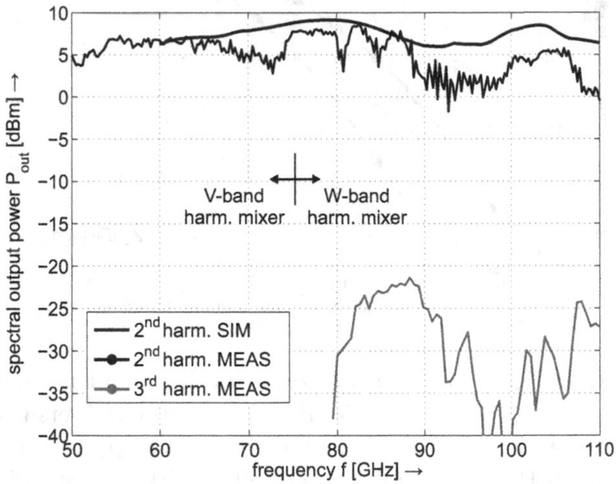

Figure 2.86: Measured spectral output power levels up to the 4th harmonic versus output frequency f_{out} at input power level P_{in} = 18 dBm. Compared to simulated 3rd harmonic output power levels with P_{in} = 18 dBm (black, solid).

Comparison and Conclusion Fig. 2.87 compares the measured output power levels P_{out} versus output frequency f_{out} at an input drive level of 18 dBm of the proposed multipliers (#1 to #5). Obviously, the frequency tripler x3_HE$_1$, frequency doublers x2_uFIN and x2_uFINring cover the largest bandwidth with the best efficiency. Conversion loss values around 18 dB from 60 to 95 GHz and below 22 dB from 60 to 110 GHz are achieved with frequency tripler #1. Frequency doubler #4 achieves conversion loss values below 16 dB from 50 to 88 GHz and below 23 dB from 50 to 110 GHz. Frequency doubler #5 operates with conversion loss values below 15 dB from 50 to 89 GHz and below 22 dB from 50 to 110 GHz. The signal generators, available for characterization, restrict the input power level to a maximum of P_{in} = 18 dBm[43]. As it can be seen from Fig. 2.57, Fig. 2.65, Fig. 2.73, Fig. 2.79, and Fig. 2.84 the multipliers are not entirely saturated at P_{in} = 18 dBm. Hence, for applications where maximum output power instead of best efficiency is preferred, the reported maximum output power levels can be increased with $P_{in} > 18$ dBm.

The presented measurement data in subsection 2.6.1 to subsection 2.6.5 deviate from the predicted co-simulation results. The main reasons for deviations are the applied Schottky junction modelling and underestimation of dielectric and conductor loss. Furthermore, the influence of the coaxial connectors is not entirely included within simulations. More sophisticated Schottky junction modelling approaches have been reported [53], but can hardly be utilized with commercial diodes due to insufficient information about material and geometry parameters of the diode.

A comprehensive comparison of the proposed frequency multipliers with reported work is given in Table 2.6. There are only a few reported multipliers [44, 70] that operate over comparable fractional bandwidths as the proposed multipliers (column four of Table 2.6). The reader should keep in mind the results of MIC multipliers are

[43] In fact, the actual generator output power level of the driving signal generator is lower than assumed (18 dBm). Hence, the measured conversion loss values of the proposed multipliers are slightly better than it is shown.

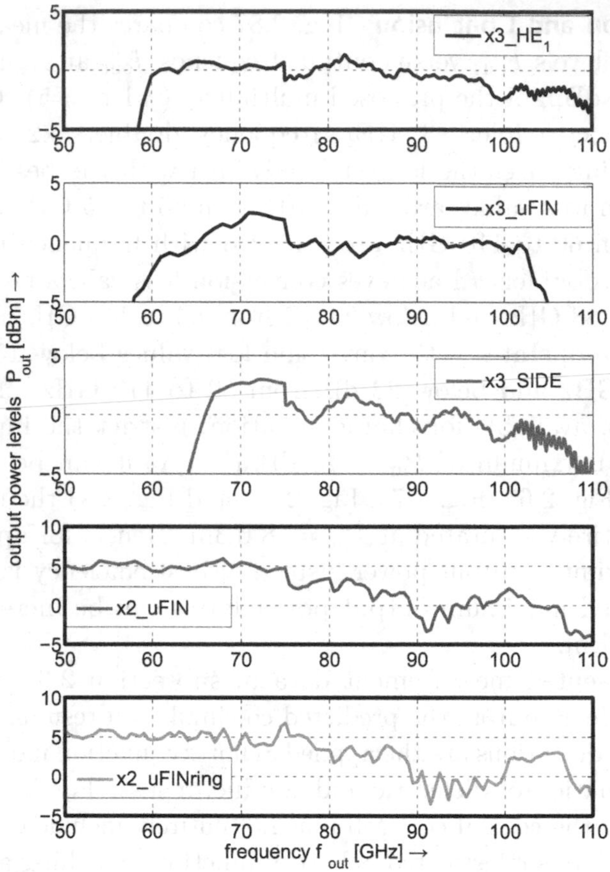

Figure 2.87: Comparison of the measured output power levels versus output frequency of the proposed multipliers (#1 to #5) with an input drive level of 18 dBm.

from on-wafer measurements of the bare die and not from a fully connectorized module, as it is with the proposed multipliers. The necessary interconnect solutions add loss at the in- and output. Considering this fact, the proposed multipliers achieve similar conversion loss values compared to MIC multipliers, but operate over greater bandwidth. Consequently, the actually necessary input power level

P_{in} of an MIC multiplier within front end modules is higher than indicated in Table 2.6.

In this section, innovative planar frequency multiplier designs dedicated to system design of VNA and signal generator frequency extension modules, as well as front end modules of ATS are presented. For cost reasons, it is preferred to manufacture the planar devices required for these modules on a single ceramic wafer with a certain set of design rules. Furthermore, restrictions from the mechanical housing and the intended level of integration make it necessary to individually choose the right multiplier architecture for a certain module. For example, multiplier #2, #4, and #5 can be directly connected to other uFIN components, if manufactured without the transition to MSL at the output, whereas multiplier #1 and #3 are best used with other MSL components.

The proposed frequency multipliers are based on cost effective commercially available GaAs Schottky diodes and conventional thin-film technology for high permittivity 5 mil Al_2O_3 instead of expensive, fragile SiO_2. The balanced multipliers do not require DC biasing or ferrite absorber material. Adequate operation is fully defined by the planar structure. Requirements on the mechanical housing are moderate, and the frequency responses are flat enough to be handled by automatic level control loops. In conjunction with commercial gallium nitride (GaN), indium phosphide (InP) or GaAs power amplifiers the achieved broadband operation and efficiency of the proposed hybrid frequency multipliers are highly suitable for the aforementioned fields of application. The presented planar architectures are not restricted to the focussed frequency ranges and substrate material Al_2O_3. In combination with appropriate THz diodes and reduced substrate dimensions, the designs are promising candidates for higher operating frequencies.

Table 2.6: Comprehensive Comparison of Reported Frequency Multipliers

references		output freq. range [GHz]	FBW [%]	diode process	mult. fact.	recom. P_{in} [dBm]	conv. loss min/max [dB]	P_{outmax} [dBm]	bias	size [mm]
[71]	m	87 to 102	16	TRW 0.15 µm GaAs pHEMT*	3	10 to 16	18/20	< −3	no	1.5/1
[43]	m	75 to 110	38	UMS BES GaAs Schottky diode*	3	10 to 18	17.2/20.6	< 3	yes	2/0.74/0.1
[44]	m	60 to 110	**59**	WIN 0.15 µm GaAs mHEMT*	3	15	13.2/20	< 5	no	1/1
[72]	h	93 to 99	6	WIN 0.15 µm GaAs pHEMT*	3	5	19/23	< −10	yes	2.5/1.5
[73]	m	94 to 100	6	0.25 µm GaAs pHEMT	3	–	–	5.5	yes	1.5/1.5
[74]	m	90 to 100	11	InAlGaAs/InP HBV◇	3	20 to 26	7.4/17	19.3	no	n/a
[75]	m	67.5 to 84	22	0.15 µm GaAs pHEMT	3	10	4.3/7	7.5	yes	1/1.5
[45]	m	72 to 90	22	GaAs Schottky diode*	3	13	18.5/21.2	< −5	no	1.1/1.4/0.1
[76]	h	75 to 110	38	n/a*	3	16	24•	n/a	no	19.1/19.1/25.4
[77]	h	70 to 110	44	n/a*	3	18 to 25	13/15	n/a	no	25.4/25.4/20.3
[50–52]	m	75 to 100	29	0.13 µm GaAs pHEMT*	2	13	−3/2	15	yes	2.25/3
[46–49]	m	70 to 78	11	GaAs Schottky diode◇	2	10 to 27	5.3/8.6	> 15	yes	1.97°/2.6°
[46–49]	m	70 to 79	12	GaAs Schottky diode◇	2	10 to 27	4.4/7.3	> 15	yes	1.97°/2.6°
[70]	m	50 to 128	**88**	WIN 0.15 µm GaAs pHEMT*	2	12.5	12.5/15	−2.5	no	0.56/0.42
[75]	m	67 to 84	23	0.15 µm GaAs pHEMT	2	16	4/13	10	yes	n/a
[78]	h	94 to 116	21	GaAs Schottky diode*	2	7 to 13	15.2/23	−2.2	no	n/a
[79]	h	64 to 78	20	JPL GaAs Schottky diode◇	2	23 to 27	8.4/12	> 18.5	yes	n/a
[80]	h	75 to 110	38	n/a*	2	16	22•	n/a	no	28.6/28.6/15.2
[81]	h	75 to 110	38	n/a*	2	24 to 27	9.6/11.6	n/a	yes	30.5/30.5/20.3
Fig. 2.58 (#1)	h	60 to 110	**59**	UMS BES GaAs Schottky diode*	3	≥ 12	16/22	< 1	no	0.9/6/0.14
Fig. 2.66 (#2)	h	60 to 110	**59**	UMS BES GaAs Schottky diode*	3	≥ 12	15/27	< 2	no	0.9/6/0.14
Fig. 2.74 (#3)	h	67 to 110	**49**	UMS BES GaAs Schottky diode*	3	≥ 14	15/24	< 2	no	0.9/6/0.14
Fig. 2.80 (#4)	h	50 to 110	**75**	UMS BES GaAs Schottky diode*	2	≥ 14	11/23	< 5	no	0.9/6/0.14
Fig. 2.85 (#5)	h	50 to 110	**75**	UMS BES GaAs Schottky diode*	2	≥ 14	11/22	< 6.8	no	0.9/6/0.14

Min. and max. conversion loss values do not necessarily correspond to the same input power level as in the case of the max. output power level P_{outmax}. ⋆ / ◇ = Varistor / Varactor mode. m / h = monolithic / hybrid realization. ∘ Estimated from die photography in [46]. • Typical values from datasheet.

2.7 D-Band Frequency Doubler and Y-Band Frequency Tripler

The presented frequency multipliers for 50 / 60 to 110 GHz of the preceding section are based on the UMS DBES105a diode. The physical dimensions of this discrete diode, especially the distance of both Schottky junctions, do not suggest operation at frequencies above 110 GHz[44]. Anyway, frequency multipliers for D-band and Y-band frequency ranges based on the DBES105a diode are valuable for extending the frequency range of cost driven industrial radar modules. At the expense of conversion loss performance, these millimeter-wave components are built up at very low cost utilizing PCB processing.

This section describes a zero bias D-band frequency doubler and Y-band frequency tripler based on pure PTFE material, split block construction and the DBES105a diode. At an input drive level range of 16 to 18 dBm the multipliers [I7] deliver output power levels $P_{\text{out}} \in [-5, 5]$ dBm from 100 to 160 GHz and $P_{\text{out}} \in [-20, -11]$ dBm from 180 to 230 GHz. As an outlook, further innovative multiplier designs are shown. Several transitions from planar waveguide to rectangular waveguides and the analytical design procedure of Dolph-Chebyshev tapers (subsection 2.7.1) are presented. A brief uncertainty analysis of spectral measurements with harmonic mixers in front of spectrum analyzers is given in subsection 2.7.2.

Introduction, Manufacturing and Assembly Successful planar Schottky diode frequency multipliers for D-band (110 to 170 GHz) and Y-band (170 to 260 GHz) output frequency ranges have been reported. These are hybrid circuits using thin SiO_2 substrates with proprietary discrete diodes or monolithic millimeter-wave integrated circuits (MMICs). Industrial short-range radar modules for various applications like distance or liquid level measurements and material

[44] The single Schottky junction of UMS DBES105a performs well up to 300 GHz and beyond. In [38], page 104, a single junction of the discrete diode is singularized in a further dicing step and used for a power detector design. In [82], a 150 to 151 GHz frequency doubler utilizing DBES105a diodes is presented.

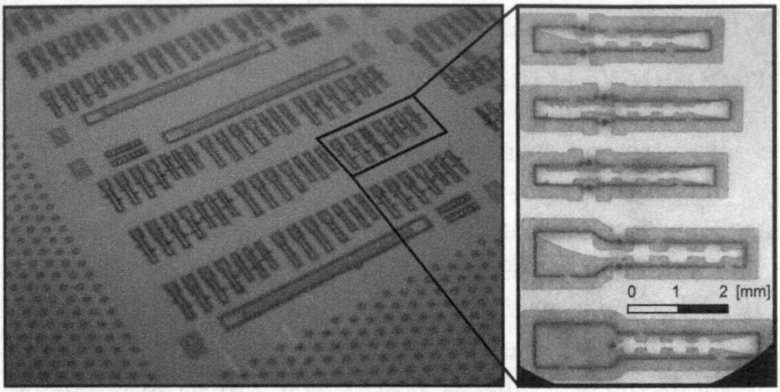

Figure 2.88: Photograph of manufactured Fastfilm 27 laminate with close up view of the proposed frequency multipliers.

parameter monitoring are narrow band, highly integrated and therefore bound to a certain technology. Frequency extension of existing modules utilizing cost effective PCB technology requires millimeter-wave frequency multipliers based on the same technology. Although the aforementioned radar modules operate at fractional bandwidths below 10 %, it is desirable to develop broadband multipliers that can be used within many different modules. Broadband cost effective D- and Y-band multipliers are further interesting for several imaging applications, where many multipliers arranged in an array are required.

The proposed multipliers are built on pure PTFE material (Fastfilm 27 from Taconic Farms, Inc.) with thickness of 50 µm and 17 µm copper cladding, processed by commercially available PCB technology. To improve lateral displacement between top and bottom layer etching, the entire PCB panel (430 × 260) mm^2 is partitioned into smaller regions (60 × 60) mm^2 that are equipped with individual alignment marks. Fig. 2.88 illustrates such a partition and close up view of the proposed multipliers.

D-Band Frequency Doubler In compliance with these design rules, an input lowpass filter (LPF) of 9th order with cut-off frequency

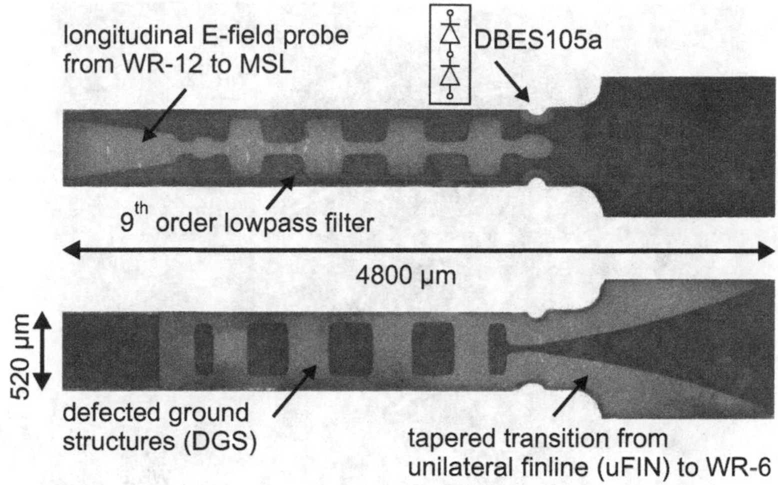

Figure 2.89: Photographs of top and bottom layer of the proposed D-band (WR-6) frequency doubler.

Table 2.7: 9th order stepped impedance MSL lowpass filter

filter elements	strip widths [µm]	lengths [µm]	inductance and capacitance values at f_c
1 and 9	75	128	5.75 pH
2 and 8	370	216	42.26 fF
3 and 7	75	283	12.72 pH
4 and 6	370	259	50.67 fF
5	75	299	13.44 pH

filter order $N = 9$, cut-off frequency $f_{\mathrm{c\,1\,dB}} = 90.7$ GHz, $f_{\mathrm{c\,3\,dB}} = 94$ GHz, passband ripple $r_{\mathrm{dB}} = 0.01$ dB, stopband rejection of 25 dB at $1.2 \times f_{\mathrm{c\,3\,dB}}$.

$f_{\mathrm{c\,1\,dB}} = 91$ GHz and passband ripple of 0.01 dB has been designed. Table 2.7 summarizes the filter parameters.

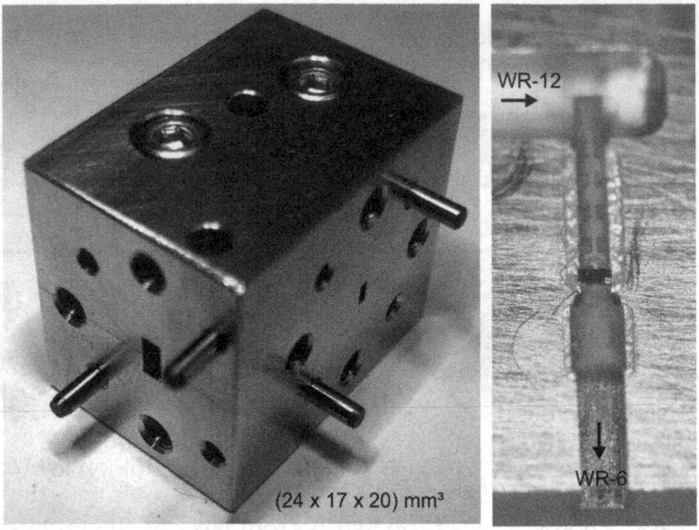

Figure 2.90: Photographs of mechanical housing and laminate assembly of the proposed D-band (WR-6) frequency doubler.

To achieve the required inductance values, the high impedance LPF sections are equipped with defected ground structures (DGS). The first and last filter elements are inductive to ensure open circuit stopband behavior for the 2nd harmonic output signal at the input. Cross-sectional views of the low (C) and high (L) impedance MSL, magnitude of the electric field strengths at 95 GHz and real parts of the characteristic impedances $\mathrm{Re}\,(Z_{\mathrm{MSL}})$ of 50 Ω MSL, high and low MSL are given in Fig. 2.91. Simulation results of Fig. 2.92 predict 20 dB rejection at 110 GHz.

In cascade to the LPF, the DBES105a diode with two Schottky junctions in series tee configuration is soldered to PTFE laminate at the position indicated in Fig. 2.89. The generated 2nd harmonic signal excites an unilateral finline (uFIN) mode (bottom layer) that exits the planar circuit by a tapered transition to WR-6 waveguide. The last filter DGS structure serves as quarterwave backshort of the MSL to uFIN transition.

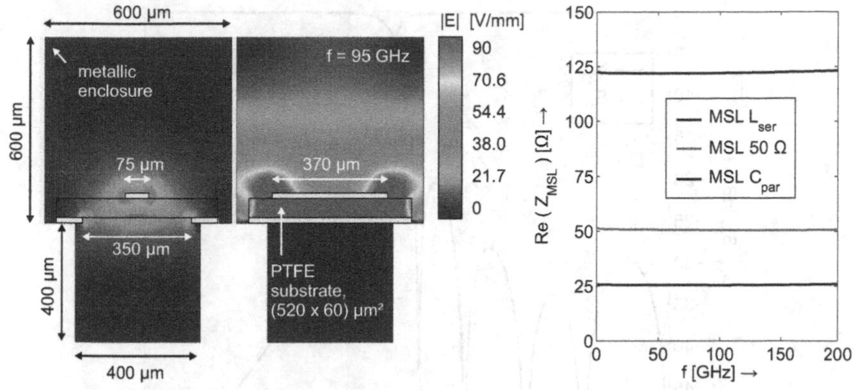

Figure 2.91: Cross-sectional views of the low (C) and high (L) impedance MSL, magnitude of the electric field strengths at 95 GHz and real parts of the characteristic impedances $\mathrm{Re}\,(Z_{\mathrm{MSL}})$ of 50 Ω MSL, high and low MSL versus frequency. $\mathrm{Re}\,(Z_{\mathrm{MSL-L}}) = 121.98\ \Omega, \mathrm{Re}\,(Z_{\mathrm{MSL-C}}) = 25.5\ \Omega$ and the corresponding effective permittivities are $\epsilon_{\mathrm{reff-L}} = 1.49, \epsilon_{\mathrm{reff-C}} = 2.24$ at $f = 95$ GHz.

A cross-sectional view of the uFIN, magnitude of the electric field strengths at 170 GHz and real parts of the characteristic impedances $\mathrm{Re}\,(Z_{\mathrm{uFIN}})$ of the first three hybrid modes are shown in Fig. 2.93. The taper contour of the transition from uFIN to WR-6 waveguide follows the Dolph-Chebyshev function, as it is outlined in subsection 2.7.1. Fig. 2.94 includes simulated scattering parameters, which also show propagation of the next higher order TE_{20} like mode. The higher order unilateral finline modes, which have even lower cut-off frequencies according to Fig. 2.93 are not excited by the transition. The milled channel of the MSL part ensure mono mode operation across the full D-band.

Fig. 2.95 shows the measured input P_{in} (red) and output power levels P_{out} versus input f_{in} (red) and output frequencies f_{out}. Precise scalar P_{out} measurements have been performed with VDI Erickson PM4 calorimeter (black) according to the procedure outlined in [83]. Scalar power measurements do not allow for distinguishing spectral components of the harmonic content.

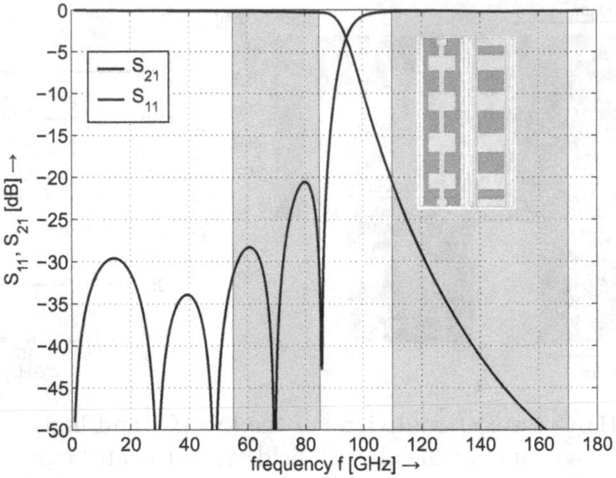

Figure 2.92: Simulated transmission (blue) and reflection coefficient (black) of the 9th order stepped impedance MSL lowpass filter versus frequency.

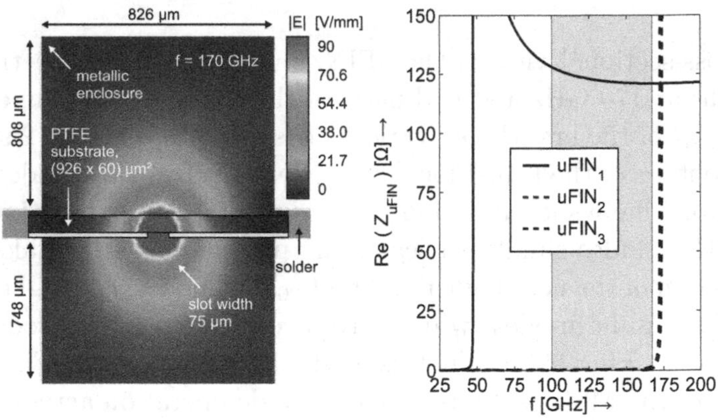

Figure 2.93: Cross-sectional view of the unilateral finline, magnitude of the electric field strengths at 170 GHz and real parts of the characteristic impedances $\mathrm{Re}\,(Z_{\mathrm{uFIN}})$ of the first three hybrid modes versus frequency.

Harmonic mixer measurements (blue) with rather poor accuracy[45] (at least 6 dBm, subsection 2.7.2) have been applied to verify the

[45]The accuracy of harmonic mixer measurements is drastically decreased if the device under test is poorly matched, which is the case with the proposed multi-

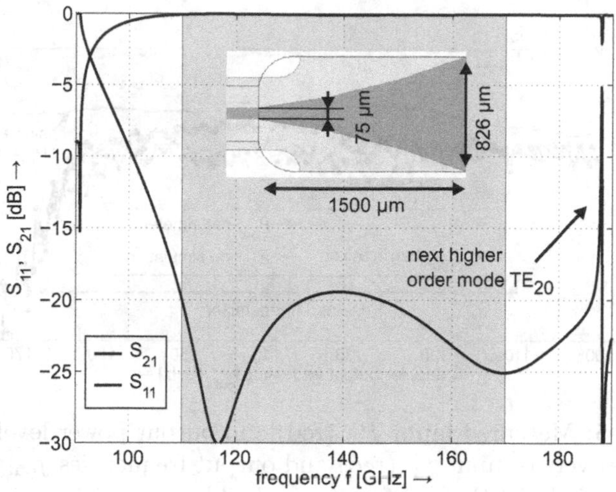

Figure 2.94: Simulated transmission (blue) and reflection coefficients (black) of the tapered (Dolph-Chebyshev) transition from unilateral finline (uFIN) to WR-6 versus frequency.

calorimetric measurements consider the envisaged spectral component (2nd harmonic).

The gray shaded area in Fig. 2.95 indicates the frequency range with sufficient input power level. An input power level of 16 dBm could be provided up to 65 GHz by commercial signal generator. From 65 to 81 GHz an amplified[46] E-band source with power levels from 16 to 19 dBm has been used in the test setup. The black curve in Fig. 2.95 without dots show P_{out} results from the synthesis procedure, that correspond to a constant input power level of 18 dBm. The measurement results are not corrected for insertion loss of interconnecting waveguides at the output port. The calorimetric P_{out} measurements of the D-Band doubler illustrate multiplier operation with 15 to 20 dB conversion loss from 100 to 160 GHz, which means output power levels in the range of -5 to 5 dBm. Considering the already mentioned

pliers. Coarse conversion loss versus frequency look-up tables and deviations from the required LO power level are further sources of errors.

[46] Amplifier RPG E-MPA-65-85-24-20 from Radiometer Physics GmbH.

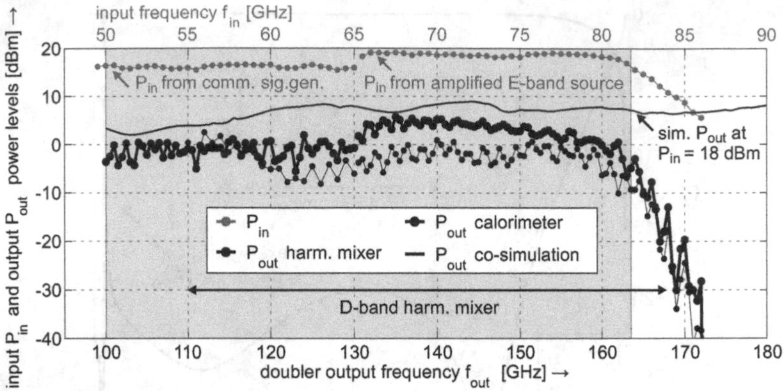

Figure 2.95: Measured input P_{in} (red) and output power levels P_{out} (black, blue) versus input f_{in} (red) and output frequencies f_{out} (black) of the proposed D-band (WR-6) frequency doubler.

problems with PCB manufacturing technology, this is in reasonable agreement to the predicted conversion loss of 10 to 15 dB. Due to insufficient input power levels no statement can be made about the doubler's performance above 160 GHz.

Y-Band Frequency Tripler The input signal of the Y-band frequency tripler of Fig. 2.96 and Fig. 2.97 enters a 7th order lowpass filter (LPF) with cut-off frequency $f_{\text{c}\,1\,\text{dB}} = 109$ GHz and passband ripple of 0.01 dB. Table 2.8 summarizes the filter parameters.

The first and last filter elements are inductive to ensure open circuit stopband behavior for the 3rd harmonic output signal at the input. Cross-sectional views of the low (C) and high (L) impedance MSL, magnitude of the electric field strengths at 116 GHz and real parts of the characteristic impedances $\text{Re}\,(Z_{\text{MSL}})$ of 50 Ω MSL, high and low MSL are given in Fig. 2.98.

To achieve the required inductance values and bandwidth of stopband, the high impedance LPF sections are equipped with defected ground structures (DGS). Simulated scattering parameters of the filter are shown in Fig. 2.99. Greater than 25 dB rejection is predicted at

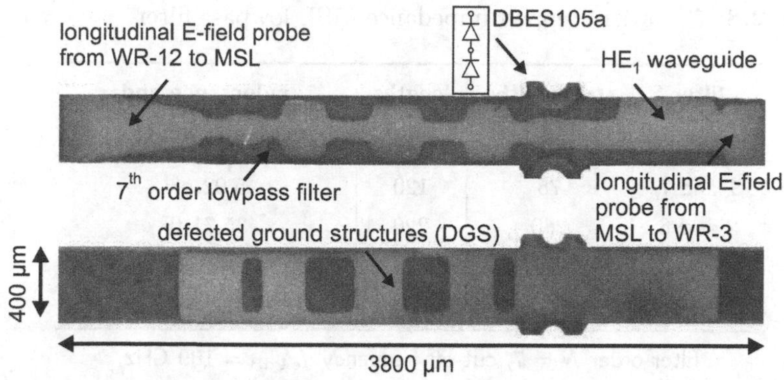

Figure 2.96: Photographs of top and bottom layer of the proposed Y-band frequency tripler with WR-3 interface.

Figure 2.97: Photographs of mechanical housing and laminate assembly of the proposed Y-band frequency tripler with WR-3 interface.

170 GHz. The dashed curve corresponds to the filter's transmission coefficient without utilization of DGS technique. Less bandwidth of stopband and a second passband at 250 GHz are visible.

In cascade to the LPF, the DBES105a diode with two Schottky junctions in series tee configuration is soldered to PTFE laminate at the position indicated in Fig. 2.96. The generated 3rd harmonic signal of the Y-band multiplier propagates along a HE_1 waveguide [60], which acts as a highpass filter (HPF). Fig. 2.100 illustrates a cross-sectional view of the HE_1 waveguide, magnitude of the electric field strengths at 240 GHz and real parts of the characteristic impedances $\mathrm{Re}\,(Z_{\mathrm{HE}})$ of the first two hybrid modes. Mono mode operation across Y-band (170 to 260 GHz) is achieved. A compromise between fundamental signal

Table 2.8: 7th order stepped impedance MSL lowpass filter

filter elements	strip widths [μm]	lengths [μm]	inductance and capacitance values at f_c
1 and 7	75	120	5.02 pH
2 and 6	280	220	35.71 fF
3 and 5	75	263	11.01 pH
4	280	258	41.88 fF

filter order $N = 7$, cut-off frequency $f_{c\,1\,dB} = 109$ GHz, $f_{c\,3\,dB} = 116$ GHz, passband ripple $r_{dB} = 0.01$ dB, stopband rejection of 15 dB at $1.2 \times f_{c\,3\,dB}$.

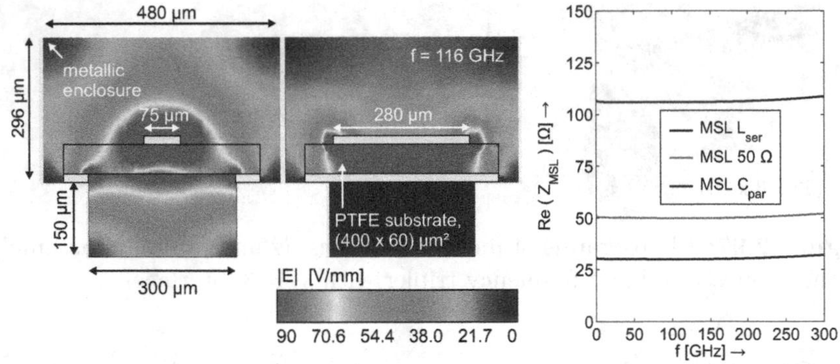

Figure 2.98: Cross-sectional views of the low (C) and high (L) impedance MSL, magnitude of the electric field strengths at 116 GHz and real parts of the characteristic impedances $\mathrm{Re}\,(Z_{\mathrm{MSL}})$ of 50 Ω MSL, high and low MSL versus frequency. $\mathrm{Re}\,(Z_{\mathrm{MSL-L}}) = 105.91\ \Omega, \mathrm{Re}\,(Z_{\mathrm{MSL-C}}) = 29.97\ \Omega$ and the corresponding effective permittivities are $\epsilon_{\mathrm{reff-L}} = 1.49, \epsilon_{\mathrm{reff-C}} = 2.13$ at $f = 116$ GHz.

rejection and additional insertion loss caused by HE_1 propagation is found by an HE_1 waveguide length of 0.5 mm. The solid curves in Fig. 2.101 correspond to simulation results of the highpass filter. Using the design procedure for distributed LPF based on quarterwave

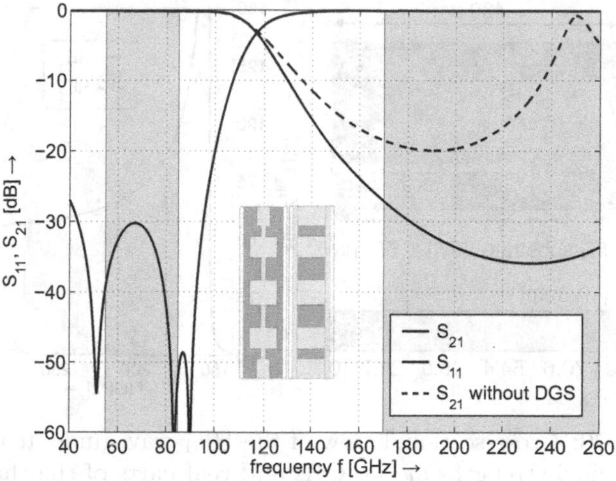

Figure 2.99: Simulated transmission (blue) and reflection coefficient (black) of the 7th order stepped impedance MSL lowpass filter versus frequency. The dashed curve corresponds to a filter design without defected ground structures.

prototypes from the appendix A, a notch LPF is developed. It is based on uFIN notches placed in quarterwave distance [84, 85] and constitutes an idler circuit for the 5th harmonic. The filter integration leads to an unintended altering of the HE_1 waveguide, which shifts the cut-off frequency to higher frequencies and therefore limits the operational bandwidth of the tripler (dashed curves in Fig. 2.101). A manufactured tripler using the integrated notch LPF is illustrated in the outlook of this section (Fig. 2.108).

The HE_1 waveguide is coupled to a WR-3 longitudinal E-field probe by a short section of 50 Ω MSL. Simulation results of the E-field probe are shown in Fig. 2.102. The milled channel of the Y-band MSL ensures mono mode operation up to 250 GHz (part of Y-band) only. Hence, the next higher order mode of the MSL channel restricts the output bandwidth of the tripler (180 to 270 GHz) and not the next higher order mode TE_{20} of the waveguide, which is also visible in Fig. 2.102 at about 345 GHz. Although Y-band frequency range

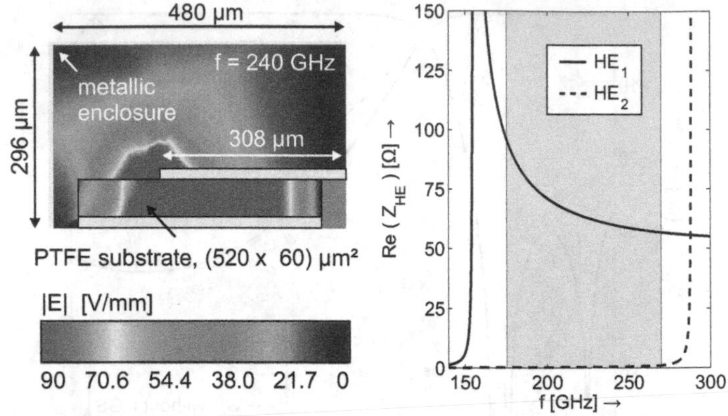

Figure 2.100: Cross-sectional view of the HE_1 waveguide, magnitude of the electric field strengths at 240 GHz and real parts of the characteristic impedances $\mathrm{Re}\,(Z_{\mathrm{HE}})$ of the first two hybrid modes versus frequency.

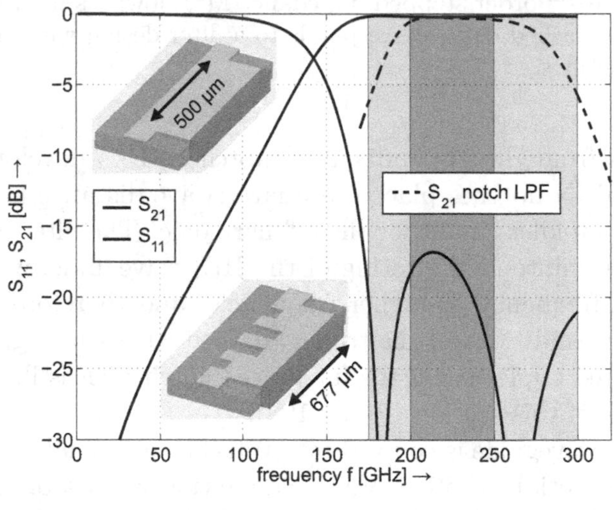

Figure 2.101: Simulated transmission (blue) and reflection coefficients (black) of the HE_1 waveguide as HPF with (dashed) and without (solid) integrated notch LPF.

(WR-4) is envisaged, a WR-3 waveguide is manufactured and used at Y-band frequencies (lower end of WR-3 frequency range).

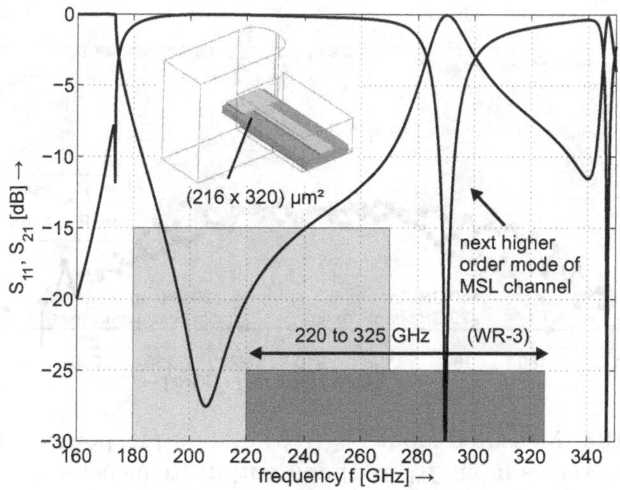

Figure 2.102: Simulated transmission (blue) and reflection coefficient (red) of a longitudinal E-field probe transition from MSL to WR-3.

Fig. 2.103 shows the measured input P_{in} (red) and output power levels P_{out} versus input f_{in} (red) and output frequencies f_{out}. The measurement setup is the same as it is with the D-band frequency doubler of the preceding paragraph.

Precise scalar P_{out} measurements have been performed with VDI Erickson PM4 calorimeter (black) according to the procedure outlined in [83]. Scalar power measurements do not allow for distinguishing spectral components of the harmonic content. Harmonic mixer measurements (blue) with rather poor accuracy (at least 6 dBm, subsection 2.7.2) have been applied to verify the calorimetric measurements consider the envisaged spectral component (3rd harmonic).

The gray shaded area in Fig. 2.103 indicates the frequency range with sufficient input power level. An input power level of 16 dBm could be provided up to 65 GHz by commercial signal generator. From 65 to 81 GHz an amplified[47] E-band source with power levels from 16 to 19 dBm has been used in the test setup. The black curve in

[47]Amplifier RPG E-MPA-65-85-24-20 from Radiometer Physics GmbH.

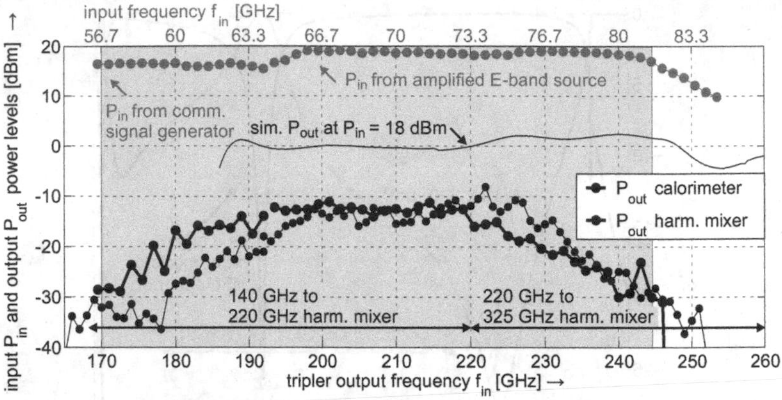

Figure 2.103: Measured input P_{in} (red) and output power levels P_{out} (black, blue) versus input f_{in} (red) and output frequencies f_{out} (black) of the proposed Y-band frequency tripler with WR-3 interface.

Fig. 2.103 without dots show P_{out} results from the synthesis procedure, that correspond to a constant input power level of 18 dBm. The measurement results are not corrected for insertion loss of interconnecting waveguides at the output ports.

The applied synthesis procedure for the Y-band tripler predicts output power levels in the range of -5 to 2 dBm from 186 to 249 GHz at 18 dBm input drivel level. The measurement results of Fig. 2.103 show output power levels $P_{\text{out}} \in [-11, -20]$ dBm from 180 to 230 GHz, which is a lot less output power than predicted. Assembly problems can be almost excluded as these results have been verified with two separate structures. Decreased output power levels at Y-band frequencies are believed to belong to the diode package and the electrically large distance of the two Schottky junctions. The frequency shift of the maximum output power level to lower frequencies, visible in Fig. 2.103, might be best explained by deteriorated performance of the longitudinal E-field probe to WR-3 waveguide. Careful inspection of Fig. 2.96 shows the separation of HE_1 waveguide and WR-3 probe is affected by the PCB manufacturing technology.

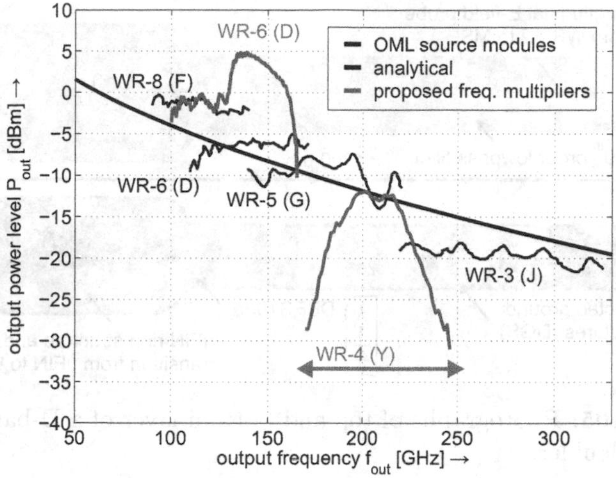

Figure 2.104: Comparison of measured output power levels P_{out} (green) of the D-band doubler and Y-band tripler with output power levels (blue) of the commercial signal generator series of OML Inc. versus frequency.

However, the reader should keep in mind that commercially available signal sources at Y-band frequencies deliver only -20 dBm of unleveled output power. Fig. 2.104 compares the measured output power levels of the proposed frequency multipliers with OML's signal generator frequency extension modules.

Efficient gallium arsenide (GaAs), gallium nitride (GaN) and indium phosphide (InP) E- and W-band MMIC power amplifiers recently became commercially available, therefore establishing 18 to 20 dBm input drive levels for the rather narrowband (10 %) radar modules can easily be realized. Whereas 15 dBm is the minimum necessary input power level. If MMIC driver amplifiers are used, bondwires to MSL instead of a WR-12 input waveguide can be utilized.

Conclusion and Outlook The developed D- and Y-band multipliers are capable for extending the frequency range of existing industrial short-range radar modules and might also be interesting for several imaging applications, where many cost effective multipliers arranged

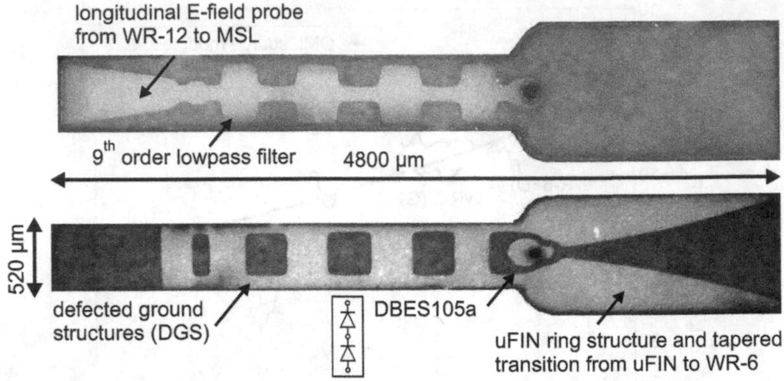

Figure 2.105: Photographs of top and bottom layer of a D-band (WR-6) frequency doubler.

in an array are required. Targeted improvements of the PCB manufacturing process, especially the realization of plated-through via holes, could further enhance the multipliers' performance.

Within the scope of this study, four more frequency multipliers are designed. Measurement results are not available, but the basic design idea is explained in the following.

The frequency doubler of Fig. 2.105 is similar to the doubler of subsection 2.6.5, which includes an explanation of the uFIN ring structure. The doubler utilizes the input LPF from Fig. 2.92. At the output the tapered transition from uFIN to WR-6 waveguide of Fig. 2.94 is used.

A D-band frequency tripler is depicted in Fig. 2.106. It utilizes the LPF of Fig. 2.99 in conjunction with the tapered (Dolph-Chebyshev) antipodal transition from MSL to WR-6 from Fig. 2.107. A transition from MSL to 1.85 mm coaxial connector is sufficient to cover the input frequency range.

Fig. 2.108 illustrates a Y-band frequency tripler, which is identical to the tripler of the preceding paragraph, but utilizes an integrated notch lowpass filter [84, 85]. The pitch of the notches is 159 µm and the individual lengths of the notches are $\ell_1 = \ell_4 = 63$ µm,

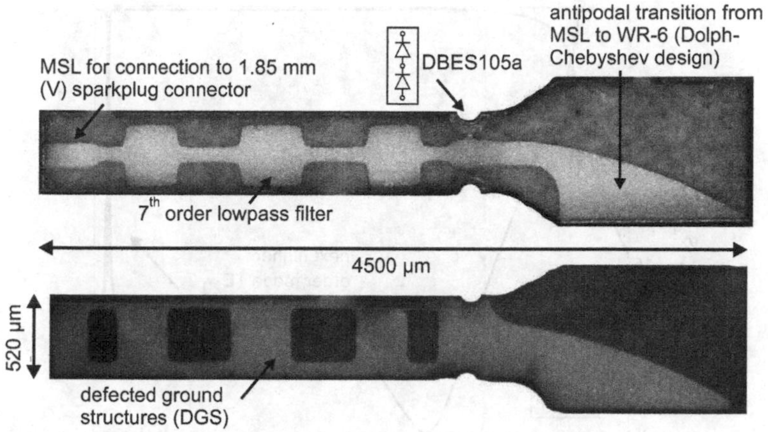

Figure 2.106: Photographs of top and bottom layer of a D-band (WR-6) frequency tripler.

$\ell_2 = \ell_3 = 139$ µm. The design procedure of distributed LPF based on quarterwave prototypes is given in the appendix A. The LPF constitutes an idler circuit for the 5th harmonic. Simulated scattering parameters are shown in Fig. 2.101. The filter integration leads to an unintended altering of the HE_1 waveguide, which shifts the cut-off frequency to higher frequencies and therefore limits the operational bandwidth of the tripler.

A Y-band frequency tripler is depicted in Fig. 2.109. It utilizes the LPF of Fig. 2.99 in conjunction with a modified version of the tapered (linear) antipodal transition from MSL to WR-3 from Fig. 2.110.

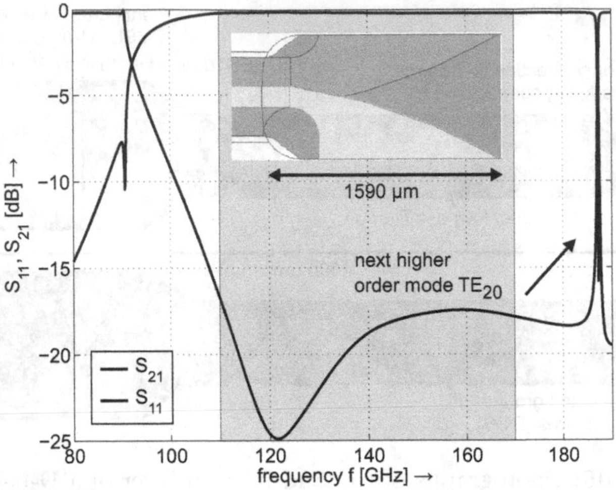

Figure 2.107: Simulated transmission (blue) and reflection coefficients (black) of the tapered (Dolph-Chebyshev) antipodal transition from MSL to WR-6 versus frequency.

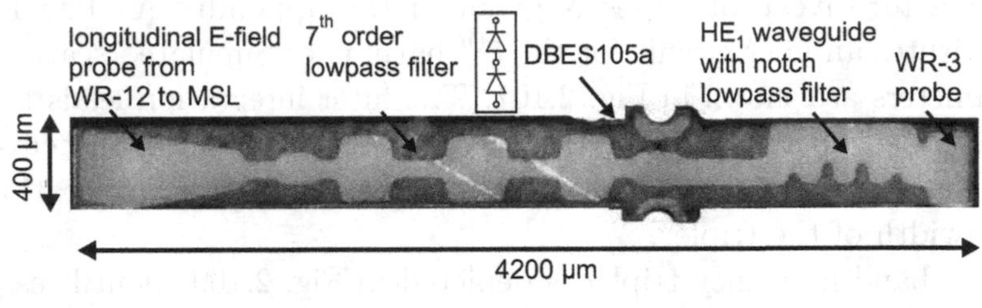

Figure 2.108: Photographs of top and bottom layer of the a Y-band (WR-3!) frequency tripler with integrated notch LPF.

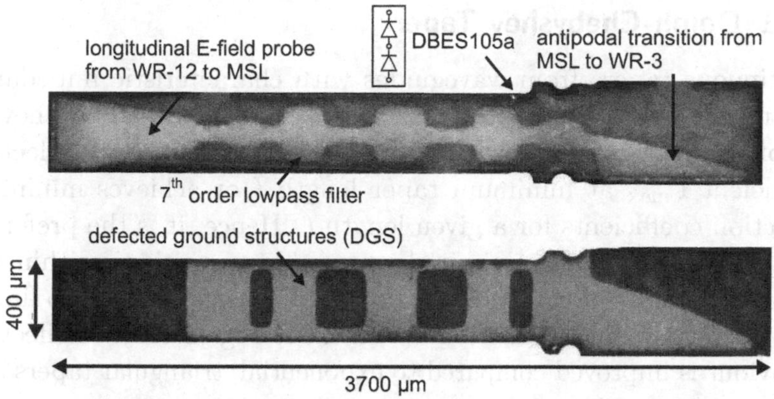

Figure 2.109: Photographs of top and bottom layer of a Y-band (WR-3!) frequency tripler.

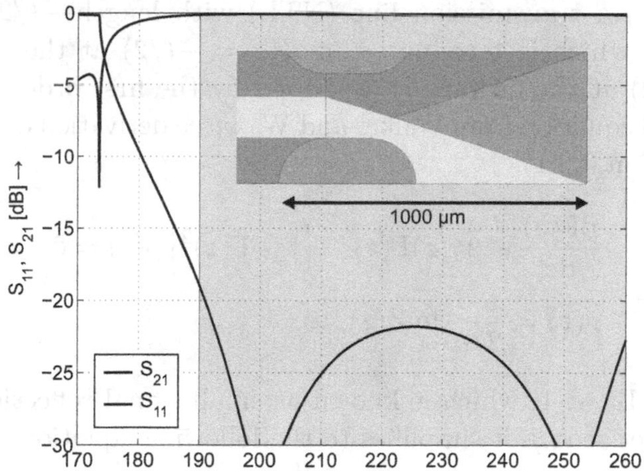

Figure 2.110: Simulated transmission (blue) and reflection coefficients (black) of the tapered (linear) antipodal transition from MSL to WR-3 versus frequency.

2.7.1 Dolph-Chebyshev Taper

Continuous tapers from waveguides with characteristic impedance Z_1 and Z_2, respectively, are considered. The Dolph-Chebyshev or Klopfenstein taper [86–88] provides a specified passband reflection coefficient $\Gamma_{\max}$ at minimum taper length ℓ or achieves minimum reflection coefficients for a given length ℓ. Hence, it is the preferred design solution for the required waveguide transitions within the proposed circuits of this thesis, especially hollow waveguide to finline and microstrip line transitions (Fig. 2.94, Fig. 2.107). Reflection behaviour is improved compared to exponential, triangular tapers and discrete multisection quarterwave transformers[48].

Derivation of the required characteristic impedance contour $Z(z)$ is given in the following. The overall reflection coefficient $\Gamma(z)$ of a nonuniform transmission line (NTL) with length $-\ell/2, \ldots, z = 0, \ldots, \ell/2$, which is terminated in $Z(z = -\ell/2)$ at the input and $Z(z = \ell/2)$ at the output, is described by the first order nonlinear differential equation from Walker and Wax (see derivation of Eq. (3.51) in subsection 3.2.1).

$$\begin{aligned} &\frac{\mathrm{d}\Gamma(z)}{\mathrm{d}z} - 2\gamma(z)\Gamma(z) + \left[1 - \Gamma(z)^2\right] p(z) = 0 \\ &p(z) = \frac{1}{2}\frac{\mathrm{d}}{\mathrm{d}z} \ln Z(z) \end{aligned} \tag{2.91}$$

In case of $\Gamma^2 \ll 1$, which is known as small signal reflection theory (page 193 et seqq.), it simplifies to the following equation.

$$\frac{\mathrm{d}\Gamma(z)}{\mathrm{d}z} - 2\gamma(z)\Gamma(z) + p(z) = 0 \tag{2.92}$$

Applying the boundary condition $\Gamma(\ell/2) = 0$ and assuming lossless transversal electromagnetic wave (TEM) propagation $\gamma = j\beta$, $\Gamma(z)$ is given by Eq. (2.93).

[48] In fact, Dolph-Chebyshev tapers are degenerated multisection quarterwave transformers with Chebyshev passband characteristics, infinite order and fixed length.

$$\begin{aligned}\Gamma(z) &= \int\limits_{-\ell/2}^{\ell/2} p(\zeta) \exp\left(-2\int\limits_{-\ell/2}^{\zeta} \gamma(\xi)\mathrm{d}\xi\right)\mathrm{d}\zeta \\ \Gamma \exp(j\beta\ell) &= \int\limits_{-\ell/2}^{\ell/2} p(\zeta)\exp(-2j\beta\zeta)\,\mathrm{d}\zeta\end{aligned} \tag{2.93}$$

Inspecting Eq. (2.93), $p(z)$ and $\Gamma e^{j\beta\ell}$ are recognized as a pair of Fourier transform. This allows for calculation of $p(z)$ from the taper's specification.

$$p(z) = \frac{1}{\pi}\int\limits_{-\infty}^{\infty} [\Gamma \exp(j\beta\ell)] \exp(j2\beta z)\,\mathrm{d}\beta \tag{2.94}$$

In case of a Chebyshev taper with infinite order and fixed length ℓ, Γ is given by Eq. (2.95) [86].

$$\Gamma \exp(j\beta\ell) = \Gamma_0 \frac{\cos\left((\beta\ell)^2 - A^2\right)}{\cosh(A)} \tag{2.95}$$

Whereas Γ_0 is the reflection coefficient without taper or taper length equal to zero $\Gamma_0 = \frac{Z_2 - Z_1}{Z_2 + Z_1}$. In [86] it is suggested to use $\Gamma_0 = \frac{1}{2}\ln\left(\frac{Z_2}{Z_1}\right)$ as an initial value for numerical calculation. To determine the constant value of A, the specified Γ_{max} and ℓ/λ are inserted into Eq. (2.95). A is known from Eq. (2.96).

$$0 = \Gamma_0 \frac{\cos\left((\beta\ell)^2 - A^2\right)}{\cosh(A)} - \Gamma_{\text{max}} \tag{2.96}$$

Inserting Eq. (2.95) into Eq. (2.94) and solving the integral leads to Eq. (2.97) with the modified Bessel function I_1 of the first kind with order one, and the unit impulse function δ.

$$p(z) = \frac{\Gamma_0}{\cosh(A)} \left(\frac{A^2}{\ell} \frac{I_1\left(A\sqrt{1-(2z/\ell)^2}\right)}{A\sqrt{1-(2z/\ell)^2}} + \frac{\delta(z-\ell/2)+\delta(z+\ell/2)}{2} \right)$$

$$p(z) = 0 \qquad |z| \leq \ell/2 \wedge |z| > \ell/2 \tag{2.97}$$

From direct integration of $p(z)$ (Eq. (2.91)), the Dolph-Chebyshev impedance contour $Z(z)$ is given by Eq. (2.98).

$$Z(z) = \exp\left(\frac{1}{2}\ln(Z_1 Z_2) + \frac{\Gamma_0}{\cosh(A)}\left(A^2\phi(2z/\ell, A) + H(z-\ell/2) + H(z+\ell/2)\right)\right) \qquad -\ell/2 \geq z \leq \ell/2 \tag{2.98}$$

The taper starts and ends with a uniform transmission line.

$$Z(z) = Z_2 \quad z > \ell/2 \qquad\qquad Z(z) = Z_1 \quad z < -\ell/2 \tag{2.99}$$

H denotes the Heaviside[49] or unit step function (Eq. (2.100)).

$$H(m) = \begin{cases} 0 & m < 0 \\ 1 & m \geq 0 \end{cases} \tag{2.100}$$

ϕ is numerically calculated from Eq. (2.101).

$$\phi(k, A) = -\phi(-k, A) = \int_0^k \frac{I_1\left(A\sqrt{1-y^2}\right)}{A\sqrt{1-y^2}} \mathrm{d}y \qquad |k| \leq 1 \tag{2.101}$$

2.7.2 Uncertainty Analysis of Spectral Power Measurements above 50 GHz

Measurements with thermocouple or diode power meters constitute the most precise method to capture scalar power levels, but do not allow for distinguishing spectral components of the harmonic content.

[49]Oliver Heaviside (1850–1925), English engineer.

At the time of writing spectrum analyzers are commercially available up to 43 / 50 / 67 GHz. The maximum input frequency is limited by the utilized magnetically tunable preselection bandpass filters (Fig. 1.3), primarily based on yttrium iron garnet (YIG). More sophisticated approaches are required to further increase the base instrument's frequency, which are subject of current research [G1], [64, 65]. Spectral measurements above 43 / 50 / 67 GHz with decreased accuracy are performed with calibrated hollow waveguide harmonic mixers (V-, E-, W-, D-, G-, J-band) in front of a spectrum analyzer to verify the power meter measurements consider the envisaged spectral components (2nd resp. 3rd harmonic) and to measure the multipliers rejection of undesired harmonics.

A brief uncertainty analysis of harmonic mixer measurements is presented in the following. Detailed information about considering several unknown systematic and random error contributions to the overall measurement error of spectrum analyzers can be found in [89–91]. These different errors, contributing to the overall measurement uncertainty of the spectrum analyzer are

- the absolute level error,
- frequency response error,
- attenuator error,
- IF gain error and
- linearity error.

To simplify the uncertainty analysis we neglect all inherent error sources of the spectrum analyzer base instrument and focus on errors due to conversion loss look-up table inaccuracies of the harmonic mixers and errors from mismatch of the harmonic mixers Γ_{MIXER} and the device under test (DUT) Γ_{DUT}.

Nonlinear devices, likewise the proposed frequency multipliers, are poorly matched to the measurement system impedance (50 Ω or characteristic impedance of the corresponding hollow waveguide).

Neglecting the errors during the calibration measurement of the harmonic mixers, performed by the manufacturer, the main source of errors is deviation from the harmonic mixer's specified local oscillator power level and temperature drift[50]. The manufacturer specifies a maximum level error of 3 dB at 25° and 4.5 dB at 5° to 40° with confidence level of 95 %. Assuming Gaussian distribution this corresponds to the variance value in Eq. (2.102), with 1.96 being the coverage factor of 95 % confidence level.

$$\sigma^2 = (3\ \mathrm{dB}/1.96)^2 = 2.34\ \mathrm{dB}^2 \tag{2.102}$$

Considering the interconnection of mismatched DUT and mismatched harmonic mixer, we build the ratio Eq. (2.103) between the measured power $P_{\max}$ in case of matched conditions $\Gamma_{\mathrm{MIXER}} = \Gamma_{\mathrm{DUT}} = 0$ and P_{meas} in the arbitrary case $\Gamma_{\mathrm{MIXER}} = \Gamma_{\mathrm{DUT}} \neq 0$.

$$\frac{P_{\max}}{P_{\mathrm{meas}}} = \frac{|1 - \Gamma_{\mathrm{DUT}}\Gamma_{\mathrm{MIXER}}|^2}{1 - |\Gamma_{\mathrm{MIXER}}|^2} \tag{2.103}$$

In Eq. (2.104) the power ratio $P_{\max}/P_{\mathrm{meas}}$ is expressed in decibel (dB) units.

$$\begin{aligned}\left(\frac{P_{\max}}{P_{\mathrm{meas}}}\right)_{\mathrm{dB}} &= 10\log_{10}\left(|1 - \Gamma_{\mathrm{DUT}}\Gamma_{\mathrm{MIXER}}|^2\right) \\ &\quad - 10\log_{10}\left(1 - |\Gamma_{\mathrm{MIXER}}|^2\right)\end{aligned} \tag{2.104}$$

The denominator in Eq. (2.103) resp. the second term in Eq. (2.104) does only depend on the magnitude of the mixer's reflection coefficient $|\Gamma_{\mathrm{MIXER}}|$ and is therefore already considered in the conversion loss look-up table. Eq. (2.105) shows maximum and minimum values of the remaining term.

[50] The spurious tone output of these harmonic mixers is another source of error. General methods for identification of spurious signals are given in [92], chapter 6. In [93] a commercially implemented procedure is described, which is based on comparing screenshots (bitmap files) of spectral measurements with different local oscillator frequencies.

$$20\log_{10}\left(|1-\Gamma_{\mathrm{DUT}}\Gamma_{\mathrm{MIXER}}|\right) = \begin{cases} \text{max.} & 20\cdot\log_{10}\left(1-|\Gamma_{\mathrm{DUT}}|\,|\Gamma_{\mathrm{LOAD}}|\right) \\ \text{min.} & 20\cdot\log_{10}\left(1+|\Gamma_{\mathrm{DUT}}|\,|\Gamma_{\mathrm{LOAD}}|\right) \end{cases} \tag{2.105}$$

The magnitude of the minimum limit in Eq. (2.105) will always be greater than the magnitude of the maximum limit in Eq. (2.105) and is therefore used as mismatch uncertainty [94]. Assuming U distribution [95–97] for errors due to mismatch, this corresponds to the variance expression in Eq. (2.106).

$$\sigma^2 = 1/2\cdot\left[20\cdot\log_{10}\left(1-|\Gamma_{\mathrm{DUT}}|\,|\Gamma_{\mathrm{LOAD}}|\right)\right]^2 \tag{2.106}$$

Although the individual errors have different error distributions, the total error distribution is assumed to be Gaussian. Strictly speaking this holds true only if there are many individual contributions that all have comparable magnitudes (central limit theorem). For Gaussian distribution the individual standard deviations are summed up in Pythagorean way to the total combined standard deviation σ_{total}, that is scaled to total error levels with certain confidence levels using the appropriate coverage factors.

$$\sigma_{\mathrm{total}} = \sqrt{\sum \sigma_k^2} \quad k \in \mathbb{N} \tag{2.107}$$

In case of 95 % confidence level the total measurement uncertainty is $\sigma_{\mathrm{total}}\cdot 1.96$. Table 2.9 summarizes the calculated total measurement errors under certain DUT and harmonic mixer matching conditions and confidence levels. In case of $|\Gamma_{\mathrm{DUT}}| = -4$ dB and $|\Gamma_{\mathrm{MIXER}}| = -5.11$ dB the total measurement error is ±3.1 dB at 68 % confidence level. The error is reduced to ±1.7 dB when using mixers with $|\Gamma_{\mathrm{MIXER}}| = -13.99$ dB. Within the scope of this work, harmonic mixers with both return loss values have been used.

The presented uncertainty analysis makes clear, well-matched harmonic mixers, utilizing isolators at the input [98, 99], can significantly improve measurement accuracy. The probability of the calculated er-

Table 2.9: Calculations of Total Measurement Error with Harmonic Mixers for Different DUT Reflection Coefficients

Γ_{DUT} [dB]	$\Gamma_{LOAD} = \Gamma_{MIXER}$ [dB]	confidence level [%]	total meas. error [dB]
−2	−5.11/ − 13.99	68	±3.9/ ± 1.9
−4	**−5.11 / −13.99**	**68**	**±3.1 / ±1.7**
−6	−5.11/ − 13.99	68	±2.5/ ± 1.7
−2	−5.11/ − 13.99	95	±7.6/ ± 3.7
−4	−5.11/ − 13.99	95	±6.0/ ± 3.4
−6	−5.11/ − 13.99	95	±4.9/ ± 3.3
−2	−5.11/ − 13.99	99	±10.0/ ± 4.8
−4	−5.11/ − 13.99	99	±7.9/ ± 4.5
−6	−5.11/ − 13.99	99	±6.5/ ± 4.3

Uncertainty of conversion loss look-up table is fixed at 3 dB according to datasheet information with confidence level of 95 %.

rors seems to be higher at the lower and upper bandwidth limits of the corresponding hollow waveguides than it is at center frequency. This is observed when using a calibrated E-band harmonic mixer, which has an overlapping frequency range with both, the V- and W-band mixers. The measured output power levels then show better agreement with the results from power meter measurements around 75 GHz. In consequence, the reader should keep in mind the presented spectral measurement data in subsections 2.6.1 to 2.6.5 and section 2.7 underlie the outlined restrictions. Precise output power level and conversion efficiency measurements are performed with power meters.

References

[1] A. A. M. Saleh. *Theory of Resistive Mixers.* Cambridge, Massachusetts: MIT Press, 1971. ISBN 0262190931.

[2] R. H. Pantell. *General Power Relationships for Positive and Negative Resistive Elements.* Proc. of the IRE, Vol. 46, No. 12, pp. 1910-1913, 1958.

[3] C. H. Page. *Frequency Conversion with Positive Nonlinear Resistors.* Journal of Reseach of the National Bureau of Standards, Vol. 56, No. 4, pp. 179-182, 1956.

[4] M. Reisch. *Halbleiter-Bauelemente.* Berlin: Springer, 2007. ISBN 3540731997.

[5] S. M. Sze. *Physics of Semiconductor Devices.* New York: John Wiley & Sons, 2nd Edition, 1981. ISBN 047109837X.

[6] Agilent Technologies. *Agilent ADS Nonlinear Devices, PN-Junction Diode Model.* Agilent ADS Manual, 2005. http://cp.literature.agilent.com/litweb/pdf/ads2005a/pdf/ccnld.pdf.

[7] M. T. Faber, J. Chramiec, and M. E. Adamski. *Microwave and Millimeter-Wave Diode Frequency Multipliers.* Boston: Artech House, 1995. ISBN 0890066116.

[8] S. A. Maas. *Nonlinear Microwave and RF Circuits.* Boston: Artech House, 2nd Edition, 2003. ISBN 1580534848.

[9] H. A. Bethe. *Theory of the Boundary Layer of Crystal Rectifiers.* MIT Radiation Laboratory Report, Vol. 43, p.12, 1942.

[10] R. E. Hummel. *Electronic Properties of Materials.* Berlin: Springer, 2010. ISBN 9781441981639.

[11] W. Schottky. *Halbleitertheorie der Sperrschicht.* Die Naturwissenschaften, Vol. 26, No. 52, pp. 843-843, 1938.

[12] S. A. Maas. *Microwave Mixers.* Boston: Artech House, 1986. ISBN 0890061718.

[13] F. Thuselt. *Physik der Halbleiterbauelemente.* Boston: Artech House, 2nd Edition, 2003. ISBN 1580534848.

[14] A. Simon. *Konzeption und Technologieentwicklung von Schottkydioden für Anwendungen im Terahertzbereich.* Darmstadt: Shaker Verlag, 1997. ISBN 3826524187.

[15] O. Cojocari. *Schottky Technology for THz-Electronics.* Ph.D. Thesis, Technical University Darmstadt, Germany, 2006.

[16] R. Williams. *Modern GaAs Processing Methods.* Boston: Artech House, 1990. ISBN 0890063435.

[17] L. E. Dickens. *An All-Solid-State Broad-Band Frequency Multiplier Chain at 1500 GHz.* IEEE Transactions on Microwave Theory & Techniques, Vol. 52, No. 5, pp. 1538-1547, 2004.

[18] M. Sotoodeh, A. H. Khalid, and A. A. Rezazadeh. *Empirical Low-Field Mobility Model for III–V Compounds Applicable in Device Simulation Codes.* Journal of Applied Physics, Vol. 87, No. 6, pp. 2890-2900, 2000.

[19] E. L. T. Kollberg, T. J. Tolmunen, M. A. Frerking, and J. R. East. *Current Saturation in Submillimeter Wave Varactors.* IEEE Transactions on Microwave Theory & Techniques, Vol. 40, No. 5, pp. 831-838, 1992.

[20] P. Drude. *Zur Elektronentheorie der Metalle.* Die Annalen der Physik, Vol. 306, No. 3, pp. 566-613, 1900.

[21] A. Sommerfeld. *Zur Elektronentheorie der Metalle.* Die Naturwissenschaften, Vol. 21, pp. 374-381, 1928.

[22] E. Episkopou, S. Papantonis, W. J. Otter, and S. Lucyszyn. *Defining Material Parameters in Commercial EM Solvers for Arbitrary Metal-Based THz Structures.* IEEE Transactions on Terahertz Science and Technology, Vol. 2, No. 5, pp. 513–524, 2012.

[23] Y. Zhou and S. Lucyszyn. *HFSS^{TM} Modelling Anomalies with THz Metal-Pipe Rectangular Waveguide Structures at Room Temperature.* PIERS Online Journal, Vol. 5, No. 3, pp. 201-211, 2009.

[24] S. Lucyszyn and Y. Zhou. *Defining Material Parameters in Commercial EM Solvers for Arbitrary Metal-Based THz Structures.* PIERS Online Journal, Vol. 101, pp. 257-275, 2010.

[25] T. W. Crowe. *GaAs Schottky Barrier Mixer Diodes for the Frequency Range from 1-10 THz.* Journal of Infrared, Millimeter, and Terahertz Waves, Vol. 10, No. 7, pp. 765-777, 1989.

[26] L. E. Dickens. *Spreading Resistance as a Function of Frequency.* IEEE Transactions on Microwave Theory & Techniques, Vol. 15, No. 2, pp. 101-109, 1967.

[27] K. S. Champlin, D. B. Armstrong, and P. D. Gunderson. *Charge Carrier Inertia in Semiconductors.* Proc. of the IEEE, Vol. 52, No. 6, pp. 677-685, 1964.

[28] K. S. Champlin and G. Eisenstein. *Cutoff Frequency of Submillimeter Schottky-Barrier Diodes.* IEEE Transactions on Microwave Theory & Techniques, Vol. 26, No. 1, pp. 31-34, 1978.

[29] W. M. Kelly and G. T. Wrixon. *Conversion Losses in Schottky-Barrier Diode Mixers in the Submillimeter Region.* IEEE Transactions on Microwave Theory & Techniques, Vol. 27, No. 7, pp. 665-672, 1979.

[30] K. J. Button. *Infrared and Millimeter Waves, Vol. 3 Submillimeter Techniques.* New York: Academic Press, 1980. ISBN 0121477037.

[31] R. H. Cox and H. Strack. *Ohmic Contacts for GaAs Devices.* Elsevier, Solid-State Electronics, Vol. 10, No. 12, pp. 1213-1214, 1967.

[32] A. G. Baca, F. Ren, J. C. Zolper, R. D. Briggs, and S. J. Pearton. *A Survey of Ohmic Contacts to III-V Compound Semiconductors.* Elsevier, Thin Solid Films, Vol. 308-309, pp. 599-606, 1997.

[33] P. Antognetti and G. Massobrio. *Semiconductor Device Modeling with SPICE.* New York: McGraw-Hill, 2nd Edition, 1993. ISBN 0070024693.

[34] Modelithics Inc. *Simulation Models for Commercial EDA Tools.* www.modelithics.com.

[35] F. Maiwald, E. Schlecht, G. Chattopadhyay, A. Maestrini, J. Gill, and I. Mehdi. *Planar GaAs Schottky Diode Frequency Multiplier Chains up to 3 THz.* Technical Discussion at Technical University of Darmstadt, Germany, May, 2002.

[36] A. Maestrini, B. Thomas, H. Wang, C. Jung, J. Treuttel, Y. Jin, G. Chattopadhyay, I. Mehdi, and G. Beaudin. *Schottky Diode based Terahertz Frequency Multipliers and Mixers.* C. R. Physique 11, 2010.

[37] E. Schlecht, G. Chattopadhyay, A. Maestrini, D. Pukala, J. Gill, and I. Mehdi. *Harmonic Balance Optimization of Terahertz Schottky Diode Multipliers using an Advanced Device Model.* Proc. of the 13th Int. Symp. on Space Terahertz Technology, pp. 187-196, March, 2002.

[38] I. Ederra. *Electromagnetic Band Gap Technology for Millimetre Wave Applications.* Ph.D. Thesis, Universidad Publica de Navarra, Spain, 2004.

[39] Agilent Technologies. *Guide to Harmonic Balance Simulation in ADS.* http://cp.literature.agilent.com.

[40] R. D. Albin. *Milllimeter-Wave Source Modules.* Hewlett Packard Journal, Vol. 39, pp 18-25, 1988.

[41] Arlon Inc. *PTFE/Woven Fiberglass/Ceramic AD600 Laminate.* www.arlon-med.com.

[42] TriQuint Semiconductor. *Frequency Tripler TGC1430G.* www.triquint.com.

[43] M. Morgan and S. Weinreb. *A Full Waveguide Band MMIC Tripler for 75-110 GHz.* IEEE-MTT-S Int. Microwave Symposium Digest, Vol. 1, pp. 103-106, 2001.

[44] G.-Y. Chen, Y.-S. Wu, H.-Y. Chang, Y.-M. Hsin, and C.-C. Chiong. *A 60-110 GHz Low Conversion Loss Tripler in 0.15-um mHEMT Process.* Proc. of the Asia Pacific Microwave Conference, pp. 377-380, 2009.

[45] Hittite Microwave Corporation. *Frequency Tripler HMC-XTB110.* www.hittite.com.

[46] F. Brauchler, J. Papapolymerou, J. East, and L. Katehi. *W Band Monolithic Multipliers.* IEEE-MTT-S Int. Microwave Symposium Digest, pp. 1225-1228, 1997.

[47] J. Papapolymerou, J. East, L. Katehi, M. Kim, and I. Mehdi. *Millimeter-Wave GaAs Monolithic Multipliers.* IEEE-MTT-S Int. Microwave Symposium Digest, pp. 395-398, 1998.

[48] J. Papapolymerou, F. Brauchler, J. East, and L. Katehi. *W-Band Finite Ground Coplanar Monolothic Multipliers.* IEEE Transactions on Microwave Theory & Techniques, Vol. 47, No. 5, pp. 614-619, 1999.

[49] Y. Lee, J. East, and L. Katehi. *High-Efficiency W-Band GaAs Monolithic Frequency Multipliers.* IEEE Transactions on Microwave Theory & Techniques, Vol. 52, No. 2, pp. 529-535, 2004.

[50] J. Lynch, E. Entchev, B. Lyons, A. Tessmann, H. Massler, A. Leuther, and M. Schlechtweg. *A Broadband 75-100 GHz MMIC Doubler.* Proc. of the 34th European Microwave Conference, pp. 1017-1020, 2004.

[51] J. Lynch, E. Entchev, B. Lyons, A. Tessmann, H. Massler, A. Leuther, and M. Schlechtweg. *Design and Analysis of a W-Band Multiplier Chipset.* IEEE-MTT-S Int. Microwave Symposium Digest, Vol. 1, pp. 227-230, 2004.

[52] J. Lynch, B. Lyons, E. Entchev, A. Tessmann, H. Massler, A. Leuther, and M. Schlechtweg. *W-Band Multiplier Chipset.* Electronics Letters, Vol. 40, No. 2, pp. 130-132, 2004.

[53] A. Y. Tang, V. Drakinskiy, K. Yhland, J. Stenarson, T. Bryllert, and J. Stake. *Analytical Extraction of a Schottky Diode Model From Broadband S-Parameters.* IEEE Transactions on Microwave Theory & Techniques, Vol. 61, No. 5, pp. 1870-1878, 2013.

[54] J. Chramiec. *Effect of Embedding Impedances on The Efficiency of Varistor Schottky Diode Frequency Multipliers.* Proc. of the 20th European Microwave Conference, pp. 1724-1729, 1990.

[55] L. Hayden. *An Enhanced Line-Reflect-Reflect-Match Calibration.* Proc. of the 67th ARFTG Conference, pp. 143-149, 2006.

[56] Y. C. Leong and S. Weinreb. *Full Band Waveguide-To-Microstrip Probe Transitions.* IEEE-MTT-S Int. Microwave Symposium Digest, pp. 1435-1438, 1999.

[57] G. L. Matthaei, L. Young, and E. M. T. Jones. *Microwave Filters, Impedance Matching Networks, and Coupling Structures.* New York: McGraw-Hill, 1964.

[58] G. F. Engen and C. A. Hoer. *Thru-Reflect-Line: An Improved Technique for Calibrating the Dual Six-Port Automatic Network Analyzer.* IEEE Transactions on Microwave Theory & Techniques, Vol. 27, No. 12, pp. 987-993, 1979.

[59] J. Chramiec. *A High-Pass Waveguide Applicable to Microwave Integrated Circuits.* Proc. of the 26th Int. Sci. Coll., Vol. A3, pp. 91-98, 1981.

[60] J. Chramiec, M. Kitlinski, and B. J. Janiczak. *Microstrip Line with Short-Circuited Edge as a High-Pass Filter Operating up to Millimeter-Wave Frequencies.* Microwaves and Optical Technical Letters, Vol. 49, No. 9, pp. 2178-2180, 2007.

[61] J. Chramiec, M. Kitlinski, A. Bochenek, and B. Janiczak. *Design of MM-Wave MIC Frequency Multipliers and Mixers using Simple Microstrip High-Pass Filters.* Proc. of the 16th Int. Conference on Microwave, Radar and Wireless Communications MIKON, pp. 893-896, 2006.

[62] K. Noujeim. *High-Pass Filtering Characteristics of Transmission-Line Combs.* IEEE Transactions on Microwave Theory & Techniques, Vol. 57, No. 11, pp. 2743-2752, 2009.

[63] K. C. Gupta, R. Garg, I. Bahl, and P. Bhartia. *Microstrip Lines and Slotlines.* Boston: Artech House, 1979. ISBN 089006766X.

[64] M. Sterns, R. Rehner, D. Schneiderbanger, S. Martius, and L.-P. Schmidt. *Novel Tunable Hexaferrite Bandpass Filter Based on Open-Ended Finlines.* IEEE-MTT-S Internatinal Microwave Symposium Digest, pp. 637-640, 2009.

[65] M. Sterns, R. Rehner, D. Schneiderbanger, S. Martius, and L.-P. Schmidt. *Magnetically Tunable Hexaferrite Bandpass Filter from 43 GHz to 83.5 GHz.* Proc. of the 39th European Microwave Conference, pp. 129-132, 2009.

[66] R. Rehner, D. Schneiderbanger, M. Sterns, S. Martius, and L.-P. Schmidt. *Novel Coupled Microstrip Wideband Filters with Spurious Response Suppression.* Proc. of the 37th European Microwave Conference, pp. 858-861, 2007.

[67] M. Sterns, R. Rehner, and D. Schneiderbanger. *Optimized Design of a Broadside-Coupled Suspended Stripline Filter.* De Gruyter Frequenz, Journal of RF-Engineering and Telecommunications, Vol. 63, Bo. 5-6, pp. 102-105, 2009. ISSN: 0016-1136.

[68] W. Bischof, W. Ehrlinger, R. Lohrmann, and M. Naehring. *Baugruppen für den Mikrowellen- und Millimeterwellenbereich.* ANT Nachrichtentechnische Berichte, Vol. 4, pp. 48-56, 1987.

[69] M. Sterns, R. Rehner, D. Schneiderbanger, S. Martius, and L.-P. Schmidt. *Broadband, Highly Integrated Receiver Frontend up to 67 GHz.* Proc. of the 41th European Microwave Conference, pp. 1063-1066, 2011.

[70] C.-S. Lin, P.-S. Wu, M.-C. Yeh, J.-S. Fu, H.-Y. Chang, K.-Y. Lin, and H. Wang. *Analysis of Multiconductor Coupled-Line Marchand Baluns for Miniature MMIC Design.* IEEE Transactions on Microwave Theory & Techniques, Vol. 55, No. 6, pp. 1190-1199, 2007.

[71] K.-Y. Lin, H. Wang, M. Morgan, T. Gaier, and S. Weinreb. *A W-Band GCPW MMIC Diode Tripler.* Proc. of the 32th European Microwave Conference, 2002.

[72] C. S. Yoo, S. Song, and K. Seo. *A W-Band Tripler with a Novel Band Pass Filter on Thin-Film Substrate.* Proc. of the Asia Pacific Microwave Conference, 2008.

[73] H. Fudem and E. Niehenke. *Novel Millimeter Wave Active MMIC Triplers.* IEEE-MTT-S Int. Microwave Symposium Digest, pp. 387-390, 1998.

[74] J. Vukusic, T. Bryllert, A. Olsen, and J. Stake. *High Power W-Band Monolithically Integrated Tripler.* Proc. of the 34th Int. Conference on IRMMW-THz, pp. 1-2, 2009.

[75] Y. Campos-Roca, L. Verweyen, M. Fernandez-Barciela, M. C. Curras-Francos, W. Sanchez, A. Huelsmann, and M. Schlechtweg. *Millimeter-Wave Active MMIC Frequency Multipliers.* Proc. of the 31th European Microwave Conference, pp. 1-4, 2001.

[76] Millitech Inc. *Frequency Tripler MUT-10.* www.millitech.com.

[77] Virginia Diodes Inc. *W-Band Frequency Tripler WR10X3.* www.vadiodes.com.

[78] G. P. Ermak and P. V. Kuprijanov. *Development of a Planar Multiplier Circuit for Millimeter-Wave Frequency Multipliers.* 4th Int. Phys. Eng. Millimeter and Sub-Millimeter Waves Symposium, Vol. 2, pp. 696-698, 2001.

[79] G.-L. Tan and G. M. Rebeiz. *High-Power Millimeter-Wave Planar Doublers.* IEEE-MTT-S Int. Microwave Symposium Digest, Vol. 3, pp. 1601-1604, 2000.

[80] Millitech Inc. *Frequency Doubler MUD-10.* www.millitech.com.

[81] Virginia Diodes Inc. *E-Band Frqeuency Doubler WR10X2.* www.vadiodes.com.

[82] C. Yao and J. Xu. *A D-Band Frequency Doubler with GaAs Schottky Varistor Diodes.* Int. Journal of Electronics,, Vol. 97, No. 12, pp. 1449–1457, 2010.

[83] Y. S. Lau, T. Denning, and C. Oleson. *Millimeter Wave Power Measurement above 110 GHz.* Proc. of the 67th ARFTG Conference, pp. 97-102, 2006.

[84] C. Nguyen and K. Chang. *Millimeter-Wave Low-Loss Finline Lowpass Filters.* Electronic Letters, Vol. 20, No. 24, pp. 1010-1011, 1984.

[85] C. Nguyen and K. Chang. *On the Design and Performance of Printed-Circuit Filters and Diplexers for Millimeter-Wave Integrated Circuits.* Int. Journal of Infrared and Millimeter Waves, Vol. 7, No. 7, pp. 971-998, 1986.

[86] R. W. Klopfenstein. *A Transmission Line Taper of Improved Design.* Proc. of the IRE, Vol. 44, No. 1, pp. 31-35, 1956.

[87] G. Razmafrouz, G. R. Branner, and B. P. Kumar. *Formulation of the Klopfenstein Tapered Line Analysis from Generalized Nonuniform Line Theory.* Proc. of the 39th IEEE Midwest Symposium on Circuits and Systems, Vol. 3, pp. 1177-1180, 1996.

[88] D. Kajfez and J. O. Prewitt. *Correction to A Transmission Line Taper of Improved Design.* IEEE Transactions on Microwave Theory & Techniques, Vol. 21, No. 5, p. 364, 1973.

[89] J. Wolf. *Application Note: 1EF36_0E, Level Error Calculation for Spectrum Analyzers.* Rohde & Schwarz GmbH & Co. KG. www.rohde-schwarz.com.

[90] Rohde & Schwarz GmbH & Co. KG. *Operating Manual of FS-Z60/-Z75/-Z90/-Z110 Harmonic Mixers.* www.rohde-schwarz.com.

[91] C. Rauscher, V. Janssen, and R. Minihold. *Fundamentals of Spectrum Analysis.* Rohde & Schwarz GmbH & Co. KG, 2001. ASIN B000R5OL26.

[92] M. Engelson and F. Telewski. *Spectrum Analyzer Theory and Applications.* Boston: Artech House, 1974. ISBN 089006024X.

[93] C. Rauscher. *Application Note: 1EF43_0E, Frequency Range Extension of Spectrum Analyzers with Harmonic Mixers.* Rohde & Schwarz GmbH & Co. KG. www.rohde-schwarz.com.

[94] Agilent Technologies. *Application Note: 1449-3, Fundamentals of RF and Microwave Power Measurements (Part 3).* http://cp.literature.agilent.com.

[95] D. Carpenter. *A Demystification of the U-Shaped Probability Distribution.* Int. Symposium on Electromagnetic Compatibility, pp. 521-525, 2003.

[96] D. Carpenter. *A Further Demystification of the U-shaped Probability Distribution.* Int. Symposium on Electromagnetic Compatibility, pp. 519-524, 2005.

[97] T. Guldbrandsen. *Uncertainty Contributions from Mismatch in Microwave Measurements.* IEE Proc. of Microwaves, Antennas and Propagation, Vol. 148, No. 6, pp. 393-397, 2001.

[98] H. J. E. Gibson, A. Walber, R. Zimmermann, B. Alderman, and O. Cojocari. *Improvements in Schottky Harmonic and Sub-Harmonic Mixers for use up to 900 GHz.* Proc. of the 40th European Microwave Conference, pp. 230-231, 2010.

[99] Agilent Technologies. *Technical Overview (5990-7718EN): Smart Waveguide Harmonic Mixers M1970E/V/W.* http://cp.literature.agilent.com.

[100] H. C. Torrey and C. A. Whitmer. *Crystal Rectifiers.* New York: McGraw-Hill, MIT Radiation Laboratory Series, Vol. 15, 1948.

[101] R. V. Pound. *Microwave Mixers.* New York: McGraw-Hill, MIT Radiation Laboratory Series, Vol. 16, 1948.

[102] C. Dragone. *Amplitude and Phase Modulations in Resistive Diode Mixers.* Bell System Technical Journal, Vol. 48, No. 6, pp. 1967-1998, 1969.

[103] E. W. Herold, R. R. Bush, and W. R. Ferris. *Conversion Loss of Diode Mixers having Image-Frequency Impedance.* Proc. of the IRE, Vol. 33, No. 9, pp. 603-609, 1945.

[104] J. A. Dobrowolski. *Microwave Network Design using the Scattering Matrix.* Boston: Artech House, 2010. ISBN 1608071294.

[105] W. Shockley. *The Theory of p-n Junctions in Semiconductors and p-n Junction Transistors.* Bell Syst. Tech. Journal, Vol. 28, pp. 435-489, 1949.

[106] H. K. Henisch. *Rectifying Semiconductor Contacts.* Oxford: Clarendon, 1957.

[107] A. R. V. Hippel. *Dielectric Materials and Applications.* MIT Press, 1954.

[108] D. Steup. *Terahertz-Verbundvervielfacher.* Erlangen: DiDacta, Wissenschaftliche Fachbuchreihe, 1997. ISBN 3519004313.

[109] L. W. Nagel. *SPICE2: A Computer Program to Simulate Semiconductor Circuits.* Thesis, EECS Department, University of California, Berkeley, USA, 1975. www.eecs.berkeley.edu/Pubs/TechRpts/1975/9602.html.

[110] S. D. Vogel. *Computer-Aided Design of a Novel Subharmonically Pumped Planar Diode Mixer for Millimeter-Wave Applications.* Ph.D. Thesis, University of Bern, Switzerland, 1994.

[111] A. Y. Tang. *Modelling of Terahertz Planar Schottky Diodes.* Ph.D. Thesis, Chalmers University of Technology, Goeteborg, Sweden, 2011.

[112] S. A. Maas. *The RF and Microwave Circuit Design Cookbook.* Boston: Artech House, 1998. ISBN 0890069735.

[113] M. Reisch. *Elektronische Bauelemente.* Berlin: Springer, 1998. ISBN 3540609911.

[114] D. A. Neamen. *Semiconductor Physics and Devices: Basic Principles.* New York: McGraw-Hill, 4th Edition, 2011. ISBN 0073529583.

[115] B. Bhat and S. K. Koul. *Analysis, Design and Applications of Fin Lines.* Boston: Artech House, 1987. ISBN 0890061955.

[116] R. G. Hicks and P. J. Khan. *Numerical Analysis of Nonlinear Solid-State Device Excitation in Microwave Circuits.* IEEE Transactions on Microwave Theory & Techniques, Vol. 30, No. 3, pp. 251-259, 1982.

[117] A. R. Kerr. *A Technique for Determining the Local Oscillator Waveforms in a Microwave Mixer.* IEEE Transactions on Microwave Theory & Techniques, Vol. 23, No. 10, pp. 828-831, 1975.

[118] K. Zeljami, J. Gutiierrez, J. P. Pascual, T. Fernandez, A. Tazon, and M. Boussouis. *Characterization and Modeling of Schottky Diodes up to 110 GHz for use in both Flip-Chip and Wire-Bonded Assembled Environments.* Progress In Electromagnetics Research, Vol. 131, pp. 457-475, 2012.

3 Planar Directional Couplers and Filters

Planar directional couplers allow for separation of forward and backward travelling waves within scattering parameter testsets of vector network analyzers (VNA). In combination with power detectors (chapter 5), couplers are used for overload protection and automatic level control of signal generators (chapter 6).

Section 3.1 introduces the modal and nodal scattering and impedance network parameters of coupled transmission lines according to the general waveguide circuit theory (GWCT) [1]. This analysis is based on the results of 2D EM Eigenmode analysis.

Backward wave directional couplers achieve directivity at frequencies as low as 10 MHz and 70 kHz, what is difficult to realize together with high maximum operating frequencies like 50 GHz and 70 GHz. Directivity is directly related to equalized phase velocities of the propagating modes. A synthesis procedure for nonuniform transmission line couplers is included. Theoretical and experimental investigations on backward wave couplers with dielectric overlay technique (stripline) and wiggly-line technique to equalize the even and odd mode phase velocities are presented.

If directional coupler operation at low frequencies is not required, codirectional couplers are the method of choice. Section 3.3 includes a synthesis procedure and simulated and measured results of various codirectional couplers on thin-film processed alumina, including the couplers used within the modules of chapter 6.

Although commercially available directional couplers with coaxial or hollow waveguide interface make use of external terminations, couplers for planar integration require internal nonreflective impedance terminations. In section 3.4, several 50 Ω terminations based on nickel chrome (NiCr) sheet resistors are presented. It further includes simulation and measurement results of the author's DC to 110 GHz attenuator series.

Beside directional coupler designs, equalized phase velocities are also beneficial for edge and broadside coupled line bandpass filters (BPF) to suppress the parasitic second passband. This is demonstrated in section 3.5 by applying wiggly-line technique to the first and last filter element of an edge coupled line BPF on 10 mil alumina.

3.1 Theoretical Foundations of Cascaded Coupled Waveguides

With the results from 2D EM Eigenmode analysis as a starting point, the modal impedance $\mathbf{Z}_{\mathrm{mt}} = \mathbf{Z}_{\mathrm{m}}$ and scattering $\mathbf{S}_{\mathrm{mt}} = \mathbf{S}_{\mathrm{m}}$ matrices of uniformly[1] coupled asymmetric lines are derived [1, 2]. The nonuniform case is covered by cascading a sufficiently high number of short uniform lines. Beside the modal description, power normalized conductor impedance $\mathbf{Z}_{\mathrm{ct}} = \mathbf{Z}_{\mathrm{n}}$ and scattering $\mathbf{S}_{\mathrm{ct}} = \mathbf{S}_{\mathrm{n}}$ matrices are introduced. Such nodal matrices are required to connect transmission lines with arbitrary lumped elements or discrete components like diodes (compare chapter 2). Nodal scattering matrices are accessible to conventional measurement. Modal scattering parameter measurements require sophisticated calibration algorithms, like TRL. Following this approach, leads to the results from Tripathi [3–5], the conductor impedance matrix of asymmetric coupled lines, but maintains a clear connection to the results from 2D EM Eigenmode analysis.

2D EM Eigenmode Analysis Fig. 3.1 shows a cross-sectional view of coupled asymmetric lines, consisting of thin conductors, substrate material and metallic enclosure. All material properties are assumed

[1]The term uniform means cross section does not change in the direction of wave propagation (z-axis).

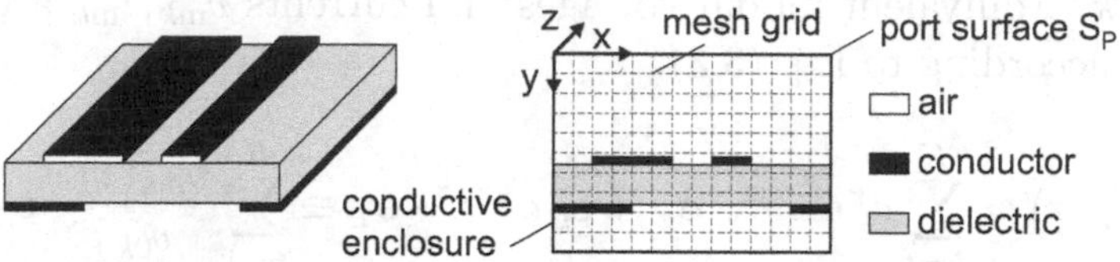

Figure 3.1: 3D sketch and cross-sectional view of uniformly coupled asymmetric lines with 2D mesh grid on port surface S_{P}.

to be known and the cross section is represented by a sufficiently fine mesh grid $\perp = \boxplus = (x, y)$. After applying numerical 2D EM Eigenmode analysis [6–8], the electric $\mathbf{E}(x, y)$ and magnetic $\mathbf{H}(x, y)$ field vectors are known at any coordinate (x, y) of the port surface S_{P}. In Eq. (3.1), the results are split into transversal $\mathbf{E}_{\perp}, \mathbf{H}_{\perp}$ and longitudinal $\mathbf{E}_{\mathrm{z}}, \mathbf{H}_{\mathrm{z}}$ components. The latter are oriented in the direction of wave propagation (z-axis).

$$\begin{aligned} \mathbf{E}(x, y) &= \mathbf{E}_{\perp} + \mathbf{E}_{\mathrm{z}} \\ \mathbf{H}(x, y) &= \mathbf{H}_{\perp} + \mathbf{H}_{\mathrm{z}} \end{aligned} \tag{3.1}$$

A $(K+1)$ conductor waveguide carries K fundamental modes (cut-off frequency equals DC). In the following a total number of K modes is assumed, which do not necessarily need to be fundamental modes. In fact, the 2D EM solvers provide separate results of every mode $k \in \{1, \ldots, K\}$ (compare Fig. 2.91, Fig. 2.93, Fig. 2.100 of chapter 2). According to Maxwell's equations[2], the transversal electromagnetic fields appear as forward and backward travelling waves with complex amplitudes c_k^+, c_k^- and exponential dependency of the z-axis $e^{-\gamma_k z}, e^{+\gamma_k z}$. Whereas γ_k is the complex propagation coefficient of

[2]The modal voltages and currents are solutions of the linear differential equation system Eq. (3.2).

$$\frac{\mathrm{d}\mathbf{v}_{\mathrm{m}}}{\mathrm{d}z} = -\gamma \mathbf{Z}_0 \ \mathbf{i}_{\mathrm{m}} \qquad \frac{\mathrm{d}\mathbf{i}_{\mathrm{m}}}{\mathrm{d}z} = -\gamma \mathbf{Z}_0^{-1} \ \mathbf{v}_{\mathrm{m}} \tag{3.2}$$

mode k. Equivalent modal voltages and currents $v_{\mathrm{m}k}, i_{\mathrm{m}k}$ are introduced according to Eq. (3.3).

$$\mathbf{E}_\perp(x,y,z) = \sum_{k=1}^{K} c_k^+ e^{-\gamma_k z} \cdot \mathbf{e}_\perp + c_k^- e^{+\gamma_k z} \cdot \mathbf{e}_\perp \equiv \sum_{k=1}^{K} \frac{v_{\mathrm{m}k}(z)}{v_{0k}} \cdot \mathbf{e}_{\perp k}(x,y)$$
$$\mathbf{H}_\perp(x,y,z) = \sum_{k=1}^{K} c_k^+ e^{-\gamma_k z} \cdot \mathbf{h}_\perp + c_k^- e^{+\gamma_k z} \cdot \mathbf{h}_\perp \equiv \sum_{k=1}^{K} \frac{i_{\mathrm{m}k}(z)}{i_{0k}} \cdot \mathbf{h}_{\perp k}(x,y) \tag{3.3}$$

The normalizing voltages and currents v_{0k}, i_{0k} are related by Eq. (3.4). This leads to $v_{\mathrm{m}k} i^\star_{\mathrm{m}k}$, representing the complex power p if only the k-th mode is travelling in forward direction.

$$v_{0k} i^\star_{0k} = p_{0k} \equiv \int_S \mathbf{e}_{\perp k} \times \mathbf{h}^\star_{\perp k} \cdot z \mathrm{d}S \tag{3.4}$$

The total complex power p is shown in Eq. (3.5).

$$p = \int_S \mathbf{E}_\perp \times \mathbf{H}^\star_\perp \cdot z\mathrm{d}S = \sum_{j,k} \frac{v_{\mathrm{m}j}(z)}{v_{0j}} \frac{i^\star_{\mathrm{m}k}(z)}{i^\star_{0k}} \int_S \mathbf{e}_\perp \times \mathbf{h}^\star_\perp \cdot z\mathrm{d}S \tag{3.5}$$

Introducing the column vectors of modal voltages $\mathbf{v}_\mathrm{m} = [v_{\mathrm{m}1}, \ldots, v_{\mathrm{m}K}]^\mathrm{T}$, currents $\mathbf{i}_\mathrm{m} = [i_{\mathrm{m}1}, \ldots, i_{\mathrm{m}K}]^\mathrm{T}$ and the cross power matrix $\mathbf{X}$, yields the compact expression in Eq. (3.6) of the total power p.

$$p = \mathbf{i}_\mathrm{m}^\mathrm{H} \; \mathbf{X} \; \mathbf{v}_\mathrm{m} = (\mathbf{i}^\star_\mathrm{m})^\mathrm{T} \; \mathbf{X} \; \mathbf{v}_\mathrm{m} \tag{3.6}$$

The cross power matrix $\mathbf{X}$ is a square matrix of size K. As a consequenc of Eq. (3.4), the diagonal elements of $\mathbf{X}$ are $X_{kj} = 1,\ k = j$. In [9], it is shown that the elements $X_{kj},\ k \neq j$ of Eq. (3.7) are nonzero, only if all or some of the K modes are degenerate[3]. Throughout this work, mode conversion is covered by separate impedance or

[3]Degenerate modes have different electromagnetic fields $\mathbf{E}_k(x,y), \mathbf{H}_k(x,y)$ but almost equal propagation coefficients $\gamma_j \approx \gamma_k, j \neq k$.

scattering matrices (e.g baluns of frequency multipliers and mixers) but ignored for simple transmission lines (planar waveguides). This is reasonable even for waveguides with cross coupling, if the waveguide's length is chosen small.

$$X_{kj} = \frac{1}{v_{0j} i_{0k}^{\star}} \int_S \mathbf{e}_\perp \times \mathbf{h}_\perp^\star \cdot z \mathrm{d}S \tag{3.7}$$

The characteristic impedance of the k-th mode is given by Eq. (3.8).

$$Z_{0k} \equiv \frac{v_{0k}}{i_{0k}} = \frac{|v_{0k}|^2}{p_{0k}} = \frac{p_{0k}}{|i_{0k}|^2} \tag{3.8}$$

Its argument $\arg(Z_{0k})$ is fixed by the surface integral of Eq. (3.4) [10]. To determine the magnitude $|Z_{0k}| = |v_{0k}|/|i_{0k}|$, either v_{0k} or i_{0k}, but not both,[4] need to be calculated by the integrals of Eq. (3.9).

$$\begin{aligned} v_{0k} &= \int_{\boldsymbol{\ell}_{\mathrm{v}_0 k}} \mathbf{e}_{\perp k} \cdot \mathrm{d}\boldsymbol{\ell} \quad \rightarrow \quad Z_{0k} = Z_{\mathrm{PV}k} = \frac{|v_{0k}|^2}{p_{0k}} \\ \text{or} \quad i_{0k} &= \oint_{\boldsymbol{\ell}_{\mathrm{i}_0 k}} \mathbf{h}_{\perp k} \cdot \mathrm{d}\boldsymbol{\ell} \quad \rightarrow \quad Z_{0k} = Z_{\mathrm{PI}k} = \frac{p_{0k}}{|i_{0k}|^2} \end{aligned} \tag{3.9}$$

This leads to the power-voltage $Z_{\mathrm{PV}k}$ or power-current $Z_{\mathrm{PI}k}$ definition of the characteristic impedance. In case of an arbitrary planar waveguide, both definitions are different and depend on the chosen integration lines $\boldsymbol{\ell}_{\mathrm{v}_0 k}, \boldsymbol{\ell}_{\mathrm{i}_0 \mathrm{k}}$.

Modal Impedance and Scattering Matrices The following derivation of the modal impedance $\mathbf{Z}_{\mathrm{mt}} = \mathbf{Z}_{\mathrm{m}}$ and scattering $\mathbf{S}_{\mathrm{mt}} = \mathbf{S}_{\mathrm{m}}$ matrices is restricted to the $(2+1)$ conductor waveguide with length

[4] If both, v_{0k} and i_{0k} are determined from integrals, Eq. (3.4) does not generally hold true and the equivalent modal voltages and currents are not related to the travelling wave power. Hence, the voltage-current definition Z_{VI} of the characteristic impedance is not an adequate choice if physical modelling is desired.

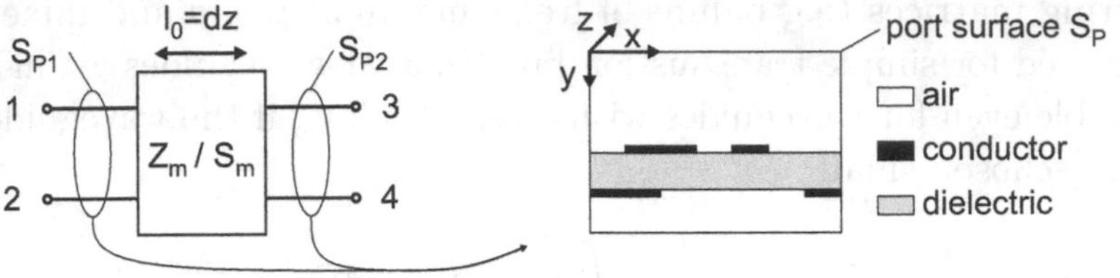

Figure 3.2: Schematic view of uniformly coupled asymmetric lines of length $\ell_0 = \mathrm{d}z$ with four ports / modes in total.

$\ell_0 = \mathrm{d}z$ of Fig. 3.2. At each physical port surface $S_{\mathrm{P1}}, S_{\mathrm{P2}}$, two modes are considered. This results in a total of four ports ($k \in \{1, \ldots, K\}$) with $\mathbf{Z}_{\mathrm{mt}} = \mathbf{Z}_{\mathrm{m}}$ and $\mathbf{S}_{\mathrm{mt}} = \mathbf{S}_{\mathrm{m}}$ being (4×4) matrices. Eq. (3.10, 3.11, 3.12) illustrate how the modal voltages and currents, propagation coefficients and characteristic impedances are summarized in vector / matrix form.

$$\begin{aligned}
\mathbf{v}_{\mathrm{mP1}} &= \begin{bmatrix} v_{\mathrm{m1}} \\ v_{\mathrm{m2}} \end{bmatrix} \quad \mathbf{i}_{\mathrm{mP1}} = \begin{bmatrix} i_{\mathrm{m1}} \\ i_{\mathrm{m2}} \end{bmatrix} \quad \rightarrow \quad \mathbf{v}_{\mathrm{m}} = \begin{bmatrix} \mathbf{v}_{\mathrm{mP1}} \\ \mathbf{v}_{\mathrm{mP2}} \end{bmatrix} \\
\mathbf{v}_{\mathrm{mP2}} &= \begin{bmatrix} v_{\mathrm{m3}} \\ v_{\mathrm{m4}} \end{bmatrix} \quad \mathbf{i}_{\mathrm{mP2}} = \begin{bmatrix} i_{\mathrm{m3}} \\ i_{\mathrm{m4}} \end{bmatrix} \quad \rightarrow \quad \mathbf{i}_{\mathrm{m}} = \begin{bmatrix} \mathbf{i}_{\mathrm{mP1}} \\ \mathbf{i}_{\mathrm{mP2}} \end{bmatrix}
\end{aligned} \tag{3.10}$$

$$\begin{aligned}
\gamma_{\mathrm{P1}} &= \operatorname{diag}\left[\gamma_1, \gamma_2\right] \quad \rightarrow \quad \gamma = \operatorname{diag}\left[\gamma_{\mathrm{P1}}, \gamma_{\mathrm{P2}}\right] \\
\gamma_{\mathrm{P2}} &= \operatorname{diag}\left[\gamma_3, \gamma_4\right]
\end{aligned} \tag{3.11}$$

$$\begin{aligned}
\mathbf{Z}_{0\mathrm{P1}} &= \operatorname{diag}\left[Z_{01}, Z_{02}\right] \quad \rightarrow \quad \mathbf{Z}_0 = \operatorname{diag}\left[\mathbf{Z}_{0\mathrm{P1}}, \mathbf{Z}_{0\mathrm{P2}}\right] \\
\mathbf{Z}_{0\mathrm{P2}} &= \operatorname{diag}\left[Z_{03}, Z_{04}\right]
\end{aligned} \tag{3.12}$$

Using this notation, the modal impedance matrix $\mathbf{Z}_{\mathrm{mt}} = \mathbf{Z}_{\mathrm{m}}$ of a section of multimode transmission line with length ℓ_0 is given by Eq. (3.13) and Eq. (3.14).

$$\mathbf{Z}_{\mathrm{mt}} = \mathbf{Z}_0 \begin{bmatrix} \coth(\gamma_{\mathrm{P1}}\ell_0) & \sinh^{-1}(\gamma_{\mathrm{P1}}\ell_0) \\ \sinh^{-1}(\gamma_{\mathrm{P2}}\ell_0) & \coth(\gamma_{\mathrm{P2}}\ell_0) \end{bmatrix} = \begin{bmatrix} \mathbf{Z}_{0\mathrm{P1}}\coth(\gamma_{\mathrm{P1}}\ell_0) & \mathbf{Z}_{0\mathrm{P1}}\sinh^{-1}(\gamma_{\mathrm{P1}}\ell_0) \\ \mathbf{Z}_{0\mathrm{P2}}\sinh^{-1}(\gamma_{\mathrm{P2}}\ell_0) & \mathbf{Z}_{0\mathrm{P2}}\coth(\gamma_{\mathrm{P2}}\ell_0) \end{bmatrix} \tag{3.13}$$

The operators $\coth(\bullet)$ and $\sinh^{-1}(\bullet) = 1/\sinh(\bullet)$ are applied component by component.

$$\mathbf{Z}_{\mathrm{mt}} = \begin{bmatrix} Z_{01}\coth(\gamma_1\ell_0) & 0 & Z_{01}\sinh^{-1}(\gamma_1\ell_0) & 0 \\ 0 & Z_{02}\coth(\gamma_2\ell_0) & 0 & Z_{01}\sinh^{-1}(\gamma_1\ell_0) \\ Z_{03}\sinh^{-1}(\gamma_3\ell_0) & 0 & Z_{03}\coth(\gamma_3\ell_0) & 0 \\ 0 & Z_{03}\sinh^{-1}(\gamma_3\ell_0) & 0 & Z_{04}\coth(\gamma_4\ell_0) \end{bmatrix} \tag{3.14}$$

The scattering matrix $\mathbf{S}$ relates the backward travelling wave amplitudes $\mathbf{b}$ with the forward travelling wave amplitudes $\mathbf{a}$.

$$\begin{aligned} &\mathbf{a} = \frac{1}{2}\mathbf{U}(\mathbf{v}_{\mathrm{m}} + \mathbf{Z}_{\mathrm{ref}}\mathbf{i}_{\mathrm{m}}) \quad \mathbf{b} = \frac{1}{2}\mathbf{U}(\mathbf{v}_{\mathrm{m}} - \mathbf{Z}_{\mathrm{ref}}\mathbf{i}_{\mathrm{m}}) \\ &\mathbf{U} = \mathrm{diag}\left(\frac{|v_{0\mathrm{k}}|}{v_{0\mathrm{k}}}\frac{\sqrt{\mathrm{Re}(\mathbf{Z}_{\mathrm{ref}})}}{|\mathbf{Z}_{\mathrm{ref}}|}\right), \quad k \in \{1, \dots, K\} \end{aligned} \tag{3.15}$$

Contrary to the impedance parameters, scattering parameters cover the behaviour of multiport networks with specified load impedances at every port. These arbitrary reference impedances are given by the diagonal matrix $\mathbf{Z}_{\mathrm{ref}}$, which has the form of Eq. (3.12). The modal scattering matrix is given by Eq. (3.16).

$$\begin{aligned} &\mathbf{S}_{\mathrm{mt}} = \mathbf{U}(\mathbf{Z}_{\mathrm{mt}} - \mathbf{Z}_{\mathrm{ref}})(\mathbf{Z}_{\mathrm{mt}} + \mathbf{Z}_{\mathrm{ref}})^{-1}\mathbf{U}^{-1} \\ &\mathbf{Z}_{\mathrm{mt}} = (\mathbf{I} - \mathbf{U}^{-1}\mathbf{S}_{\mathrm{mt}}\mathbf{U})^{-1}(\mathbf{I} + \mathbf{U}^{-1}\mathbf{S}_{\mathrm{mt}}\mathbf{U})\mathbf{Z}_{\mathrm{ref}} \\ &\mathbf{I} = \text{identity matrix} \end{aligned} \tag{3.16}$$

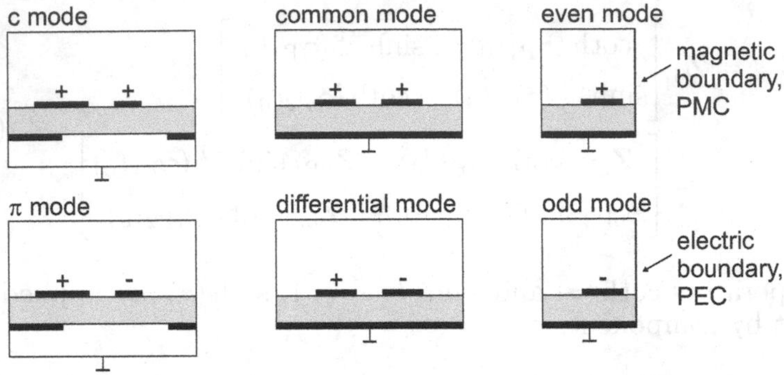

Figure 3.3: Cross-sectional views of planar waveguide carrying c and π mode (left), common and differential mode (middle) and even and odd mode (right).

If the reference impedances equal the characteristic impedances $\mathbf{Z}_{\mathrm{ref}} = \mathbf{Z}_0$ of Eq. (3.12), the scattering matrix consists of four nonzero entries only. These are transmission coefficients of the two modes, which propagate bidirectional (Eq. (3.17)). $\gamma_{\mathrm{P1}} = \gamma_{\mathrm{P2}}$ holds true in general.

$$\mathbf{S}_{\mathrm{mt}} = \left[\begin{array}{cc|cc} 0 & 0 & e^{-\gamma_1 \ell_0} & 0 \\ 0 & 0 & 0 & e^{-\gamma_2 \ell_0} \\ \hline e^{-\gamma_1 \ell_0} & 0 & 0 & 0 \\ 0 & e^{-\gamma_2 \ell_0} & 0 & 0 \end{array}\right] \tag{3.17}$$

In the case of arbitrary but equal and real positive reference impedances $\mathbf{Z}_{\mathrm{ref}} = \mathrm{diag}\left[Z_{\mathrm{ref}}, Z_{\mathrm{ref}}, Z_{\mathrm{ref}}, Z_{\mathrm{ref}}\right], Z_{\mathrm{ref}} \in \mathbb{R}^+$, the modal scattering matrix is given by Eq. (3.18, 3.19). Mode 1,3 and 2,4 are referred to as c mode and π mode. According to Fig. 3.3, c mode enforces both strips to have same potential at a time and currents through both strips flow in the same direction. Whereas π mode currents have opposite directions.

$$\mathbf{S}_{\mathrm{mt}} = \left[\begin{array}{cc|cc} S_{11\mathrm{c}} & 0 & S_{12\mathrm{c}} & 0 \\ 0 & S_{11\pi} & 0 & S_{12\pi} \\ \hline S_{21\mathrm{c}} & 0 & S_{22\mathrm{c}} & 0 \\ 0 & S_{21\pi} & 0 & S_{22\pi} \end{array}\right] \tag{3.18}$$

$$\begin{aligned}
S_{11\mathrm{c}} = S_{22\mathrm{c}} &= \frac{\left(2\frac{Z_{0\mathrm{c}}}{Z_{\mathrm{ref}}} - \frac{Z_{\mathrm{ref}}}{2Z_{0\mathrm{c}}}\right)\sinh(\gamma_{\mathrm{c}}\ell_0)}{2\cosh(\gamma_{\mathrm{c}}\ell_0) + \left(\frac{2Z_{0\mathrm{c}}}{Z_{\mathrm{ref}}} + \frac{Z_{\mathrm{ref}}}{2Z_{0\mathrm{c}}}\right)\sinh(\gamma_{\mathrm{c}}\ell_0)} \\
S_{21\mathrm{c}} = S_{12\mathrm{c}} &= \frac{2}{2\cosh(\gamma_{\mathrm{c}}\ell_0) + \left(\frac{2Z_{0\mathrm{c}}}{Z_{\mathrm{ref}}} + \frac{Z_{\mathrm{ref}}}{2Z_{0\mathrm{c}}}\right)\sinh(\gamma_{\mathrm{c}}\ell_0)} \\
S_{11\pi} = S_{22\pi} &= \frac{\left(\frac{1}{2}\frac{Z_{0\pi}}{Z_{\mathrm{ref}}} - \frac{2Z_{\mathrm{ref}}}{Z_{0\pi}}\right)\sinh(\gamma_{\pi}\ell_0)}{2\cosh(\gamma_{\pi}\ell_0) + \left(\frac{1}{2}\frac{Z_{0\pi}}{Z_{\mathrm{ref}}} + \frac{2Z_{\mathrm{ref}}}{Z_{0\pi}}\right)\sinh(\gamma_{\pi}\ell_0)} \\
S_{21\pi} = S_{12\pi} &= \frac{2}{2\cosh(\gamma_{\pi}\ell_0) + \left(\frac{1}{2}\frac{Z_{0\pi}}{Z_{\mathrm{ref}}} + \frac{2Z_{\mathrm{ref}}}{Z_{0\pi}}\right)\sinh(\gamma_{\pi}\ell_0)}
\end{aligned} \tag{3.19}$$

Eq. (3.18, 3.19) are valid also for the simplified case of uniformly coupled symmetric lines (middle of Fig. 3.3). Then, modes are usually called common and differential mode. Due to the symmetry, these are more easily characterized by analysis of only half the port surface. As it is shown on the right side of Fig. 3.3, even and odd mode analysis requires magnetic (perfect magnetic conductor, PMC) and electric (perfect electric conductor, PEC) boundaries, respectively. The common mode case is like two even mode cases in parallel. Hence, the even mode impedance is twice the common mode impedance $2 \cdot Z_{0\mathrm{common}} = Z_{0\mathrm{e}}$. Whereas, the differential mode case is like two odd mode cases connected in series $Z_{0\mathrm{differential}} = 2 \cdot Z_{0\mathrm{o}}$. Propagation coefficients are not affected by this simplification $\gamma_{\mathrm{common}} = \gamma_{\mathrm{e}}, \gamma_{\mathrm{differential}} = \gamma_{\mathrm{o}}$. Eq. (3.20) includes the corresponding scattering matrix entries, which are required for directional coupler synthesis in subsection 3.2.1.

$$
\begin{aligned}
S_{11\mathrm{e}} = S_{22\mathrm{e}} &= \frac{\left(\frac{Z_{0\mathrm{e}}}{Z_{\mathrm{ref}}} - \frac{Z_{\mathrm{ref}}}{Z_{0\mathrm{e}}}\right)\sinh(\gamma_{\mathrm{e}}\ell_0)}{2\cosh(\gamma_{\mathrm{e}}\ell_0) + \left(\frac{Z_{0\mathrm{e}}}{Z_{\mathrm{ref}}} + \frac{Z_{\mathrm{ref}}}{Z_{0\mathrm{e}}}\right)\sinh(\gamma_{\mathrm{e}}\ell_0)} \\
S_{21\mathrm{e}} = S_{12\mathrm{e}} &= \frac{2}{2\cosh(\gamma_{\mathrm{e}}\ell_0) + \left(\frac{Z_{0\mathrm{e}}}{Z_{\mathrm{ref}}} + \frac{Z_{\mathrm{ref}}}{Z_{0\mathrm{e}}}\right)\sinh(\gamma_{\mathrm{e}}\ell_0)} \\
S_{11\mathrm{o}} = S_{22\mathrm{o}} &= \frac{\left(\frac{Z_{0\mathrm{o}}}{Z_{\mathrm{ref}}} - \frac{Z_{\mathrm{ref}}}{Z_{0\mathrm{o}}}\right)\sinh(\gamma_{\mathrm{o}}\ell_0)}{2\cosh(\gamma_{\mathrm{o}}\ell_0) + \left(\frac{Z_{0\mathrm{o}}}{Z_{\mathrm{ref}}} + \frac{Z_{\mathrm{ref}}}{Z_{0\mathrm{o}}}\right)\sinh(\gamma_{\mathrm{o}}\ell_0)} \\
S_{21\mathrm{o}} = S_{12\mathrm{o}} &= \frac{2}{2\cosh(\gamma_{\mathrm{o}}\ell_0) + \left(\frac{Z_{0\mathrm{o}}}{Z_{\mathrm{ref}}} + \frac{Z_{\mathrm{ref}}}{Z_{0\mathrm{o}}}\right)\sinh(\gamma_{\mathrm{o}}\ell_0)}
\end{aligned}
\tag{3.20}
$$

Modal impedance and scattering matrices of multimode transmission lines are presented. The reader should keep in mind, results from 2D EM Eigenmode analysis $\mathbf{E}(x,y), \mathbf{H}(x,y)$ are used as excitation (varying boundary conditions) in 3D EM fullwave simulations. In this sense, results from 3D EM simulation, $\mathbf{S}$ or $\mathbf{Z}$, are inherently modal.

If different components within a front end module, like filters, couplers, mixers, multipliers, have the same waveguide interface, capturing the module performance by connecting the individual modal impedance or scattering matrices is the method of choice. These matrices cover the modal transmission and reflection behaviour as well as the amount of power, which is converted from one mode to another.

In case of different waveguide interfaces or if lumped elements or discrete components (e.g. diodes) are mounted to the waveguide, interactions with the entire electromagnetic field (all modes) take place. This depends on the exact mounting position[5]. Nodal impedance and scattering matrices are adequate parameters in such cases and are presented in the next paragraph.

[5]To accurately cover the interactions between waveguides and other components, like diodes, precise 3D EM models of the entire assembly are required (compare the co-simulation procedures of chapters 2, 4 and 5).

Power Normalized Conductor Impedance and Scattering Matrices
In full analogy to the modal equivalent voltages and currents of Eq. (3.10), nodal (conductor based) equivalent quantities are introduced in Eq. (3.21).

$$\begin{aligned}\mathbf{v}_{\mathrm{cP1}} &= \begin{bmatrix} v_{\mathrm{c1}} \\ v_{\mathrm{c2}} \end{bmatrix} \quad \mathbf{v}_{\mathrm{cP2}} = \begin{bmatrix} v_{\mathrm{c3}} \\ v_{\mathrm{c4}} \end{bmatrix} \quad \rightarrow \quad \mathbf{v}_{\mathrm{c}} = \begin{bmatrix} \mathbf{v}_{\mathrm{cP1}} \\ \mathbf{v}_{\mathrm{cP2}} \end{bmatrix} \\ \mathbf{i}_{\mathrm{cP1}} &= \begin{bmatrix} i_{\mathrm{c1}} \\ i_{\mathrm{c2}} \end{bmatrix} \quad \mathbf{i}_{\mathrm{cP2}} = \begin{bmatrix} i_{\mathrm{c3}} \\ i_{\mathrm{c4}} \end{bmatrix} \quad \rightarrow \quad \mathbf{i}_{\mathrm{c}} = \begin{bmatrix} \mathbf{i}_{\mathrm{cP1}} \\ \mathbf{i}_{\mathrm{cP2}} \end{bmatrix}\end{aligned} \tag{3.21}$$

The unitless matrices $\mathbf{M}_{\mathrm{v}}, \mathbf{M}_{\mathrm{i}}$ map the modal quantities $\mathbf{v}_{\mathrm{m}}, \mathbf{i}_{\mathrm{m}}$ to the nodal ones $\mathbf{v}_{\mathrm{c}}, \mathbf{i}_{\mathrm{c}}$.

$$\mathbf{v}_{\mathrm{c}} = \mathbf{M}_{\mathrm{v}}\mathbf{v}_{\mathrm{m}} \qquad \mathbf{i}_{\mathrm{c}} = \mathbf{M}_{\mathrm{i}}\mathbf{i}_{\mathrm{m}} \tag{3.22}$$

$$\begin{aligned}\mathbf{v}_{\mathrm{cP1}} &= \mathbf{M}_{\mathrm{vP1}}\mathbf{v}_{\mathrm{mP1}} & \mathbf{v}_{\mathrm{cP2}} &= \mathbf{M}_{\mathrm{vP2}}\mathbf{v}_{\mathrm{mP2}} \\ \mathbf{M}_{\mathrm{v}} &= \mathrm{diag}\left(\mathbf{M}_{\mathrm{vP1}}, \mathbf{M}_{\mathrm{vP2}}\right) & \mathbf{M}_{\mathrm{i}} &= \mathrm{diag}\left(\mathbf{M}_{\mathrm{iP1}}, \mathbf{M}_{\mathrm{iP2}}\right)\end{aligned} \tag{3.23}$$

Consequently, conductor impedance matrix is related to the modal impedance matrix by Eq. (3.24).

$$\mathbf{Z}_{\mathrm{ct}} = \mathbf{M}_{\mathrm{v}}\mathbf{Z}_{\mathrm{mt}}\mathbf{M}_{\mathrm{i}}^{-1} \tag{3.24}$$

The conductor voltages and currents are integral quantities of the total transversal electromagnetic field from 2D EM analysis $\mathbf{E}_{\perp}(x, y)$, $\mathbf{H}_{\perp}(x, y)$, including all modes. The integration lines $\boldsymbol{\ell}_k$ are chosen to match the lumped element or component, which is intended to be connected to the impedance matrix. According to Eq. (3.25), either all conductor voltages $v_{\mathrm{c}k}$ or all conductor currents $i_{\mathrm{c}k}$ may be fixed, but not both.

$$v_{\mathrm{c}k} = -\int_{\boldsymbol{\ell}_k} \mathbf{E}_{\perp} \cdot \mathrm{d}\boldsymbol{\ell} \qquad \text{or} \qquad i_{\mathrm{c}k} = \oint_{\boldsymbol{\ell}_k} \mathbf{H}_{\perp} \cdot \mathrm{d}\boldsymbol{\ell} \tag{3.25}$$

To ensure $v_{ck} \cdot i_{ck}^{\star}$ corresponds to the complex power at node k within circuit simulations, the power normalization of Eq. (3.26) is required [2].

$$\mathbf{M}_{\mathrm{i}}^{\mathrm{T}} \mathbf{M}_{\mathrm{v}} = \mathbf{X} \tag{3.26}$$

In Eq. (3.25) the conductor quantities are expressed as field integrals. With the help of $\mathbf{M}_{\mathrm{v}}$ and $\mathbf{M}_{\mathrm{i}}$, the conductor quantities are given as a weighted sum of the modal quantities with the matrix elements $M_{\mathrm{v}kj}, M_{\mathrm{i}kj}$ as weighting factors. Each element is obtained from electromagnetic field integration of a single mode only (Eq. (3.27)).

$$M_{\mathrm{v}kj} = \frac{-1}{v_{0j}} \int_{\ell_k} \mathbf{e}_{\perp j} \cdot \mathrm{d}\boldsymbol{\ell} \quad \forall\ j \quad \text{or} \quad M_{\mathrm{i}kj} = \frac{1}{i_{0j}} \oint_{\ell_k} \mathbf{h}_{\perp j} \cdot \mathrm{d}\boldsymbol{\ell} \quad \forall\ j \tag{3.27}$$

If either $\mathbf{M}_{\mathrm{v}}$ or $\mathbf{M}_{\mathrm{i}}$ is known, applying Eq. (3.26) ensures power normalization and determines the other matrix. Inserting Eq. (3.22) into Eq. (3.6), the total complex power in the conductor representation is then given by Eq. (3.28).

$$p = \mathbf{i}_{\mathrm{c}}^{\mathrm{H}} (\mathbf{M}_{\mathrm{i}}^{-1})^{\mathrm{H}} \mathbf{X} \mathbf{M}_{\mathrm{v}}^{-1} \mathbf{v}_{\mathrm{c}} \tag{3.28}$$

Fig. 3.4 illustrates a specific choice of modal and nodal / conductor voltages. Integration lines of the conductor voltages are oriented from left top conductor to ground and right top conductor to ground, respectively. Hence, the conductor impedance and scattering matrices cover the interaction of all modes with these four new nodes (terminals, ports).

In order to derive analytical expressions, the matrix $\mathbf{M}_{\mathrm{v}}$ is assumed as shown in Eq. (3.29, 3.30). The parameters $a, R_{\mathrm{c}}, R_{\pi}$ are complex in general and account for the contribution of all modes to the nodal voltages.

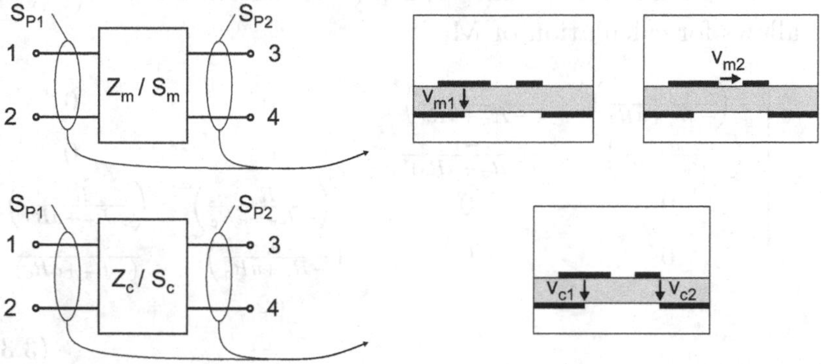

Figure 3.4: Specific modal (top) and nodal / conductor voltages (bottom) of uniformly coupled asymmetric lines.

$$\mathbf{v}_{\mathrm{cP1}} = \begin{bmatrix} v_{\mathrm{c1}} \\ v_{\mathrm{c2}} \end{bmatrix} = \begin{bmatrix} 1 & a \\ R_{\mathrm{c}} & aR_{\pi} \end{bmatrix} \begin{bmatrix} v_{\mathrm{m1}} \\ v_{\mathrm{m2}} \end{bmatrix} \tag{3.29}$$

$$\begin{aligned} \mathbf{M}_{\mathrm{vP1}} &= \begin{bmatrix} v_{\mathrm{c1}} \\ v_{\mathrm{c2}} \end{bmatrix} = \begin{bmatrix} 1 & a \\ R_{\mathrm{c}} & aR_{\pi} \end{bmatrix} \begin{bmatrix} v_{\mathrm{m1}} \\ v_{\mathrm{m2}} \end{bmatrix} \\ \mathbf{M}_{\mathrm{vP2}} &= \begin{bmatrix} v_{\mathrm{c3}} \\ v_{\mathrm{c4}} \end{bmatrix} = \begin{bmatrix} 1 & a \\ R_{\mathrm{c}} & aR_{\pi} \end{bmatrix} \begin{bmatrix} v_{\mathrm{m3}} \\ v_{\mathrm{m4}} \end{bmatrix} \end{aligned} \tag{3.30}$$

If cross coupling is zero [9] $\mathbf{X} = \mathbf{I}$, Eq. (3.26) becomes $\mathbf{M}_\mathrm{i} = \left(\mathbf{M}_\mathrm{V}^{-1}\right)^\mathrm{H}$ and allows for calculation of $\mathbf{M}_\mathrm{i}$.

$$\mathbf{M}_\mathrm{i} = \begin{bmatrix} -\left(\frac{R_\pi}{-R_\pi+aR_\mathrm{c}}\right)^\star & \left(\frac{R_\mathrm{c}}{-R_\pi+aR_\mathrm{c}}\right)^\star & 0 & 0 \\ \left(\frac{a}{-R_\pi+aR_\mathrm{c}}\right)^\star & \frac{-1}{(-R_\pi+aR_\mathrm{c})^\star} & 0 & 0 \\ 0 & 0 & -\left(\frac{R_\pi}{-R_\pi+aR_\mathrm{c}}\right)^\star & \left(\frac{R_\mathrm{c}}{-R_\pi+aR_\mathrm{c}}\right)^\star \\ 0 & 0 & \left(\frac{a}{-R_\pi+aR_\mathrm{c}}\right)^\star & \frac{-1}{(-R_\pi+aR_\mathrm{c})^\star} \end{bmatrix} \tag{3.31}$$

With Eq. (3.14) and Eq. (3.24) the resulting power normalized, nodal impedance matrix is known. The results of Eq. (3.33, 3.34) are a generalized form of the Tripathi[6] equations [3–5] with clear connection to results from 2D EM Eigenmode analysis and general waveguide circuit theory [1].

$$\mathbf{Z}_\mathrm{ct} = \left[\begin{array}{c|c} \mathbf{Z}_\mathrm{ct11} & \mathbf{Z}_\mathrm{ct12} \\ \hline \mathbf{Z}_\mathrm{ct21} & \mathbf{Z}_\mathrm{ct22} \end{array}\right] \tag{3.33}$$

[6]Eq. (3.33, 3.34) become equal to the equations in [4], if $a = 1/2$ and the modal characteristic impedances $Z_{01} = Z_{03}, Z_{02} = Z_{04}$ are replaced according to Eq. (3.32).

$$Z_{01} = Z_{03} = \frac{Z_{\mathrm{c}1}^2}{Z_{\mathrm{c}1} + Z_{\mathrm{c}2}/R_\pi^2} \qquad Z_{02} = Z_{04} = \frac{Z_{\pi 1}^2}{Z_{\pi 1} + Z_{\pi 2}/R_\mathrm{c}^2} \tag{3.32}$$

$$\gamma_{01} = \gamma_{03} = \gamma_\mathrm{c} \qquad \gamma_{02} = \gamma_{04} = \gamma_\pi$$

$$\mathbf{Z}_{\mathrm{ct11}} =$$
$$\begin{bmatrix} Z_{01}\coth(\gamma_{01}\ell_0) + a^2 Z_{02}\coth(\gamma_{02}\ell_0) & R_c Z_{01}\coth(\gamma_{01}\ell_0) + aR_\pi Z_{02}\coth(\gamma_{02}\ell_0) \\ R_c Z_{01}\coth(\gamma_{01}\ell_0) + aR_\pi Z_{02}\coth(\gamma_{02}\ell_0) & R_c^2 Z_{01}\coth(\gamma_{01}\ell_0) + R_\pi^2 Z_{02}\coth(\gamma_{02}\ell_0) \end{bmatrix}$$

$$\mathbf{Z}_{\mathrm{ct12}} = \begin{bmatrix} \frac{Z_{01}}{\sinh(\gamma_{01}\ell_0)} + \frac{a^2 Z_{02}}{\sinh(\gamma_{02}\ell_0)} & \frac{R_c Z_{01}}{\sinh(\gamma_{01}\ell_0)} + \frac{aR_\pi Z_{02}}{\sinh(\gamma_{02}\ell_0)} \\ \frac{R_c Z_{01}}{\sinh(\gamma_{01}\ell_0)} + \frac{aR_\pi Z_{02}}{\sinh(\gamma_{02}\ell_0)} & \frac{R_c^2 Z_{01}}{\sinh(\gamma_{01}\ell_0)} + \frac{R_\pi^2 Z_{02}}{\sinh(\gamma_{02}\ell_0)} \end{bmatrix}$$

$$\mathbf{Z}_{\mathrm{ct21}} = \begin{bmatrix} \frac{Z_{01}}{\sinh(\gamma_{01}\ell_0)} + \frac{a^2 Z_{02}}{\sinh(\gamma_{02}\ell_0)} & \frac{R_c Z_{01}}{\sinh(\gamma_{01}\ell_0)} + \frac{aR_\pi Z_{02}}{\sinh(\gamma_{02}\ell_0)} \\ \frac{R_c Z_{01}}{\sinh(\gamma_{01}\ell_0)} + \frac{aR_\pi Z_{02}}{\sinh(\gamma_{02}\ell_0)} & \frac{R_c^2 Z_{01}}{\sinh(\gamma_{01}\ell_0)} + \frac{R_\pi^2 Z_{02}}{\sinh(\gamma_{02}\ell_0)} \end{bmatrix}$$

$$\mathbf{Z}_{\mathrm{ct22}} =$$
$$\begin{bmatrix} Z_{01}\coth(\gamma_{01}\ell_0) + a^2 Z_{02}\coth(\gamma_{02}\ell_0) & R_c Z_{01}\coth(\gamma_{01}\ell_0) + aR_\pi Z_{02}\coth(\gamma_{02}\ell_0) \\ R_c Z_{01}\coth(\gamma_{01}\ell_0) + aR_\pi Z_{02}\coth(\gamma_{02}\ell_0) & R_c^2 Z_{01}\coth(\gamma_{01}\ell_0) + R_\pi^2 Z_{02}\coth(\gamma_{02}\ell_0) \end{bmatrix}$$

(3.34)

In the symmetric case (middle and right side of Fig. 3.3), $R_c = 1$, $R_\pi = -1$ and at least at low frequencies $a = 1/2$ hold true. The resulting impedance parameters Eq. (3.36, 3.37) simplify to the conventional textbook expressions[7] (e.g. [11]), if the modal characteristic impedances are replaced by the even $Z_{01} = 1/2 \cdot Z_{0e}$ and odd mode impedances $Z_{02} = 2 \cdot Z_{0o}$.

$$\mathbf{Z}_{\mathrm{ct}} = \left[\begin{array}{c|c} \mathbf{Z}_{\mathrm{ct11}} & \mathbf{Z}_{\mathrm{ct12}} \\ \hline \mathbf{Z}_{\mathrm{ct21}} & \mathbf{Z}_{\mathrm{ct22}} \end{array}\right] \tag{3.36}$$

[7] Conventional impedance parameters of uniformly coupled symmetric transmission lines with length ℓ_0.

$$\begin{aligned} Z_{11} = Z_{22} = Z_{33} = Z_{44} &= \frac{1}{2}\left(Z_{0e}\coth\gamma_e\ell_0\right) + \frac{1}{2}\left(Z_{0o}\coth\gamma_o\ell_0\right) \\ Z_{13} = Z_{31} = Z_{24} = Z_{41} &= \frac{1}{2}\left(Z_{0e}\coth\gamma_e\ell_0\right) - \frac{1}{2}\left(Z_{0o}\coth\gamma_o\ell_0\right) \\ Z_{14} = Z_{41} = Z_{32} = Z_{23} &= \frac{1}{2}\left(\frac{Z_{0e}}{\sinh(\gamma_e\ell_0)}\right) - \frac{1}{2}\left(\frac{Z_{0o}}{\sinh(\gamma_o\ell_0)}\right) \\ Z_{12} = Z_{21} = Z_{43} = Z_{34} &= \frac{1}{2}\left(\frac{Z_{0e}}{\sinh(\gamma_e\ell_0)}\right) + \frac{1}{2}\left(\frac{Z_{0o}}{\sinh(\gamma_o\ell_0)}\right) \end{aligned} \tag{3.35}$$

$$\mathbf{Z}_{\mathrm{ct11}} = \begin{bmatrix} Z_{01}\coth(\gamma_1\ell_0) + aa^\star Z_{02}\coth(\gamma_2\ell_0) & Z_{01}\coth(\gamma_1\ell_0) - aa^\star Z_{02}\coth(\gamma_2\ell_0) \\ Z_{01}\coth(\gamma_1\ell_0) - aa^\star Z_{02}\coth(\gamma_2\ell_0) & Z_{01}\coth(\gamma_1\ell_0) + aa^\star Z_{02}\coth(\gamma_2\ell_0) \end{bmatrix}$$

$$\mathbf{Z}_{\mathrm{ct12}} = \begin{bmatrix} \frac{Z_{01}}{\sinh(\gamma_1\ell_0)} + aa^\star \frac{Z_{02}}{\sinh(\gamma_2\ell_0)} & \frac{Z_{01}}{\sinh(\gamma_1\ell_0)} - aa^\star \frac{Z_{02}}{\sinh(\gamma_2\ell_0)} \\ \frac{Z_{01}}{\sinh(\gamma_1\ell_0)} - aa^\star \frac{Z_{02}}{\sinh(\gamma_2\ell_0)} & \frac{Z_{01}}{\sinh(\gamma_1\ell_0)} + aa^\star \frac{Z_{02}}{\sinh(\gamma_2\ell_0)} \end{bmatrix}$$

$$\mathbf{Z}_{\mathrm{ct21}} = \begin{bmatrix} \frac{Z_{01}}{\sinh(\gamma_1\ell_0)} + aa^\star \frac{Z_{02}}{\sinh(\gamma_2\ell_0)} & \frac{Z_{01}}{\sinh(\gamma_1\ell_0)} - aa^\star \frac{Z_{02}}{\sinh(\gamma_2\ell_0)} \\ \frac{Z_{01}}{\sinh(\gamma_1\ell_0)} - aa^\star \frac{Z_{02}}{\sinh(\gamma_2\ell_0)} & \frac{Z_{01}}{\sinh(\gamma_1\ell_0)} + aa^\star \frac{Z_{02}}{\sinh(\gamma_2\ell_0)} \end{bmatrix}$$

$$\mathbf{Z}_{\mathrm{ct22}} = \begin{bmatrix} Z_{01}\coth(\gamma_1\ell_0) + aa^\star Z_{02}\coth(\gamma_2\ell_0) & Z_{01}\coth(\gamma_1\ell_0) - aa^\star Z_{02}\coth(\gamma_2\ell_0) \\ Z_{01}\coth(\gamma_1\ell_0) - aa^\star Z_{02}\coth(\gamma_2\ell_0) & Z_{01}\coth(\gamma_1\ell_0) + aa^\star Z_{02}\coth(\gamma_2\ell_0) \end{bmatrix} \tag{3.37}$$

The nodal scattering parameters are related to the impedance matrix by Eq. (3.38).

$$\begin{aligned} \mathbf{S}_{\mathrm{ct}} &= \mathbf{U}\left(\mathbf{M}_{\mathrm{v}}\mathbf{Z}_{\mathrm{mt}}\mathbf{M}_{\mathrm{i}}^{-1} - \mathbf{Z}_{\mathrm{ref}}\right)\left(\mathbf{M}_{\mathrm{v}}\mathbf{Z}_{\mathrm{mt}}\mathbf{M}_{\mathrm{i}}^{-1} + \mathbf{Z}_{\mathrm{ref}}\right)^{-1}\mathbf{U}^{-1} \\ \mathbf{S}_{\mathrm{ct}} &= \mathbf{U}\left(\mathbf{Z}_{\mathrm{ct}} - \mathbf{Z}_{\mathrm{ref}}\right)\left(\mathbf{Z}_{\mathrm{ct}} + \mathbf{Z}_{\mathrm{ref}}\right)^{-1}\mathbf{U}^{-1} \end{aligned} \tag{3.38}$$

Choosing $\mathbf{Z}_{\mathrm{ref}} = \mathrm{diag}\left(Z_{\mathrm{ref}}, Z_{\mathrm{ref}}, Z_{\mathrm{ref}}, Z_{\mathrm{ref}}\right), Z_{\mathrm{ref}} \in \mathbb{R}$, the modal scattering parameters are the same as in Eq. (3.20). Their relation to the nodal scattering parameters is shown in Eq. (3.39).

$$\begin{aligned} S_{\mathrm{ct11}} = S_{\mathrm{ct22}} = S_{\mathrm{ct33}} = S_{\mathrm{ct44}} &= \frac{1}{2}\left(S_{11\mathrm{e}} + S_{11\mathrm{o}}\right) \\ S_{\mathrm{ct12}} = S_{\mathrm{ct21}} = S_{\mathrm{ct34}} = S_{\mathrm{ct43}} &= \frac{1}{2}\left(S_{21\mathrm{e}} + S_{21\mathrm{o}}\right) \\ S_{\mathrm{ct13}} = S_{\mathrm{ct24}} = S_{\mathrm{ct31}} = S_{\mathrm{ct42}} &= \frac{1}{2}\left(S_{11\mathrm{e}} - S_{11\mathrm{o}}\right) \\ S_{\mathrm{ct14}} = S_{\mathrm{ct23}} = S_{\mathrm{ct32}} = S_{\mathrm{ct41}} &= \frac{1}{2}\left(S_{21\mathrm{e}} - S_{21\mathrm{o}}\right) \end{aligned} \tag{3.39}$$

3.2 Backward Wave Directional Couplers

The backward wave directional couplers or contradirectional couplers, discussed throughout this work, are reciprocal four port devices, which exhibit double symmetry[8] (Fig. 3.5).

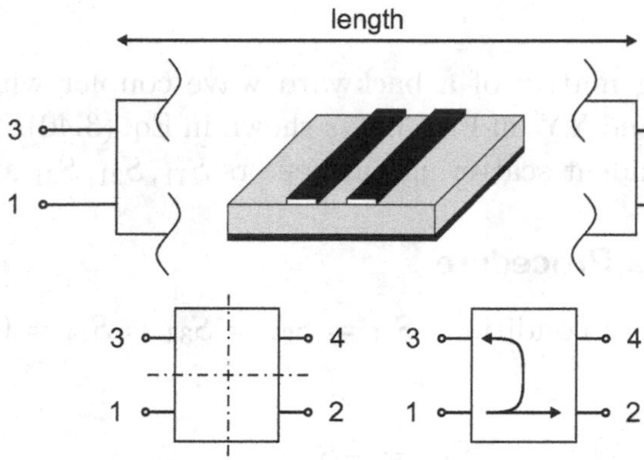

Figure 3.5: Block diagrams of backward wave directional couplers.

Synthesis is based on coupled transmission lines. To allow for ultrawide bandwidths, the focus is on nonuniform couplers rather than discrete section couplers. Uniformly coupled asymmetric lines are discussed in the preceding section. The nonuniform case is covered by cascading many short sections of uniform coupled lines. The synthesis procedure of the next subsection, is restricted to symmtric coupled lines. Hence, the modal and nodal scattering parameters of Eq. (3.20) and Eq. (3.39) are valid.

[8]Strictly speaking, only the part with coupled transmission lines exhibits double symmetry. Couplers with applied wiggly line technique do only have vertical symmetry, but within synthesis double symmetry is assumed anyway.

$$\underbrace{\left[\begin{array}{cc|cc} S_{11} & S_{12} & S_{13} & S_{14} \\ S_{21} & S_{22} & S_{23} & S_{24} \\ \hline S_{31} & S_{32} & S_{33} & S_{34} \\ S_{41} & S_{42} & S_{43} & S_{44} \end{array}\right]}_{\mathbf{S}} = \left[\begin{array}{cc|cc} S_{11} & S_{21} & S_{31} & S_{41} \\ S_{21} & S_{11} & S_{41} & S_{31} \\ \hline S_{31} & S_{41} & S_{11} & S_{21} \\ S_{41} & S_{31} & S_{21} & S_{11} \end{array}\right] = \left[\begin{array}{cc|cc} \bigcirc & \blacksquare & \square & 0 \\ \blacksquare & \bigcirc & 0 & \square \\ \hline \square & 0 & \bigcirc & \blacksquare \\ 0 & \square & \blacksquare & \bigcirc \end{array}\right] \tag{3.40}$$

The scattering matrix of a backward wave coupler with double symmetry (XX$'$ and YY$'$ in Fig. 3.6) is shown in Eq. (3.40). There are only four independent scattering parameters S_{11}, S_{21}, S_{31} and S_{41}.

3.2.1 Synthesis Procedure

Assuming matched conditions $S_{11} = S_{22} = S_{33} = S_{44} = 0$ leads to Eq. (3.41).

$$S_{11\mathrm{e}} = -S_{11\mathrm{o}} \tag{3.41}$$

Comparing the first and third line of Eq. (3.20), Eq. (3.41) holds true if the even and odd mode propagation coefficients or phase velocities are equal and the reference impedance equals the geometric average of the even and odd mode impedances.

$$\begin{aligned} \frac{Z_{0\mathrm{e}}}{Z_{\mathrm{ref}}} = \frac{Z_{\mathrm{ref}}}{Z_{0\mathrm{o}}} \quad &\rightarrow \quad Z_{\mathrm{ref}}^2 = Z_0^2 = Z_{0\mathrm{e}} Z_{0\mathrm{o}} \\ \gamma_{\mathrm{e}} = \gamma_{\mathrm{o}} \quad &\rightarrow \quad v_{\mathrm{e}} = v_{\mathrm{o}} \end{aligned} \tag{3.42}$$

Negligible forward wave coupling $S_{14} = S_{23} = S_{32} = S_{41} = 0$ results in Eq. (3.43).

$$S_{21\mathrm{e}} = S_{21\mathrm{o}} \tag{3.43}$$

Inserting Eq. (3.41) into Eq. (3.39), the backward wave coupling $S_{13} = S_{24} = S_{31} = S_{42}$ is given by the even mode reflection coefficient

S_{11e}. If losses are neglected, the propagation coefficients have zero real parts and S_{11e} in Eq. (3.20) simplifies[9] to Eq. (3.44).

$$S_{13} = S_{11e} = \frac{j\left(\frac{Z_{0e}}{Z_{ref}} - \frac{Z_{ref}}{Z_{0e}}\right)\sin(\beta_e \ell_0)}{2\cos(\beta_e \ell_0) + j\left(\frac{Z_{0e}}{Z_{ref}} + \frac{Z_{ref}}{Z_{0e}}\right)\sin(\beta_e \ell_0)} \tag{3.44}$$

The coupling coefficient's magnitude $|S_{13}|$ reaches a maximum, if $\beta_e \ell_0 = \pi/2$. This occurs at $\ell_0 = \lambda/4$. The maximum value $c = S_{13}$ is called voltage coupling factor and given by Eq. (3.45).

$$c = S_{13} = S_{24} = S_{31} = S_{42}\Big|_{\ell_0=\lambda/4} = \frac{Z_{0e} - Z_{0o}}{Z_{0e} + Z_{0o}} \tag{3.45}$$

Ultrawideband directional couplers[10] are obtained by cascading several (N) quarterwave coupler sections with different voltage coupling factors $c[k]$, $k \in \{1, \ldots, N\}$, as it is shown on the right side of Fig. 3.6.

According to Eq. (3.45), the coupled arm response S_{31} equals the even mode reflection coefficient S_{11e} versus frequency. Hence, it is sufficient to design a single ended prototype waveguide with $S_{11\,\mathrm{prototype}}$, rather than the more complicated coupled waveguides. The even mode impedances of the coupler $Z_{0e}[k]$ are then chosen equal to the prototype's characteristic impedances $Z_{0p}[k]$ versus k (Eq. (3.46), Fig. 3.7).

$$S_{13} = S_{24} = S_{31} = S_{42} \overset{!}{=} S_{11e} \overset{!}{=} S_{11\,\mathrm{prototype}} \tag{3.46}$$

As it is with most synthesis procedures of commensurate line structures [12], the problem can be more easily be described after Richards transform [13]. Uniform coupler synthesis is then based on the method of equating the coefficients of the input impedance function of the prototype waveguide $Z_{11p}(S)$ and the desired Chebyshev polynomial $Z_{\mathrm{Chebyshev}}(S)$ with respect to the complex Richards variable $S =$

[9] $\sinh(\gamma) = \sinh(j\beta) = j\sin(\beta) \qquad \cosh(\gamma) = \cosh(j\beta) = \cos(\beta)$

[10] Couplers with a single quarterwave section (e.g. 3 dB $\pm$ 0.3 dB) operate approximately over an octave bandwidth $f_2 \approx 2f_1$.

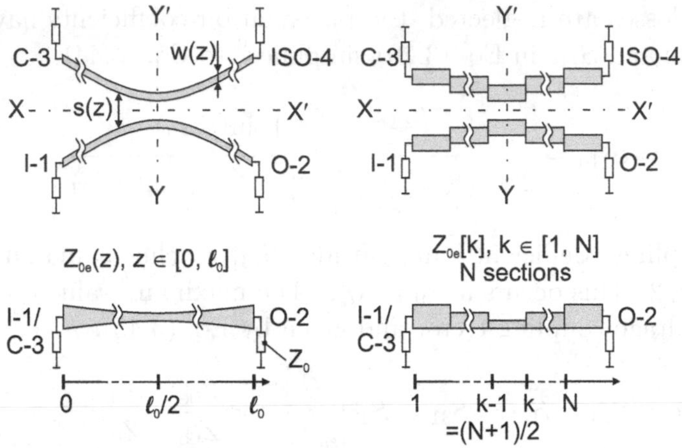

Figure 3.6: Symmetric nonuniform (left) and uniform (right) backward wave couplers together with single ended prototype waveguides to synthesize the appropriate even mode impedance distribution $Z_{0e}[k], Z_{0e}(z)$.

$j \cdot \tan(2\pi/\lambda \cdot \ell_{\lambda/4})$. Where $\ell_{\lambda/4}$ is the mechanical length of a single quarterwave coupler section.

$$Z_{11p}(S) \stackrel{!}{=} Z_{\text{Chebyshev}}(S) \tag{3.47}$$

As a result, the even mode impedances $Z_{0e}[k]$ and odd mode impedances $Z_{0o}[k]$ of all coupler sections which have to fulfill the specified Chebyshev behaviour are known [14].

High frequency performance of such discrete multisection couplers suffers from parasitics of the interconnections $(N-1)$ between adjacent coupler elements. Therefore, nonuniform directional couplers are discussed in the following (left side of Fig. 3.6). In this case, the voltage coupling factor and the characteristic impedances are continuous functions $c(z), Z_{0p}(z), Z_{0e}(z), Z_{0o}(z)$ $z \in [0, \ell_0]$ of the longitudinal axis z. Eq. (3.42) holds true at every position z.

The task of the synthesis procedure, is determination of $c(z)$ or $Z_{0e}(z)$ or $Z_{0o}(z)$ from the specification of the coupled arm response

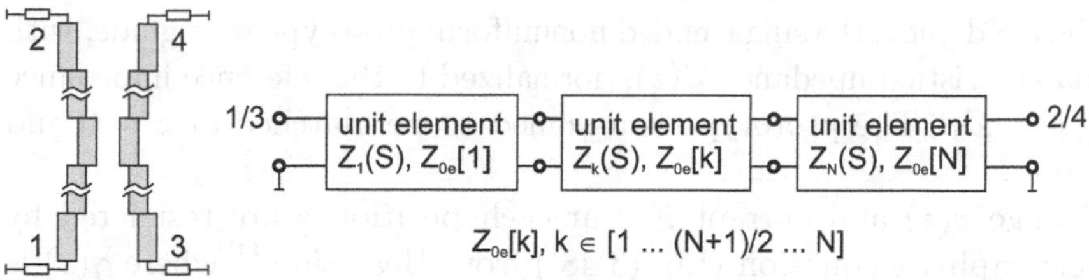

Figure 3.7: Symmetric, uniform backward wave coupler and its single ended prototype waveguide to synthesize the even mode impedances $Z_{0e}[k]$.

$S_{31}(\omega)$ versus frequency, the coupler's order N or length ℓ_0, the reference impedance Z_{ref} and the maximum realizable voltage coupling factor c_{max}. An additional step of synthesis is extraction of the necessary geometrical parameters, like strip widths $w(z)$ and strip spacings $s(z)$, for a given waveguide configuration.

The analytical design procedure is mainly based on the early works from Tresselt in 1966 [15] and Kammler in 1969 [16] together with the modifications from Uysal [17] in 1991. Whereas the design approaches in [15, 16] were based on the approximative solutions [18, 19], Uysal integrated the series solution from Bergquist [20, 21]. Although it is well documented, a short review is given to introduce nomenclature and the necessary scaling parameters E_i which are used to balance out deviations of the design assumptions in the proposed coupler structures.

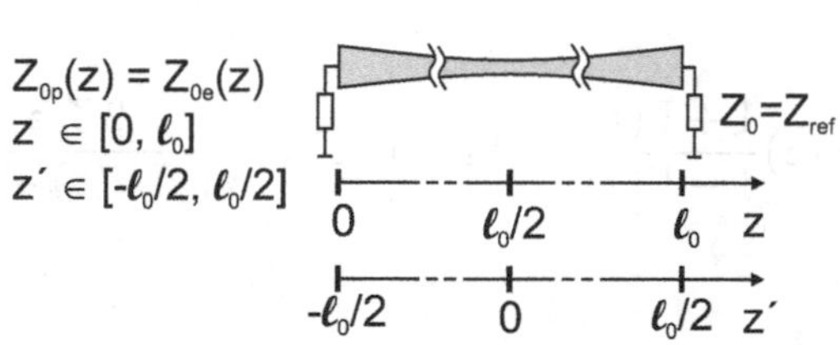

Figure 3.8: Single ended nonuniform prototype waveguide with overall reflection coefficient versus frequency $S_{11}(\omega)$, if terminated with the reference impedance $Z_{\text{ref}} = Z_0$.

Fig. 3.8 depicts the single ended nonuniform prototype waveguide, with characteristic impedance $Z(z)$, normalized to the reference impedance $Z_{\text{ref}} = Z_0$. The prototype is assumed to be matched at $z = 0$ and $z = \ell_0$.
Voltage $v(z)$ and current $i(z)$ at each position z are restricted by Telegrapher's equation (Eq. (3.48)) from Heaviside,[11] where $\gamma(z)$ is the propagation coefficient of the prototype waveguide.

$$\begin{aligned} \frac{\mathrm{d}v(z)}{\mathrm{d}z} &= -\gamma(z)Z(z)\cdot i(z) \\ \frac{\mathrm{d}i(z)}{\mathrm{d}z} &= -\gamma(z)/Z(z)\cdot v(z) \end{aligned} \tag{3.48}$$

The quantity $\Gamma(z)$ of Eq. (3.49) constitutes the overall reflection coefficient of a nonuniform transmission line (NTL[12]) with length z. In this sense, the corresponding NTL is terminated in $Z(0)$ at the input and $Z(z)$ at the output. Some authors call $\Gamma(z)$ the reflection coefficient at any point along the NTL, which is incorrect. A more detailed discussion of this misunderstanding is found in [22].

$$\Gamma(z) = \frac{\frac{v(z)/i(z)}{Z(z)} - 1}{\frac{v(z)/i(z)}{Z(z)} + 1} \tag{3.49}$$

Derivation of Eq. (3.49) with respect to z using the quotient rule, together with the Telegrapher's equation and the substitutions of Eq. (3.50)

$$v(z)/i(z) = Z(z)\frac{1+\Gamma(z)}{1-\Gamma(z)} \qquad 1-\Gamma(z) = \frac{2Z(z)}{v(z)/i(z)+Z(z)} \tag{3.50}$$

[11]Oliver Heaviside (1850–1925), English engineer.
[12]The abbreviation NTL (nonuniform transmission line) should not be mixed up with NLTL (nonlinear transmission line).

lead to the first order nonlinear differential equation from Walker and Wax [23]. This type of differential equation (Eq. (3.51)) is known as Riccati[13] equation.

$$\frac{\mathrm{d}\Gamma(z)}{\mathrm{d}z} - 2\gamma(z)\Gamma(z) + \left[1 - \Gamma(z)^2\right] p(z) = 0 \tag{3.51}$$

The term $p(z)$ is called the reflection coefficient distribution function. The argument of the natural logarithm of Eq. (3.52) is the characteristic impedance $Z_{0\mathrm{p}}(z)$ of the prototype waveguide (NTL), normalized to the reference impedance Z_{ref}.

$$p(z) = \frac{1}{2}\frac{\mathrm{d}}{\mathrm{d}z} \ln Z(z) = \frac{1}{2}\frac{\mathrm{d}}{\mathrm{d}z} \ln \frac{Z_{0\mathrm{p}}(z)}{Z_{\mathrm{ref}}} \tag{3.52}$$

Bergquist [20, 21] reported a general solution of Eq. (3.51), which is an alternative form of the results from Protonotarios [24, 25] and is given in Eq. (3.53).

$$\Gamma(z) = \frac{\phi_1 + \Gamma(0)\ \psi_2}{\phi_2 + \Gamma(0)\ \psi_1} \cdot \exp\left(2 \int_0^z \gamma(z)\mathrm{d}z\right) \tag{3.53}$$

The occurring functions $\phi_1, \phi_2, \psi_1, \psi_2$ and K_k, Q_k, f_1, f_2, with $k \in \mathbb{N}$ are recursively defined in Eq. (3.54 to 3.57).

$$\begin{aligned}
\phi_1 &= \sum_{k=1}^{\infty} K_{2k-1} = K_1 + K_3 + K_5 + \ldots \\
\phi_2 &= \sum_{k=1}^{\infty} K_{2k} = 1 + K_2 + K_4 + K_6 + \ldots \\
\psi_1 &= \sum_{k=1}^{\infty} Q_{2k-1} = Q_1 + Q_3 + Q_5 + \ldots \\
\psi_2 &= \sum_{k=1}^{\infty} Q_{2k} = 1 + Q_2 + Q_4 + Q_6 + \ldots
\end{aligned} \tag{3.54}$$

[13] Jacopo Francesco Riccati (1676–1754), Italian mathematician.

$$
\begin{aligned}
K_1 &= \int_0^z f_2 \mathrm{d}z & K_2 &= \int_0^z f_1 \cdot K_1 \mathrm{d}z \\
K_3 &= \int_0^z f_2 \cdot K_2 \mathrm{d}z & K_4 &= \int_0^z f_1 \cdot K_3 \mathrm{d}z \\
K_5 &= \dots
\end{aligned}
\tag{3.55}
$$

$$
\begin{aligned}
Q_1 &= \int_0^z f_2 \mathrm{d}z & Q_2 &= \int_0^z f_1 \cdot Q_1 \mathrm{d}z \\
Q_3 &= \int_0^z f_2 \cdot Q_2 \mathrm{d}z & Q_4 &= \int_0^z f_1 \cdot Q_3 \mathrm{d}z \\
Q_5 &= \dots
\end{aligned}
\tag{3.56}
$$

$$
\begin{aligned}
f_1 &= p(z) \cdot \exp\left(-2 \int_0^z \gamma(z) \mathrm{d}z\right) \\
f_2 &= p(z) \cdot \exp\left(+2 \int_0^z \gamma(z) \mathrm{d}z\right)
\end{aligned}
\tag{3.57}
$$

Assuming perfectly matched conditions,[14] $Z_{0\mathrm{e}} = Z_{0\mathrm{o}} = Z_{0\mathrm{p}} = Z_0 = Z_{\mathrm{ref}}$ at $z = 0$ and $z = \ell_0$, leads to $\Gamma(0) = \Gamma(\ell_0) = 0$ and Bergquist's solution simplifies to Eq. (3.58).

$$
\Gamma(z) = \frac{\phi_1}{\phi_2} \cdot \exp\left(2 \int_0^z \gamma(z) \mathrm{d}z\right) \tag{3.58}
$$

[14]This is possible if the strip spacing is infinitely (sufficiently) large $s \to \infty$.

Neglecting losses $\gamma(z) = j\beta = j\omega/v$ and introducing the function $G(z)$ of Eq. (3.59),

$$G = \left| \int_0^{\ell_0} \sin\left(\frac{2\omega}{v_g} z\right) \cdot p(z) \mathrm{d}z \right| \tag{3.59}$$

Eq. (3.54) reduces to Eq. (3.60). Where v_g is the waveguide phase velocity, that equals the even and odd mode phase velocity $v_g = v_e = v_o$.

$$\begin{aligned} \phi_1 &= \sum_{k=1}^{\infty} G^{2k-1}/(2k-1)! = G + G^3/3! + G^5/5! + \cdots = \sinh(G) \\ \phi_2 &= 1 + \sum_{k=1}^{\infty} G^{2k}/(2k)! = 1 + G^2/2! + G^4/4! \quad + \cdots = \cosh(G) \end{aligned} \tag{3.60}$$

The relationship between $p(z)$ from Eq. (3.52) and $S_{11e}(\omega) = S_{31}(\omega)|_{\Gamma(z)}$ is identified as a corresponding pair of a Fourier integral transform.

$$p(z) \quad \circ\!\!-\!\!\bullet \quad S_{31}(\omega) = S_{11e}(\omega)$$

$$S_{31(\omega)} = \tanh\left(\int_0^{\ell_0} \sin\left(\frac{2\omega z}{v_g}\right) p(z) \mathrm{d}z \right) \tag{3.61}$$

From a given desired coupling function $S_{31\mathrm{spec}}(\omega)$ the reflection coefficient distribution function $p(z)$ is extracted by inverse Fourier transform.

$$p(z) = -\frac{2}{\pi v_g} \int_{\omega_1}^{\omega_2} \sin\left(\frac{2\omega z}{v_g}\right) \operatorname{arctanh}\left(S_{31\mathrm{spec}}(\omega)\right) \mathrm{d}\omega \tag{3.62}$$

The resulting $S_{31}(\omega)$, the required even mode impedances $Z_{0p}(z) = Z_{0e}(z)$ and the corresponding voltage coupling factor $c(z)$ are then known from Eq. (3.63).

$$
\begin{aligned}
S_{31} &= \tanh\left(\int_0^{\ell_0} \sin\left(\frac{2\omega z}{v_\mathrm{g}}\right) p(z)\mathrm{d}z\right), \\
\frac{Z_{0\mathrm{e}}(z)}{Z_0} &= \exp\left(2\int_0^z p(z)\mathrm{d}z\right), \; c(z) = \frac{Z_{0\mathrm{e}}(z) - Z_{0\mathrm{o}}(z)}{Z_{0\mathrm{e}}(z) + Z_{0\mathrm{o}}(z)}.
\end{aligned} \tag{3.63}
$$

If the specified $S_{31\mathrm{spec}}(\omega)$ has rectangular dependency of ω, there will be Gibbs–Wilbraham[15] phenomenon in the synthesized coupled arm response.

$$
S_{31\mathrm{spec}}(\omega) = \begin{cases} C_\mathrm{spec} & \omega_1 \leq \omega \leq \omega_2 \\ 0 & \text{else} \end{cases} \qquad 0 \leq C_\mathrm{spec} \leq 1 \tag{3.64}
$$

To obtain quasi equal ripple response, the Fourier transform is performed at least twice. In a second run #2, the deviation of the first #1 result from the specified $S_{31\mathrm{spec}}(\omega)$ is used to calculate the new coupled arm target response (Eq. (3.65)). C_spec has to be chosen small enough to keep the maximum voltage coupling factor c_max, which occurs at the coupler's center, realizable. c_max mainly corresponds to the minimum realizable strip spacing s_min, but also to the minimum strip width w_min.

$$
\begin{aligned}
S_{31\mathrm{spec}} &\rightarrow p^{\#1}(z) \;\rightarrow\; S_{31}^{\#1} \\
|S_{31\mathrm{spec}}| + \left(|S_{31\mathrm{spec}}| - S_{31}^{\#1}\right) &\rightarrow p^{\#2}(z) \;\rightarrow\; S_{31}^{\#2}
\end{aligned} \tag{3.65}
$$

Weighting functions as introduced in [15, 16] are obsolete. This is one main advantage of using Bergquist's solution [20] instead of [18]. At the far left and right end of the symmetric coupler, the synthesized voltage coupling factor vanishes, what corresponds to infinite strip spacing $s(0), s(\ell_0) \rightarrow \infty$. As this can hardly be guaranteed, Uysal's

[15] Josiah Willard Gibbs (1839–1903), American scientist.
Henry Wilbraham (1825–1883), English mathematician.

scaling parameter E_0 [17] has shown to be highly useful in practical designs ($E_0 = 0 \ldots 0.04$).

$$\frac{Z_{0\mathrm{e}E_0}(z)}{Z_0} = \frac{Z_{0\mathrm{e}}(z)}{Z_0} \cdot \exp\left(E_0\right) \tag{3.66}$$

E_0 is increased until the voltage coupling factor c at $z = 0$ and $z = \ell_0$ becomes realizable with the chosen coupled microstrip line (cMSL) waveguides. $S_{31\mathrm{E}_0}$ in Eq. (3.67) is the unintended modification of the desired coupled arm response S_{31} caused by scaling the normalized even mode impedance, which is tolerable if E_0 is chosen sufficiently small.

$$S_{31\mathrm{E}_0}(\omega) = S_{31}(\omega) + jE_0 \sin\left(\frac{\omega \ell_0}{v_\mathrm{g}}\right) \tag{3.67}$$

After applying the presented synthesis procedure, a realizable coupler configuration is found and extraction of the required geometrical parameters, strip widths $w(z)$ and spacings $s(z)$, is possible by 2D EM simulations or analytical models.

The presented procedure requires equalized phase velocities of the even and odd mode $v_\mathrm{e} = v_\mathrm{o}$ to obtain isolation $S_{41} \approx 0$ and therefore high directivity $D = S_{31}/S_{41}$, respectively. As most practical waveguides show dispersive behaviour, additional effort is necessary. Loss mechanisms, which always differ for both modes, further complicate the scenario.

Several phase equalization methods for planar couplers have been reported. These rely on lowering the odd mode phase velocity (wiggly-line technique, capacitive loading), increasing the even mode phase velocity (defected ground structure technique) or are based on almost transversal electromagnetic (TEM) waveguides with inherent phase equalization (dielectric overlay technique).

Capacitive loading has been studied in several papers (e.g. [26]), but operation in the millimeter-wave frequency range could only be demonstrated with monolithic integrated circuit technology [27] and a quite high minimum operating frequency. The same holds true

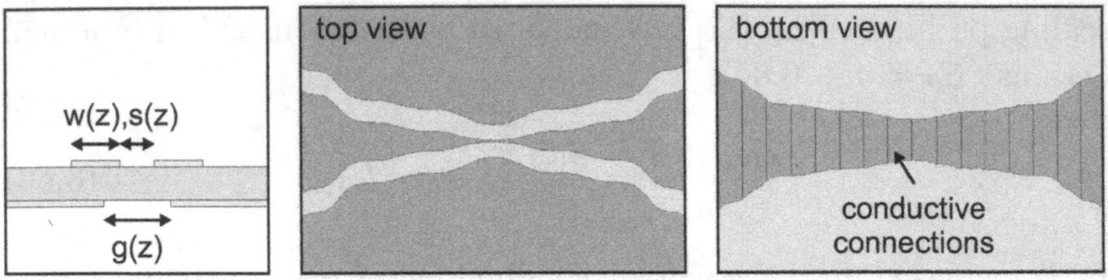

Figure 3.9: Cross section (left), top (middle) and bottom (right) view of a directional coupler on 10 mil alumina with defected ground structure (DGS) technique. dgs coupler $w(z), s(z), g(z)$, thin connections between separated ground planes to avoid higher order modes.

for meandered coupler configurations [28]. Inductive compensation techniques that increase the even mode phase velocity have also been reported [29] and are proved to work well at low operating frequencies. An asymmetric coupler with highpass frequency characteristic and resistive compensation elements has been demonstrated in [30]. The achieved broadband simulation results are promising for the aforementioned field of applications, but [30] does not contain measurement results.

Defected ground structures (DGS) are used throughout this work to compensate for shunt capacitive behaviour of transitions from coaxial waveguide to microstrip line (Fig. 2.39 of chapter 2), to increase the characteristic impedance of series inductances within lowpass filter designs (Fig. 2.44 of chapter 2) and to compensate for mounting pads of DC block series capacitors (section 6.1). The author took great effort to also realize ultrawideband nonuniform MSL directional couplers on dispersive 10 mil alumina with DGS technique, based on the proposed synthesis procedure. Fig. 3.9 shows from left to right, an exemplary coupler cross section, top and bottom view. Beside strip widths $w(z)$ and strip spacings $s(z)$ varying with z, the widths of DGS $g(z)$ constitutes another degree of freedom for coupler design. In theory, DGS allows for sufficient equalization of v_e and v_o. Experimental investigations make clear handling / avoiding higher order modes over ultrawide bandwidth is hard to realize, although conductive connections of the separated ground planes (right side of Fig. 3.9) are added. In practice, success of the synthesized directional coupler is highly sensitive to the chosen substrate material and the manufactured mechanical housing. DGS technique does not always lead to successful designs. The same holds true for applying DGS technique to design of edge coupled MSL halfwave resonator bandpass filters with high parasitic passband suppression (compare subsection 3.5). Successful coupler results could only be achieved from 1 to 20 GHz. Therefore, the following subsections 3.2.2 and 3.2.3 focus on dielectric overlay technique (stripline couplers STL) and wiggly-line technique.

The main problem with dielectric overlay technique is finding an appropriate layer stackup with proprietary materials (cores and prepregs), that exhibit equalized phase velocities $v_e = v_o$. Process technology to build multilayer printed circuit boards (PCB) constitutes a severe limitation to the design rule set. Minimum strip widths and spacings are large compared to thin-film technology. The necessity of transitions from coaxial or microstrip line to the TEM stripline is another limiting factor. These things are discussed in subsection 3.2.2 together with simulation and measurement results of manufactured stripline directional couplers.

Subsection 3.2.3 deals with microstrip line directional couplers on 10 mil alumina, processed by thin-film technology. Phase equalization is established by wiggly-line technique. The author's theoretical and experimental investigations are presented.

The following paragraph describes a simple heuristic design strategy, which is included for the sake of completeness and to distinguish the proposed method from more simple approaches.

Nonuniform Coupler Synthesis Based On Small Signal Reflection Theory There is a simple heuristic design strategy, which is found in many textbooks (e.g. [11]). It is explained in the following to show the relation to the proposed synthesis procedure.

The reflection coefficient of an infinitesimal short section $\mathrm{d}z$ of prototype waveguide $Z_{0\mathrm{p}}(z) = Z_{0\mathrm{e}}(z)$ is called $\mathrm{d}\Gamma$ in Eq. (3.68).

$$\mathrm{d}\Gamma = \frac{Z_{0\mathrm{e}} + \mathrm{d}Z_{0\mathrm{e}} - Z_{0\mathrm{e}}}{Z_{0\mathrm{e}} + \mathrm{d}Z_{0\mathrm{e}} + Z_{0\mathrm{e}}} \approx \frac{\mathrm{d}Z_{0\mathrm{e}}}{2Z_{0\mathrm{e}}} \tag{3.68}$$

Applying the simplification $2Z_{0\mathrm{e}} \gg \mathrm{d}Z_{0\mathrm{e}}$ is equivalent to neglecting multiple reflections. This is known as small signal reflection theory.

$$\frac{1}{Z_{0\mathrm{e}}} = \frac{\mathrm{d}\left(\ln Z_{0\mathrm{e}}\right)}{\mathrm{d}Z_{0\mathrm{e}}} \qquad \int \frac{1}{Z_{0\mathrm{e}}} \mathrm{d}Z_{0\mathrm{e}} = \ln Z_{0\mathrm{e}} \tag{3.69}$$

Including the integral relation of Eq. (3.69), Eq. (3.68) becomes Eq. (3.70).

$$\mathrm{d}\Gamma \approx \frac{\mathrm{d}Z_{0\mathrm{e}}}{2Z_{0\mathrm{e}}} = \frac{1}{2}\mathrm{d}\left(\ln Z_{0\mathrm{e}}\right) = \frac{1}{2}\frac{\mathrm{d}}{\mathrm{d}z}\left(\ln Z_{0\mathrm{e}}\right)\mathrm{d}z \tag{3.70}$$

Multiplying the exponential function $e^{-2j\beta z}$ to $\mathrm{d}\Gamma$ results in the corresponding infinitesimal reflection coefficient $\mathrm{d}S_{11\,\mathrm{prototype}} = \mathrm{d}S_{11\mathrm{e}}$, which is the contribution of the reflection coefficient $\mathrm{d}\Gamma$ at position z to the overall reflection coefficient $S_{11\mathrm{e}} = S_{31}$ at $z = 0$.

$$\mathrm{d}S_{11\mathrm{e}} = \mathrm{d}\Gamma e^{-2j\beta z} = \frac{1}{2}e^{-2j\beta z}\frac{\mathrm{d}}{\mathrm{d}z}\left(\ln Z_{0\mathrm{e}}\right)\mathrm{d}z \tag{3.71}$$

Summing up all contributions $\mathrm{d}S_{11\mathrm{e}}$ with correct phase over the entire coupler length from $z = 0$ to $z = \ell_0$, leads to the coupled arm response in Eq. (3.72).

$$S_{11e} = \frac{1}{2}\int_0^{\ell_0} e^{-2j\beta z}\frac{\mathrm{d}}{\mathrm{d}z}(\ln Z_{0\mathrm{e}})\mathrm{d}z = S_{31} \tag{3.72}$$

In Eq. (3.73) the coordinate transformation $z \rightarrow z' = z - \ell_0/2$ (Fig. 3.8) and the substitution $p(z') = 1/2 \cdot \mathrm{d}/\mathrm{d}z' \, (\ln Z_{0\mathrm{e}})$ are applied.

$$\begin{aligned} S_{31} = S_{11\mathrm{e}} &= \frac{1}{2} e^{-j\beta\ell_0} \int\limits_{-\ell_0/2}^{\ell_0/2} e^{-2j\beta z'} \frac{\mathrm{d}}{\mathrm{d}z'} (\ln Z_{0\mathrm{e}}) \, \mathrm{d}z' \\ S_{31} &= e^{-j\beta\ell_0} \int\limits_{-\ell_0/2}^{\ell_0/2} e^{-2j\beta z'} p(z') \mathrm{d}z' \end{aligned} \tag{3.73}$$

Symmetry considerations allow for further simplification of the integral in Eq. (3.73). $p(x)$ is an odd function, multiplied by the complex exponential function $e^{jx} = \cos(x) + j\sin(x)$, $x \in \mathbb{C}$. It is $\cos(-x) = \cos(x)$ and $\sin(-x) = -\sin(x)$. An even function (cos) multiplied by an odd function (p) results in an odd function. Integrating odd functions symmetrically around the center of the coordinate system results in zero value. With these considerations, the coupled arm response and $p(z')$ are recognized as a pair of Fourier transform.

$$\begin{aligned} S_{31} &= -je^{-j\beta\ell_0} \int\limits_{-\ell_0/2}^{\ell_0/2} \sin(2\beta z') p(z') \mathrm{d}z' \\ S_{31} e^{j\beta\ell_0} &= \int\limits_{-\infty}^{\infty} e^{-2j\beta z'} p(z') \mathrm{d}z' \quad \bullet\!\!-\!\!\circ \quad p(z') \end{aligned} \tag{3.74}$$

This heuristic solution is equivalent to the results from Kammler [16], which are solutions of the differential equation Eq. (3.75).

$$\frac{\mathrm{d}\Gamma(z)}{\mathrm{d}z} - 2\gamma(z)\Gamma(z) + p(z) = 0 \tag{3.75}$$

The latter is a simplified version ($\Gamma(z)^2 \ll 1$) of Eq. (3.51) and best known from synthesis of matched transitions (e.g. Dolph-Chebyshev tapers) between different types of waveguides (comp. subsection 2.7.1).

3.2.2 Dielectric Overlay Technique

Since Oliver published the first exact design theory for a single section transversal electromagnetic field (TEM) coupler in 1954 [31], directional couplers became one of the most important passive components with widespread fields of application up to the millimeter-wave frequency range and beyond. Especially ultrawideband, loose couplers with average coupling values in the range of 10 to 20 dB and a preferably high directivity, greater than 10 dB, play an important role in front end modules of semiconductor automatic test systems and scattering parameter testsets of vector network analyzers (VNA). Due to applied calibration methods, the aforementioned applications can deal with a quite large coupling ripple, but sufficiently high directivity values from frequencies as low as 0.003 MHz or 10 MHz up to 60/67/70 GHz have to be guaranteed.

Modern VNAs perform scattering parameter measurements in the frequency range of 0.003 MHz or 10 MHz up to 1.1 THz. The upper millimeter-wave frequencies, 60 GHz (E-band) to 1.1 THz, are accessible through VNA frequency converters. These additional testsets are driven by the VNA base instrument and consist of frequency multipliers at the RF / LO ports, as well as harmonic mixers for down-conversion and cover a full hollow waveguide band (FBW $\approx$ 40 %). Wave separation is performed by hollow waveguide directional couplers, which provide inherently high directivity and low insertion loss. If necessary, codirectional (forward wave) couplers (compare section 3.3) on thin quartz substrates can serve as planar alternative realization [32]. For use with VNA frequency converters, the base instrument needs to provide RF and LO input signals at about 20 GHz only.

VNA base instruments operate from 0.003/10 MHz up to 67/70 GHz. Within the corresponding testsets, the design of appropriate directional couplers that provide directivity values greater than zero across the whole frequency range is a challenging task. The minimum operating frequency can only be handled by backward wave couplers,[16] which are

[16]Some vendors of VNA systems use resistive directional bridges to cover the very low frequencies. Hence, low frequency requirements on the directional

advantageously designed for planar integration. The parasitic forward wave coupling S_{41} and therefore the achievable minimum directivity $D = S_{31}/S_{41}$ is related to the phase velocity difference of the occuring even and odd mode waves $v_e^{-1} - v_o^{-1}$. The author's investigations to build high directivity backward wave couplers on single layer substrates are given in subsection 3.2.3. Phase velocitiy equalization of the edge coupled microstrip lines (MSL) is based on wiggly-line technique [J1]. These designs fit best into highly integrated planar testset modules, but require comparably sophisticated manufacturing technology.

In the following, an approach based on conventional multilayer PCBs, dielectric overlay technique, is presented. Stripline based directional couplers with maximum operating frequencies of 50 and 70 GHz are presented. Two different PCB stackups are investigated. Besides an appropriate synthesis, the design of well matched transitions from STL to MSL constitutes an inevitable requirement to obtain high directivity. Two different transitions and the associated critical design parameters are introduced.

Ultrawideband stripline couplers are commercially available on the market with coaxial connector interface [33–38], which is unfavourable for planar integration. The proposed couplers can either be integrated with other STL components, like filters or connected to conventional MSL components by bondwire interconnection.

Detailed theoretical and experimental results on STL ultrawideband coupler design for six port reflection analyzers have been reported by Potter and Hjipieris [39–41]. They successfully have built up 8.34 dB couplers with very high order (coupler length), operating from 0.75 to 26.5 GHz. Within 3 dB tandem configuration, these couplers achieve greater than 15 dB directivity. A summary of available STL couplers and comparison with the proposed ones is given in the first part of Table 3.2 in subsection 3.2.3.

---

couplers are relaxed on the expense of an additional switched path and therefore system cost. Resistive directional bridges are basically Wheatstone bridges (Sir Charles Wheatstone (1802–1875), English scientist), which can easily obtain high directivity but suffer from main line insertion loss and low maximum operating frequency.

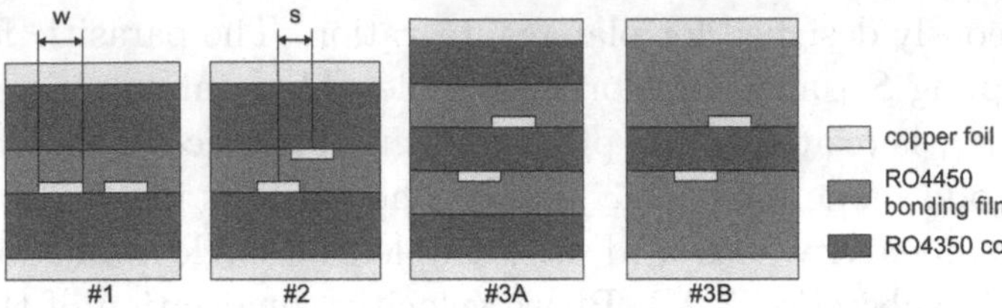

Figure 3.10: Four different multilayer PCB configurations (stackups) suitable for building up directional couplers.

Quasi TEM Waveguide Configuration Equal phase velocities of the even and odd mode are a distinct precondition to gain high isolation performance. Within a multilayer stackup we have to claim for minimum deviation of the bulk permittivities of the core and prepreg materials, while a maximum nominal permittivity value is needed to ensure a sufficiently high maximum voltage coupling factor c_{max}. To the author's knowledge, the glass-reinforced hydrocarbon / ceramic laminate RO4350 and the corresponding bonding film RO4450 constitute the best compromise (RO4350/4450, $\epsilon_{\mathrm{r}} = 3.48/3.54 \pm 0.05$). These laminates are available with 17.5 µm electrodeposited copper foils. Hydrocarbon substrates were intended for use in low cost multi purpose microwave components and show increased dielectric loss ($\tan\delta = 0.0037/0.0040$ at 10 GHz) compared with PTFE materials. Due to the carbon, laser ablation and laser cutting can hardly be utilized because the occuring heat leads to a significant increase of insertion loss.

Fig. 3.10 shows four different multilayer configurations. The first #1 is based on edge coupled STL and suffers from the realizable minimum conductor spacing of common PCB technology ($s_{\mathrm{min}} = 100$ µm $\rightarrow c_{\mathrm{max}} = 0.28$). The offset coupled STL in configurations #2 and #3B are used to build up the proposed couplers ($c_{\mathrm{max}} \approx 0.5$). The lateral displacement of both structured core materials in case of configuration #2 is a critical issue. This can be solved by using configuration #3A or #3B, which are a bit more complicated and cost efficient but the unwanted offset between both strips can be guaranteed to stay below 50 µm. In all four cases plating of blind and through vias is performed as last manufacturing step, which causes copper thicknesses of $\approx$ 35 µm at the outer layers and leads to an aspect ratio of $\varnothing_{\text{VIA}}/\text{length}_{\text{VIA}} = 1/1$.

To establish a mapping from the synthesized coupling factor $c(z)$ of the performed synthesis to the geometric data of the STL waveguides, 2D EM simulations in the frequency domain are performed. Although there are closed form expressions available, for example the ones published by Shelton in 1966 [42, 43], the 2D EM approach is necessary to account for the different dielectric sheets and their electromagnetic behaviour up to the maximum operating frequency. Fig. 3.11 shows STL design data for configuration #2. The plotted strip widths w, strip offsets s and phase velocities $v_{\text{e/o}}$ correspond to a terminating impedance $Z_0 = Z_{\text{ref}} = 50\ \Omega$. As we know $w(c), s(c)$ and $c(z)$, calculation of $w(z), s(z)$ and generation of the coupler layout is possible. The parameters used for the design of the proposed couplers with $f_{\text{max}} = 50$ GHz and $f_{\text{max}} = 70$ GHz are given by Eq. (3.76). Fig. 3.12 illustrates the corresponding layouts.

$$\begin{aligned} E_0 &= 0.020,\ f_{1/2} = 0.01/52\ \text{GHz},\ \ell_0 = 8.0\ \text{mm},\ S_{31\text{spec}} = 0.263 \\ E_0 &= 0.022,\ f_{1/2} = 0.00/72\ \text{GHz},\ \ell_0 = 9.8\ \text{mm},\ S_{31\text{spec}} = 0.231 \end{aligned} \tag{3.76}$$

Scattering parameters from 3D EM simulation versus frequency of the 70 GHz coupler with STL waveguide interfaces are shown in Fig. 3.13. Furthermore a comparison with the analytic coupled arm response and the influence of the already mentioned lateral displacement present to configuration #2 are visualized. The achievable displacement with configuration #2 depends on the overall size of the substrate dimensions, which are used for building up the multilayer stackup. The common panel size of 250 mm $\times$ 450 mm has been applied. With respect to this constraint, $\Delta y \pm 50$ µm constitutes a challenging specification. As we can see in Fig. 3.13, $\Delta y < 0$ has considerable impact on the couplers performance. Although the coupling

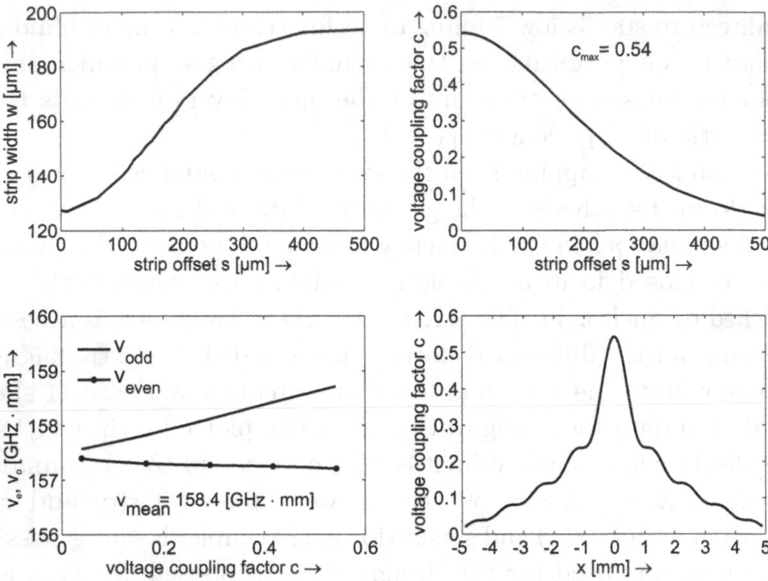

Figure 3.11: Extracted strip width w, strip offset s, even and odd mode phase velocity $v_{\mathrm{e/o}}$ values at 30 GHz from 2D EM simulation together with the synthesized voltage coupling factor $c(z)$ of the 70 GHz STL coupler.

performance significantly deteriorates, the isolation holds good. The latter depends more sensitively on equalized phase velocities $v_{\mathrm{e}} \approx v_{\mathrm{o}}$ than on matched impedances $Z_0 \approx \sqrt{Z_{0\mathrm{e}} Z_{0\mathrm{o}}}$. Within inhomogeneous stackups achieving isolation values better than 25 dB is a by far more difficult business ([J1], subsection 3.2.3).

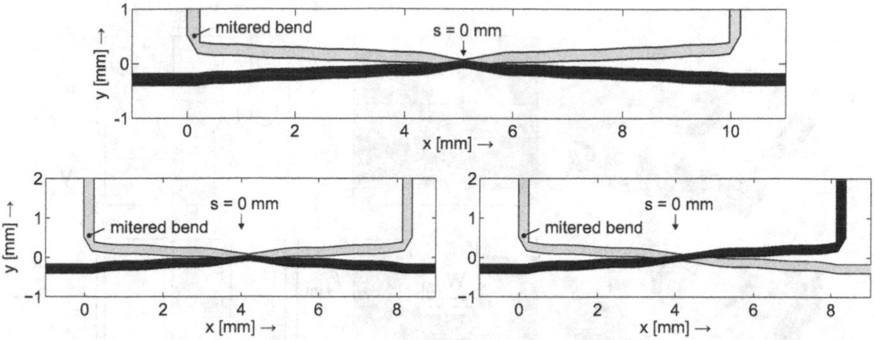

Figure 3.12: Layouts of the 70 GHz (top) and 50 GHz (bottom) stripline directional couplers with and without crossing (configuration #2 of Fig. 3.10).

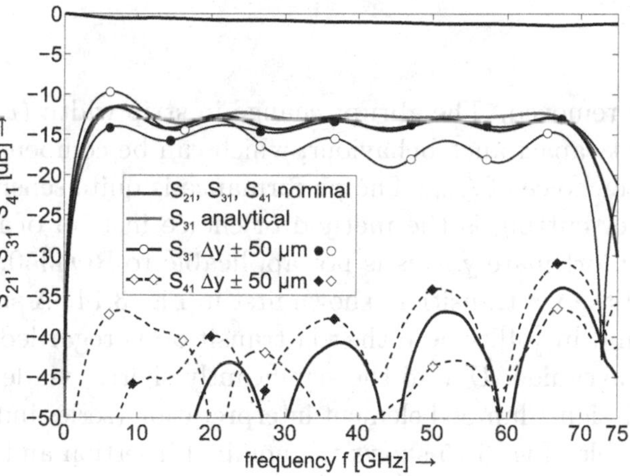

Figure 3.13: Simulated insertion loss, coupling and isolation versus frequency of the 70 GHz coupler, together with the analytically synthesized coupling function and the influence of lateral displacement in y direction.

Waveguide Transitions from Stripline to Microstrip Line Fig. 3.14 shows a 2D and 3D view of two possible transitions from STL to MSL resp. coplanar waveguide (CPW). To realize the one at the bottom of Fig. 3.14, an outer copper foil and the underlying core material

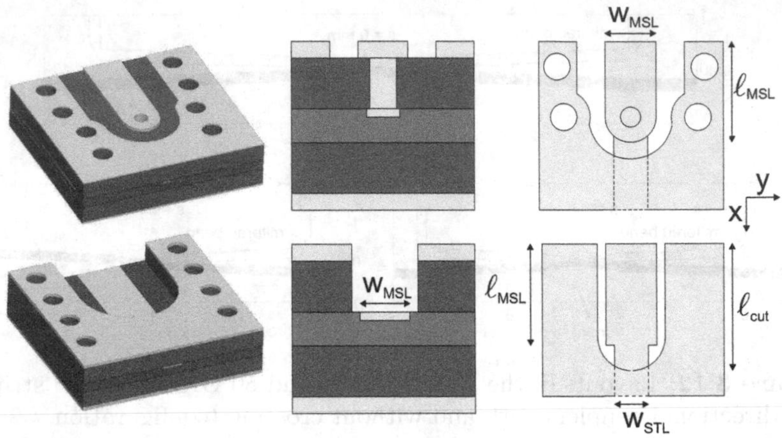

Figure 3.14: 2D and 3D view of two possible transitions from STL to MSL/CPW.

need to be removed. The abrupt change in strip width ($w_{\mathrm{MSL}}, w_{\mathrm{STL}}$) shows shunt capacitance behaviour, which can be compensated by an appropriate choice of ℓ_{cut}. The performance is quite sensitive to ℓ_{cut}. Hence, laser cutting is the method of choice instead of mechanical milling. Unfortunately, this is not applicable to RO4350. Therefore we realized the via transition, shown first in Fig. 3.14. A similar setup can be found in [44], where the via transition is regarded as a three wire line waveguide. Due to the significantly shorter via length in the proposed design a lumped element interpretation (series inductance) is still applicable. Fig. 3.15 shows the nominal insertion and return loss of the optimized transition and the influence of ± 100 µm shift of the via coordinates. Both transitions of Fig. 3.14 tend to radiate power to freespace and there is coupling to the parallel plate (PP) mode within the multilayer PCB. The first problem is solved by choosing ℓ_{MSL} sufficiently small. To prevent power conversion to PP mode, via fences have to be placed along the whole coupling structure. Ensuring proper operation up to the maximum frequency of 70 GHz the distance of the opposite fences has to be chosen smaller than $\lambda/2 \approx 1.1$ mm, otherwise there will be power conversion to a TE_{10} like mode.

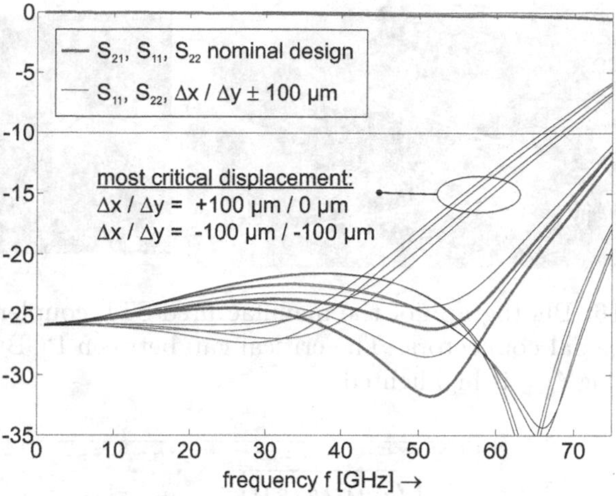

Figure 3.15: Simulated scattering parameters of the via transition from STL to MSL and the influence of ±100 μm via displacement.

Measurement Results with Configuration #2 of Fig. 3.10

Fig. 3.16 shows a photograph of one of the manufactured STL couplers. Within a highly integrated testset module, bondwire interconnects would be used instead of the 1.85 mm coaxial connectors. Scattering parameter measurements up to 70 GHz were performed using a commercial four port VNA. The coaxial interfaces build the reference planes of the measurements, established by common short-open-load-thru (SOLT) calibration.

The applied transition to coaxial waveguide is highly sensitive to ℓ_{gap} (Fig. 3.16) at frequencies greater than 20 GHz. Silver filled adhesive was used to establish sufficiently small ℓ_{gap} values (Fig. 3.19).

A comparison of the proposed couplers performance with 3D EM simulations, that partly account for the transitions to coaxial waveguide, are given in the following.

Fig. 3.17 and Fig. 3.18 include results of the 50 GHz STL coupler (lower left side of Fig. 3.12). The targeted maximum lateral displacement of $\Delta y \pm 50$ μm could not be achieved within this manufacturing

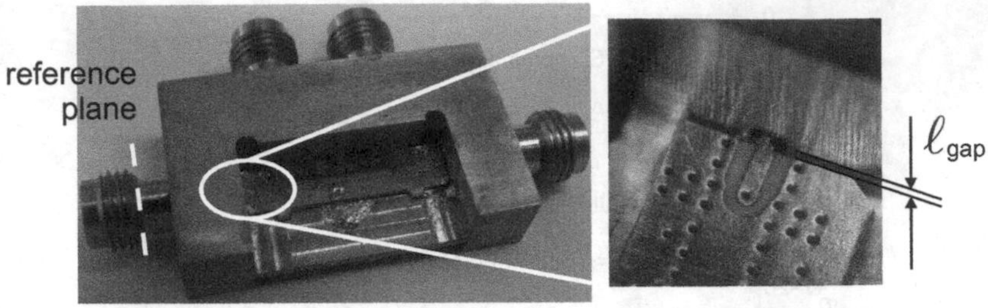

Figure 3.16: Photograph of the manufactured STL couplers with 1.85 mm coaxial connectors. The critical gap between PCB and mechanical housing ℓ_{gap} is highlighted.

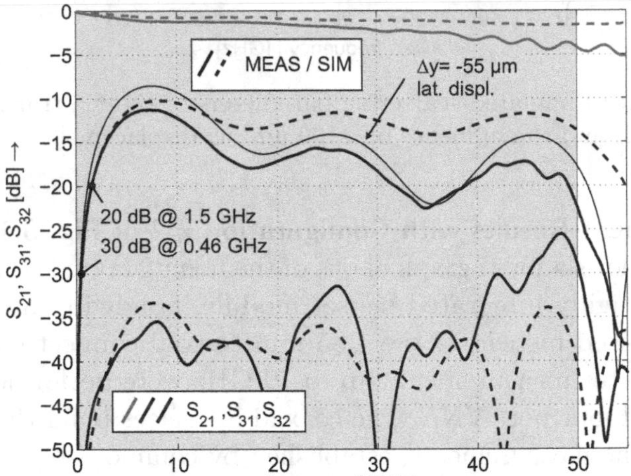

Figure 3.17: Simulated (dashed) and measured (solid) scattering parameters (transmission, coupling, isolation) versus frequency of the 50 GHz STL directional coupler. Simulated coupled arm response is also shown for Δy =-55 µm of lateral displacement.

run. The solid black line (3D EM) in Fig. 3.17 clearly illustrates the manufactured coupler is affected by $\Delta y = -55$ µm displacement. This deteriorates the coupled arm response, but directivity values greater

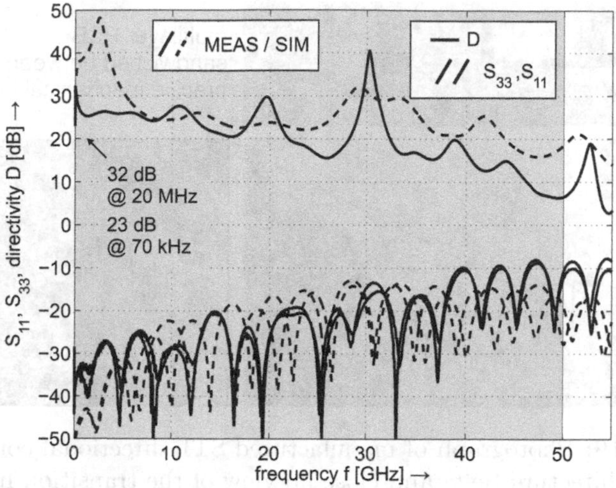

Figure 3.18: Simulated (dashed) and measured (solid) directivity and reflection coefficients of the 50 GHz STL directional coupler.

than 15 dB from 70 kHz up to 40 GHz and at least greater than 5 dB up to 50 GHz are achieved.

In addition, a variant of the 50 GHz coupler with crossed striplines (lower right side of Fig. 3.12) has been investigated. Fig. 3.19 depicts the multilayer PCB sandwiched between the mechanical housing. The corresponding simulation and measurement results are given by Fig. 3.20 and Fig. 3.21. Deterioration of the coupled arm response is similar to the straight architecture ($\Delta y = -55$ µm), but directivity stays above 10 dB up to 50 GHz. Less isolation is achieved between 20 and 40 GHz, which is believed to belong to the more complicated housing and associated assembly of the sparkplug connectors. At these frequencies, the already mentioned problems with $\ell_{\text{gap}} > 0$ could be observed in several assemblies.

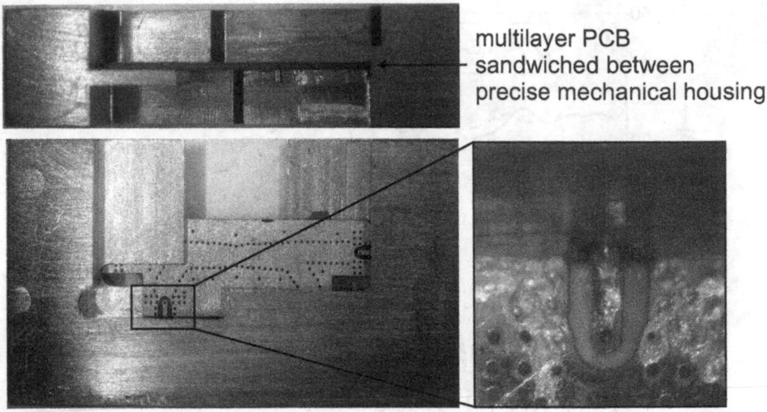

Figure 3.19: Photograph of manufactured STL directional couplers with crossed architecture (left) and close up view of the transition from PCB to coaxial waveguide (right).

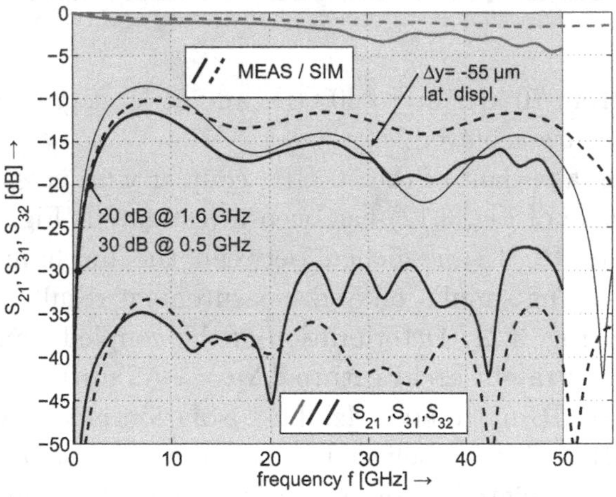

Figure 3.20: Simulated (dashed) and measured (solid) scattering parameters (transmission, coupling, isolation) versus frequency of the 50 GHz STL directional coupler with crossed architecture. Simulated coupled arm response is also shown for Δy =-55 μm of lateral displacement.

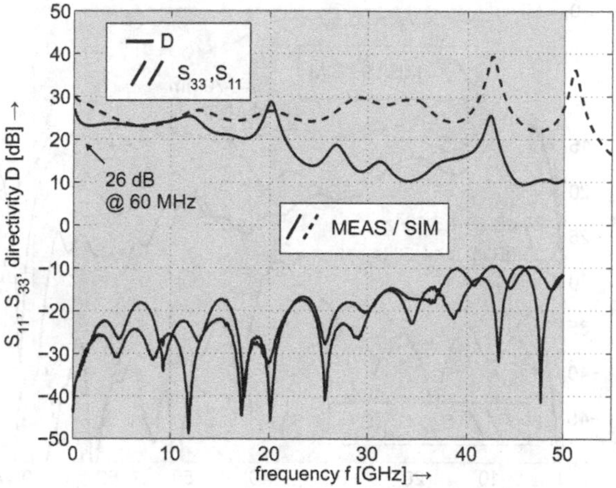

Figure 3.21: Simulated (dashed) and measured (solid) directivity and reflection coefficients of the 50 GHz STL directional coupler with crossed architecture.

Fig. 3.22 and Fig. 3.23 illustrate simulation and measurement results of the 70 GHz STL coupler (first row of Fig. 3.12). Deterioration of the measured coupled arm response is more pronounced than it is with the 50 GHz couplers. Lateral displacement is determined to be $\Delta y = -70$ µm. Decreased isolation at about 35 GHz is believed to belong to problems with $\ell_{\text{gap}} > 0$. Directivity values are greater than 10 dB from 70 kHz to 70 GHz except for frequencies around 35 GHz.

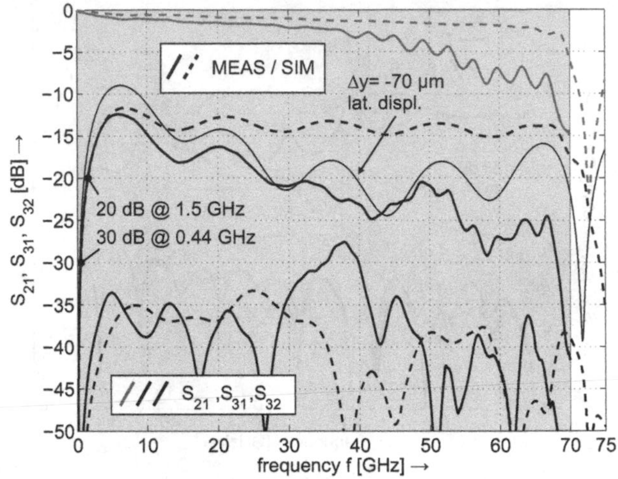

Figure 3.22: Simulated (dashed) and measured (solid) scattering parameters (transmission, coupling, isolation) versus frequency of the 70 GHz STL directional coupler. Simulated coupled arm response is also shown for Δy =-70 µm of lateral displacement.

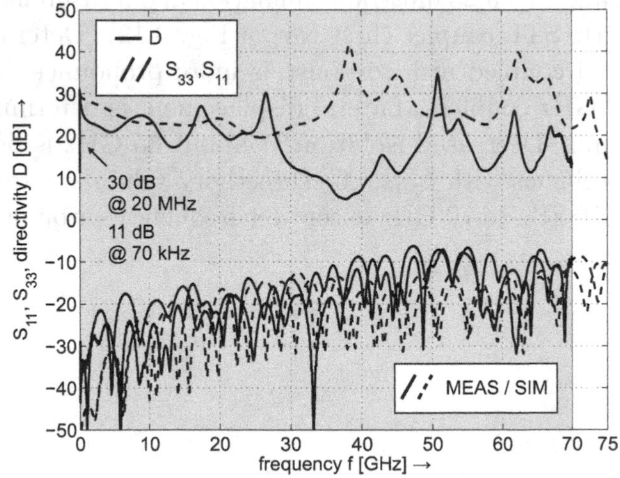

Figure 3.23: Simulated (dashed) and measured (solid) directivity and reflection coefficients of the 70 GHz STL directional coupler.

The 70 GHz STL coupler has also been manufactured in a different PCB processing run, which coincidentally experienced less lateral displacement. The coupler design is identical to the first row of Fig. 3.12 but the feeding striplines are significantly enlarged to allow for assembly of end launch connectors. Fig. 3.24 illustrates the setup. The utilized connectors from Southwest Microwave Inc. are pressed against the PCB and operate without solder. Simulation and measurement

Figure 3.24: Photograph of manufactured STL directional couplers with extended length for use of end launch connectors and close up view of the solderless transition from PCB to coaxial waveguide. Substrate dimensions (43 ×18) mm^2.

results are shown in Fig. 3.25 and Fig. 3.26. The additional stripline lengths at all four coupler ports lead to significantly increased insertion loss. Anyway, the measurement results prove STL coupler operation up to 70 GHz with cost effective hydrocarbon laminates if problems with lateral displacement are under control.

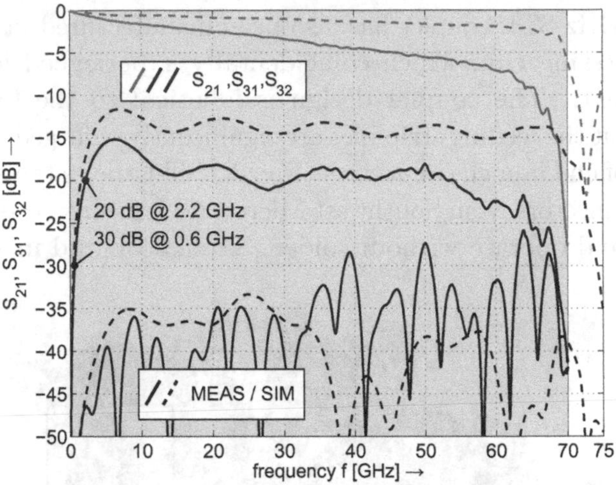

Figure 3.25: Simulated (dashed) and measured (solid) scattering parameters (transmission, coupling, isolation) versus frequency of the 70 GHz STL directional coupler with extended length for use end launch coaxial connectors.

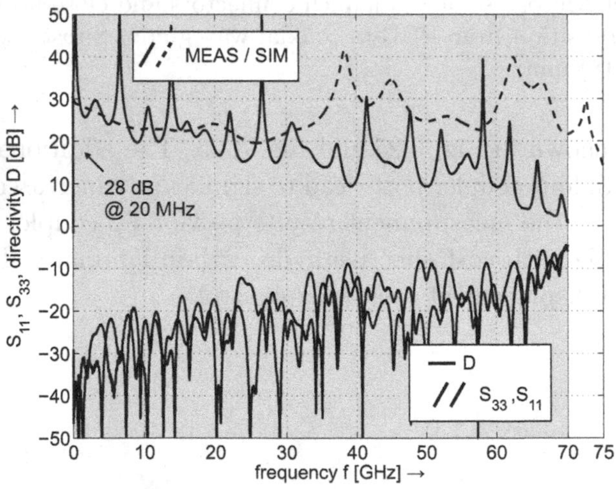

Figure 3.26: Simulated (dashed) and measured (solid) directivity and reflection coefficients of the 70 GHz STL directional coupler with extended length for use end launch coaxial connectors.

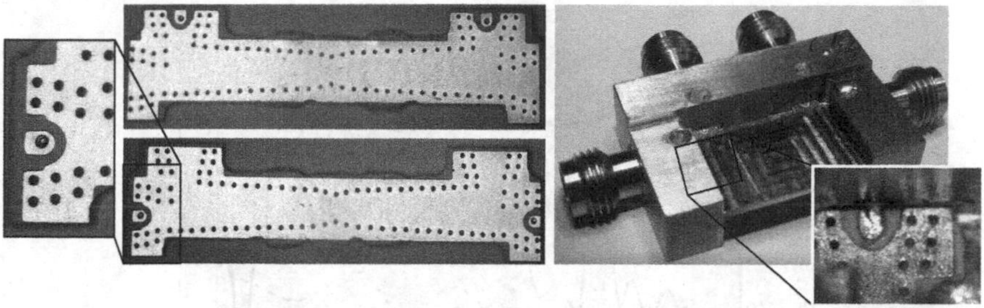

Figure 3.27: Photograph of manufactured 70 GHz STL directional couplers with stackup configuration #3B of Fig. 3.10 and close up view of the transition from PCB to coaxial waveguide. Coupler length ℓ_0 =15.94 mm.

Measurement Results with Configuration #3B of Fig. 3.10 In order to overcome the problems with lateral displacement, couplers have been built up with stackup configuration #3B of Fig. 3.10. The signal conductors are manufactured on top and bottom layer of 168 µm RO4350 hydrocarbon laminate with 17.5 µm copper cladding. The applied PCB process should allow for lateral displacement below 50 µm. Two RO4450 bonding films with overall thickness of 200 µm are bonded to both sides of the core laminate, whereas the outer copper ground planes are bonded to RO4450. Electroplated vias through the entire stackup lead to an outer ground plane thickness of 35 µm, which prohibits small spacings and strip widths on the outer layers. Based on 2D EM simulations, in analogy to Fig. 3.11, a 70 GHz STL coupler with stackup configuration #3B has been designed. The assembly is shown in Fig. 3.27.

Simulation and measurement results are given by Fig. 3.28 and Fig. 3.29. As expected, there is no deterioration due to lateral displacement. Simulated and measured coupled arm response are in reasonable agreement. Compared to the couplers from the last paragraph, isolation performance is worse. Separation of the single couplers from the large PCB panel is done by mechanical milling. Deviations of up to 200 µm between the coupler outlines and the actual milling

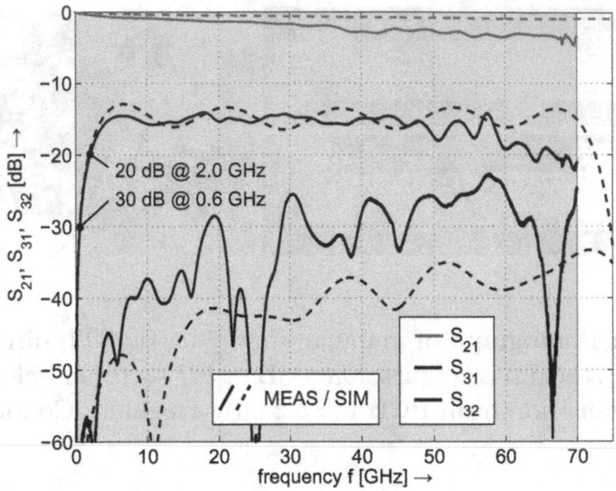

Figure 3.28: Simulated (dashed) and measured (solid) scattering parameters (transmission, coupling, isolation) versus frequency of the 70 GHz STL directional coupler with stackup configuration #3B of Fig. 3.10.

contour had to be accepted (left side of Fig. 3.27). Although manual adjustment has been applied, the measured return loss values and consequently isolation / directivity values fall short of expectations. Anyway, adding additional alignment marks will improve the milling process without adding cost. Hence, successful coupler designs based on hydrocarbon laminates with stackup configuration #3B are possible.

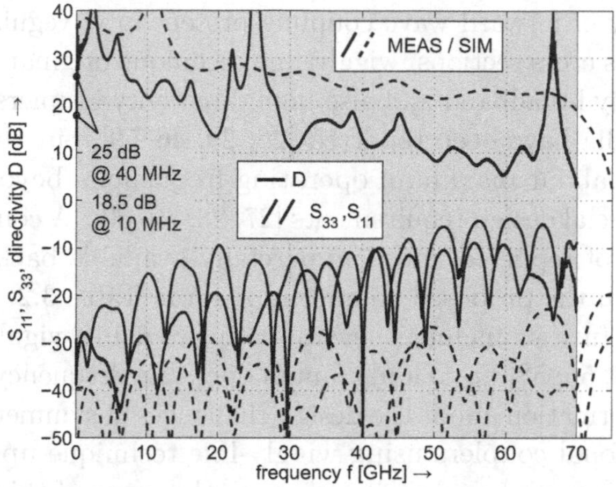

Figure 3.29: Simulated (dashed) and measured (solid) directivity and reflection coefficients of the 70 GHz STL directional coupler with stackup configuration #3B of Fig. 3.10.

3.2.3 Wiggly-Line Technique

Successful ultrawideband, loose directional coupler designs for front end modules of semiconductor automatic test systems and scattering parameter testsets of VNAs have been shown in the preceding subsection 3.2.2. These are based on cost effective hydrocarbon multilayer PCBs. Equalization of the even and odd mode phase velocities depends on the chosen core and prepreg materials. With this approach, the required minimum directivity value ($D_{\min} = 10$ dB) could be achieved over ultrawide bandwidth. However, the transitions needed for integration of STL couplers into planar front end modules degrade directivity performance and add avoidable insertion loss. Further on multilayer stackups increase manufacturing cost, if the coupler is intended to be integrated with other planar single layer structures that can all be manufactured on the same wafer in a single thin-film process.

Therefore we focus on nonuniform, symmetric backward wave couplers utilizing edge coupled microstrip lines (cMSL). To overcome the tendency of forward wave coupling present in waveguides with inhomogeneous cross sections, wiggly-line technique originated from [45] is used. Many broadband and also high directivity couplers in inhomogeneous media have been reported [26, 29, 46–59], but there is little information about maximum operating frequencies beyond 20 GHz and their special design requirements [27, 28, 30, 60]. A comprehensive comparison of reported and commercially available backward wave couplers with the proposed coupler is given in Table 3.2.

Table 3.2 further includes several couplers using wiggly-line technique [49–53, 56, 58] up to a maximum operating frequency of 20 GHz. Detailed information about the design challenges of symmetric, nonuniform directional couplers using wiggly-line technique up to 50 GHz and a measurement based verification are the scope of this subsection. In contrast to the other couplers with operating frequencies greater than 20 GHz in Table 3.2 [27, 28, 30, 60], the proposed coupler [J1] achieves slightly lower directivity values and exhibits a bit higher ripple but provides greater bandwidth and in particular a lower minimum operating frequency. Suitability to planar integration is another advantage incorporated with the proposed design. The achieved performance does not depend on the mechanical housing, as it is with other approaches like [60].

Synthesis Procedure The design procedure from subsection 3.2.1 does not account for the dispersive nature of the cMSL waveguide. Equalization of the even and odd mode phase velocity is done by Podell's wiggly-line technique [45]. Due to the high maximum operating frequency, cMSL waveguide design data is extracted from 2D EM simulation rather than analytical closed form approximations [61–64]. The top of Fig. 3.30 illustrates simulated even, odd and microstrip (MSL) mode configurations, as well as the underlying equivalent circuits to extract the frequency dependent per unit length capacitance values $C_\mathrm{p}, C_\mathrm{f}, C_\mathrm{fe}, C_\mathrm{fo}$.

$$
\begin{aligned}
C_{\mathrm{p}} &= \epsilon_0 \epsilon_{\mathrm{r}} \frac{w}{h} & C_{\mathrm{f}} &= \frac{1}{2}\left[(v_{\mathrm{MSL}} Z_{\mathrm{MSL}})^{-1} - C_{\mathrm{p}}\right] \\
C_{\mathrm{e/o}} &= \left(v_{\mathrm{e/o}} Z_{0\mathrm{e/o}}\right)^{-1} & C_{\mathrm{fe/o}} &= C_{\mathrm{e/o}} - C_{\mathrm{p}} - C_{\mathrm{f}}
\end{aligned}
\tag{3.77}
$$

The enhanced propagation way of the wiggles, shown at the bottom of Fig. 3.30, allow for lowering the odd mode phase velocity in z direction. It can easily be shown, that the desired increase of the odd mode capacitance C_o by factor $\epsilon_{\mathrm{re}}/\epsilon_{\mathrm{ro}}$ results in the wiggle depth $d(z)$

$$
d(z) = \frac{\Delta z}{2} \exp\left(E_1 \frac{C_{\mathrm{fe}}}{C_{\mathrm{fo}}}\right) \sqrt{\left(\frac{C_{\mathrm{fow}}}{C_{\mathrm{fo}}}\right)^2 - 1}\,, \tag{3.78}
$$

where Δz is the wiggle length and C_{fow} is the odd mode fringing capacitance after wiggling (Eq. (3.79)).

$$
C_{\mathrm{fow}} = (C_{\mathrm{p}} + C_{\mathrm{f}})(\epsilon_{\mathrm{re}}/\epsilon_{\mathrm{ro}} - 1) + \epsilon_{\mathrm{re}}/\epsilon_{\mathrm{ro}} C_{\mathrm{fo}} \tag{3.79}
$$

As suggested by Uysal, the effect of wiggling on the even mode is considered by the additional factor $\exp(E_1 \cdot C_{\mathrm{fe}}/C_{\mathrm{fo}})$ with E_1 as scaling parameter. The optimum value of E_1 depends on the cMSL waveguide configuration and is chosen empirically. The compensated phase velocity after wiggling is then given by $v_{\mathrm{c}} = [(C_{\mathrm{p}} + C_{\mathrm{f}} + C_{\mathrm{fow}}) Z_{0\mathrm{o}}]^{-1}$.

According to the bottom of Fig. 3.30 the overall coupler length ℓ_0 is divided into m wiggle sections $\ell_0 = m \cdot \Delta z$, $m \in \mathbb{R}^+$. The appropriate choice of Δz is determinative for successful designs. Within the family of couplers in scope of this work, the wiggles of both coupler arms intersect to a high degree at the tight coupling region. Normally the loose coupling spacing $s_{\mathrm{l}k}$ is used to slow down the odd mode phase velocity. In case of great intersection of both arms we have to calculate the effective tight coupling spacing $s_{\mathrm{t}k}$. As extensive simulations have shown a continuous transition from $s_{\mathrm{l}k}$ to $s_{\mathrm{t}k}$ has to be established. Therefore E_2 is introduced as another scaling parameter with an initial value $E_2 = 1.0$.

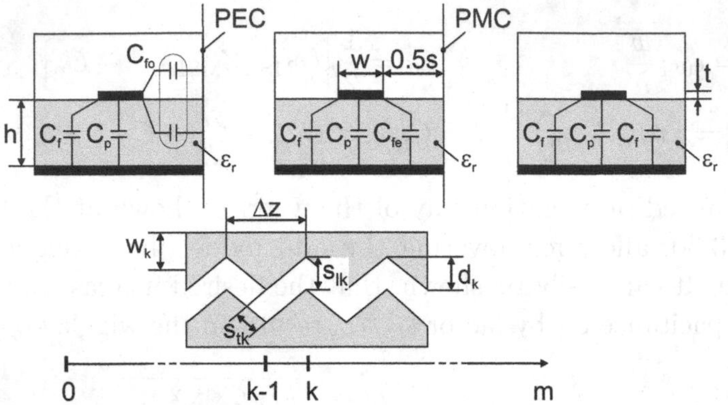

Figure 3.30: Distributed waveguide equivalent circuits used for capacitance parameter $C_{\mathrm{p}}, C_{\mathrm{f}}, C_{\mathrm{fe}}, C_{\mathrm{fo}}$ extraction and visualization of the applied wiggly-line technique.

$$s(z) = E_2 \left(\frac{(1 - s_{\mathrm{l}k}(z)/d_k(z))^2}{\max|\,(1 - s_{\mathrm{l}k}(z)/d_k(z))^2\,|} \right) \cdot (s_{\mathrm{t}k} - s_{\mathrm{l}k}) + s_{\mathrm{l}k} \tag{3.80}$$

Fig. 3.31 shows cMSL design data for 254 µm Al_2O_3 alumina substrate with bulk permittivity $\epsilon_{\mathrm{r}} = 10.2$ and $t = 5$ µm Au strip thickness. The plotted strip widths $w(s)$, spacings $s(c)$, phase velocities $v_{\mathrm{e/o/c}}(c)$ and wiggle depths $d(c)$ correspond to a reference impedance $Z_{\mathrm{ref}} = Z_0 = 50\ \Omega$. The flow chart in Fig. 3.32 summarizes the whole synthesis procedure.

The proposed coupler is synthesized with the parameters of Eq. (3.81) and Eq. (3.82).

$$\begin{aligned} E_0 &= 0.02, & E_1 &= 0.8, & E_2 &= 1.1, \\ f_0 &= 30\ \mathrm{GHz}, & \Delta z &= 300\ \mu\mathrm{m}, & \ell_0 &= 8\ \mathrm{mm} \end{aligned} \tag{3.81}$$

$$S_{31\mathrm{spec}}(\omega) = \begin{cases} 0.201 & 0.75\ \mathrm{GHz} \le \omega/(2\pi) \le 55\ \mathrm{GHz} \\ 0 & \text{else} \end{cases} \tag{3.82}$$

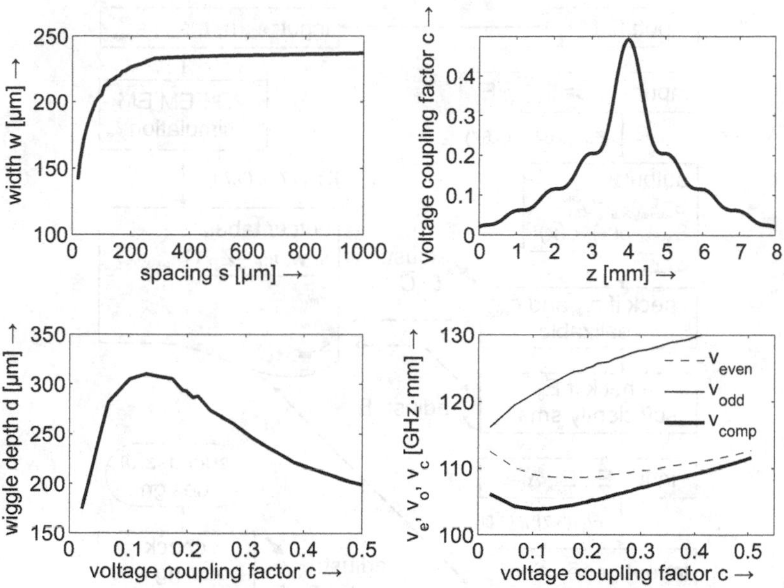

Figure 3.31: Extracted strip width w, spacing s, wiggle depth d, even, odd mode and compensated phase velocity $v_{\mathrm{e}}, v_{\mathrm{o}}, v_{\mathrm{c}}$ values from 2D EM simulation together with the synthesized voltage coupling factor $c(z)$.

For the desired bandwidth and average coupling $\overline{C} = 13.9$ dB, $\ell_0 = 8$ mm leads to the maximum realizable coupler order due to $c_{\max} = 0.45$. As we know $w(c), s(c), d(c)$ and $c(x)$ (upper right of Fig. 3.31) we can calculate $w(z), s(z), d(z)$ and the compensated phase velocity v_{c}, which is enough to generate the coupler layout (Fig. 3.33).

Besides $E_0, E_1, E_2, \Delta z, l_0$ the extraction frequency f_0 of the performed 2D EM simulation should be seen as another scaling parameter. To show the stringent necessity of appropriate scaling parameters, Fig. 3.34 compares results from 3D EM fullwave simulations of the initial and $E_k, k \in \mathbb{N}$ scaled coupler design together with the analytically synthesized coupling function. Without E_k scaling there is significant forward wave coupling due to impedance mismatch and unequal phase velocities of the initial design.

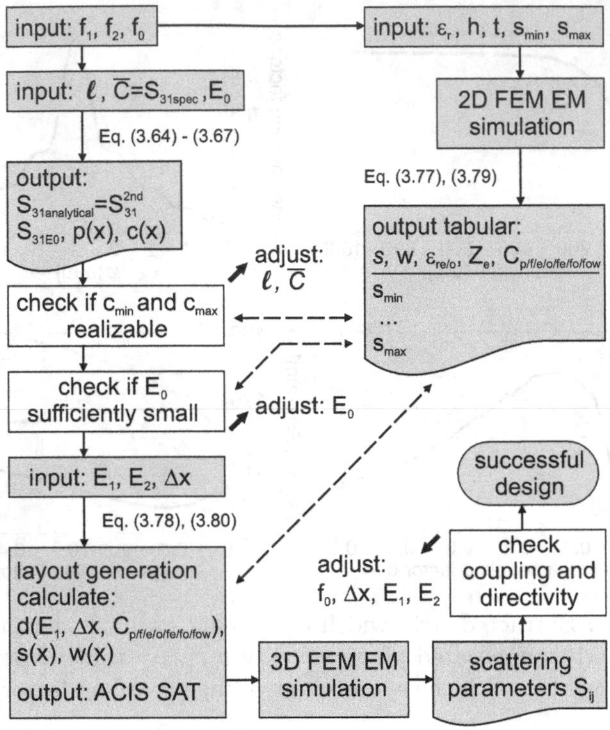

Figure 3.32: Flow chart illustrating the synthesis procedure of the proposed ultrawideband wiggly-line directional couplers.

Sensitivity Analysis In distinction from discrete section directional couplers [65, 66] there is no straightforward possibility to generally map deviations from the design assumptions to the couplers performance in the nonuniform case. The nonlinear slope of the continuous coupling function $\mathrm{d}c(z)/\mathrm{d}z$ which differs in each design, according to desired bandwidth and coupler length, determines the parasitic impact of the transitions between adjacent coupling sections. Anyway, we can study the robustness of the proposed coupler (Fig. 3.33) against common manufacturing tolerances present in thin-film and

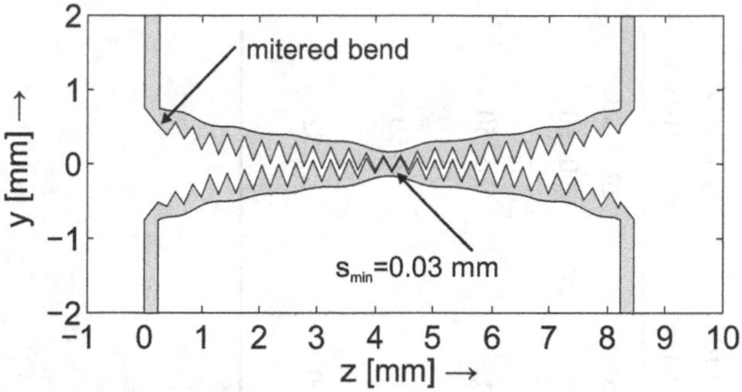

Figure 3.33: Ultrawideband wiggly-line directional coupler layout.

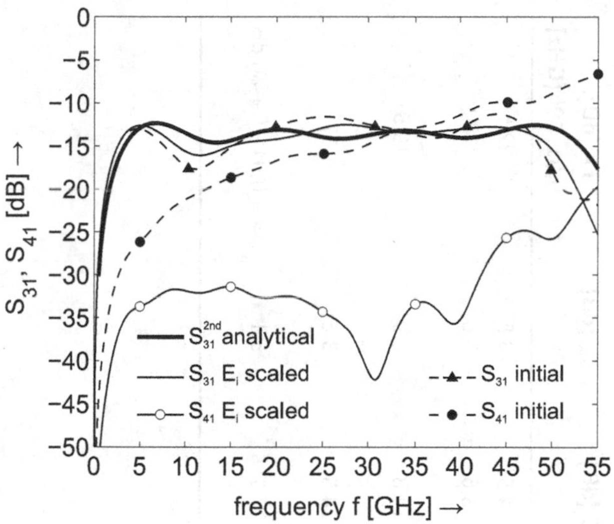

Figure 3.34: Simulated coupling and isolation versus frequency of the initial and E_k scaled coupler design, together with the analytically synthesized coupling function.

laser ablation processing by 3D EM fullwave simulations. Table 3.1 summarizes the results.

Table 3.1: The influence of manufacturing tolerances on the coupler performance

Parameter Tolerances	**Performance Influence**				
	av. coupling $\overline{\mathrm{C}}$ **[dB]**	**max. ripple** r_{max} **[dB]**	**av. directivity** $\overline{\mathrm{D}}$ **[dB],** $f \in [0.01, 50]$ **[GHz]**	**min. directivity** D_{min} **[dB] @ 40 GHz**	**min. directivity** D_{min} **[dB] @** f_{Dmin} **[GHz]**
nominal values	14.5	3.5	22.5	25	14.8, 48
bulk substrate permittivity $e_r \pm 2\,\%$	13.6	3.6	18	20.2	10, 50
substrate height $h \pm 5\,\%$	14.8	3.9	19.5	21	10, 50
strip thickness $t \pm 10\,\%$	14.5	3.6	17	17.2	9, 50
lateral displacement $s_i + 50\,\mu m\ \forall\, i \in [0, m]$	17.1	5.1	$\Rightarrow -0.94 \cdot f + 30$ dB	< 0	0, 32

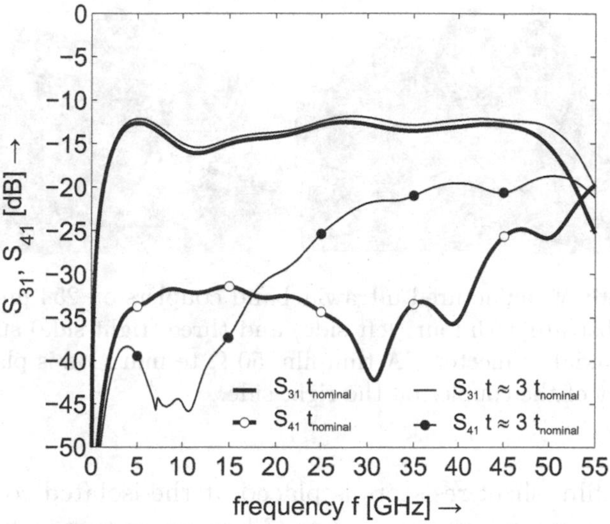

Figure 3.35: Simulated coupling and isolation versus frequency for nominal strip thicknesses $t_{\mathrm{nom}} = 5\ \mu\mathrm{m}$ and $t = 17.5\ \mu\mathrm{m} \approx 3 \cdot t_{\mathrm{nom}}$.

Although we recognize the coupler's directivity will deteriorate in consequence of all examined manufacturing tolerances, lateral displacement below 50 µm is a requirement. Variations of the nominal strip thickness t are of great importance. Fig. 3.35 illustrates, usage of approximately three times thicker strips leads to distinct forward coupling phenomenon at frequencies greater than 20 GHz due to inequality of the even and odd mode phase velocities. What impressively shows that waveguide design data from analytical closed form approximations can hardly lead to successful coupler designs in the millimeter-wave frequency range.

Measurement Results Fig. 3.36 shows the manufactured alumina ultrawideband couplers. The couplers are intended to be used within highly integrated planar front end modules, but transitions to 1.85 mm coaxial sparkplug connectors help to simplify the measurement setup. To fulfill the needs of planar integration a broadband 50 Ω termination

Figure 3.36: Manufactured ultrawideband couplers on 254 µm Al_2O_3 alumina substrate with four (left side) and three (right side) sparkplug 1.85 mm coaxial connectors. A thin-film 50 Ω termination is placed at the isolated port of the coupler on the right side.

using thin-film sheet resistors is placed at the isolated coupler port of one of the manufactured devices. The termination is discussed in chapter 3.4 (Fig. 3.57, Fig. 3.58).

Scattering parameter measurements up to 55 GHz are performed using a commercial four port VNA. The coaxial connector interfaces represent the reference planes of the measurements, established by common multiport short-open-load-thru (SOLT) calibration.

Even with complete metallic enclosure, which is not shown in Fig. 3.36, there is no power coupling phenomenon to spurious cavity modes. Although the latter can exist in the focused frequency range with the actual substrate dimensions of 20.2 mm × 11.8 mm. Within the highly integrated front end module, commercially available ferrite absorber material is placed at the top metal cover.

Fig. 3.37 shows the measured insertion loss, coupling and isolation versus frequency of the manufactured coupler, depicted on the right side of Fig. 3.36. The measurement results are compared with 3D finite element EM simulations, which also account for the transitions to the 1.85 mm coaxial waveguides but not the entire connector.

The measured coupling versus frequency is in reasonable agreement to the analytical design from Fig. 3.34. Due to additional loss in the measurement setup, the average coupling is slightly downshifted to $\overline{C}$ = 16.6 dB. We further recognize an increase in coupling ripple

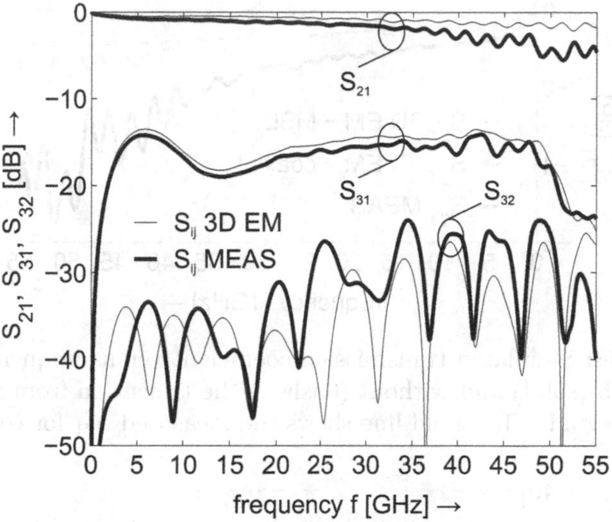

Figure 3.37: Transmission coefficient, coupling and isolation of simulation and measurement versus frequency of the coupler with planar 50 Ω thin-film termination at the isolated port.

at frequencies greater than 35 GHz, which corresponds to degraded performance of the transition from MSL to coaxial waveguide at these frequencies. The latter effect is also shown in Fig. 3.38 which compares the measured insertion loss with 3D EM simulations of the coupler with and without the transition.

The measured minimum isolation of -25.6 dB decreased the minimum directivity to 9.3 dB at 40 GHz (Fig. 3.39) compared with the nominal values of Table 3.1. The dashed line in Fig. 3.39 further illustrates the achievable directivity versus frequency without the deterioration of the waveguide transitions.

Conclusion The design and construction of an ultrawideband $\overline{C} = 16.6 \pm 2.4$ dB backward wave directional coupler based on nonuniform cMSL waveguide and wiggly-line technique to equalize the phase velocities of the even and odd mode have been presented. Directivity

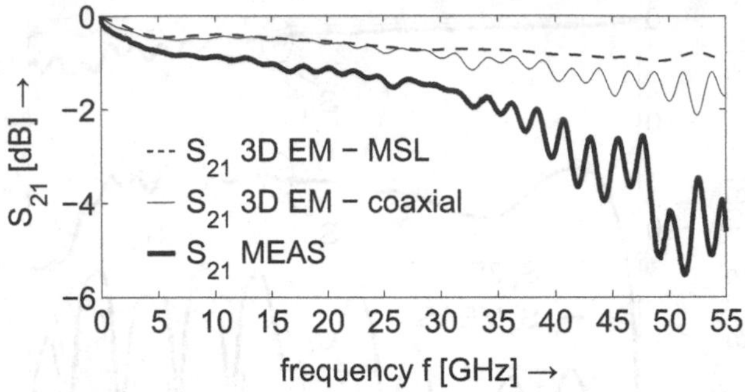

Figure 3.38: Simulated transmission coefficient versus frequency of the coupler with (solid) and without (dashed) the transition from MSL to coaxial waveguide. The bold line shows the measured S_{21} for comparison.

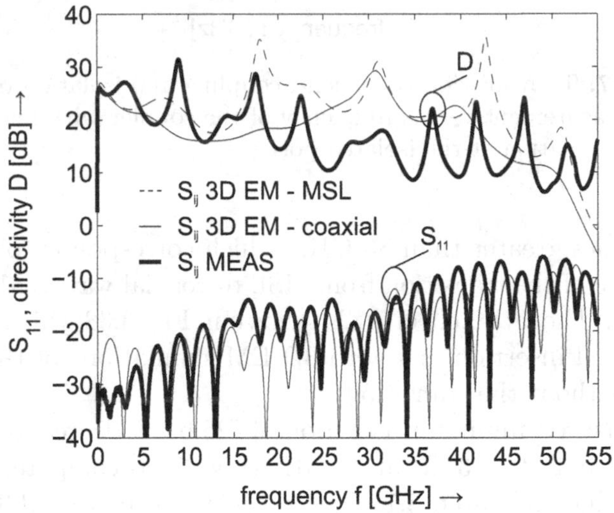

Figure 3.39: Reflection coefficient and directivity of simulation and measurement versus frequency of the coupler with planar 50 Ω thin-film termination at the isolated port. Dashed line shows simulated directivity without the transition from MSL to coaxial waveguide.

values greater than 8 dB are guaranteed within the frequency range of 10 MHz to 48.4 GHz and greater than 10 dB up to 33.8 GHz. The necessity of 2D EM simulations to extract accurate waveguide design data valid up to the maximum operating frequency has been outlined. In combination with the presented thin-film 50 Ω termination the coupler is suitable for integration to planar front end modules of modern measurement systems. The introduced scaling parameters E_0, E_1, E_2 and a careful choice of the coupler length ℓ_0, the wiggle length Δz and the waveguide parameter extraction frequency f_0 give sufficient tuning opportunities to account for the remaining discrepancy between the optimum and synthesized geometry. Robustness of the design towards most of the usual manufacturing tolerances within thin-film processing has been demonstrated.

Table 3.2: Comprehensive Comparison of Reported and Commercially Available Couplers

	av. coup. ± ripple $\overline{C} \pm r$ **[dB],** $f \in [f_1, f_2]$ **[GHz]**	**min. directivity** D_{min} **[dB],** $f \in [f_1, f_2]$ **[GHz]**	**realization type**	**size or length [mm]**
Ref.	**Multilayer Couplers / Homogeneous Media**			
[33]	10 ± 1.5, [4, 50]°	> 14 up to 50°	nonuniform, STL	40.6
[34]	13 ± 1.75, [1, 50]°	> 15 up to 34° > 9 up to 50°	nonuniform, STL	66.0
[35, 36]	11.2 ± 1.2, [1, 40]°	> 15 up to 35° 12 at 38°	STL	101.6
[37]	13.0 ± 2.0, [1, 65]°	> 10, [1, 30]° > 7.2 [1, 65]°	STL	63.5
[38]	13.0 ± 1.5, [1, 65]°	> 10, [0.04, 65]°	STL	78.74
[39–41]	4.5 ± 1.5, [0.75, 26.5]	> 15 up to 26.5	nonuniform, STL, tandem config.	81.5
[67]	4 ± 2, [2, 18]	> 9 up to 18	nonuniform, STL, tandem config.	50.0
Fig. 3.20,	15.6 ± 3.9, [1.8, 49]	> 20, [0.06, 22] > 10 up to 50	nonuniform, STL	8.0
Fig. 3.25	19.8 ± 4.6, [1.26,65]	> 20, [0.02, 19.5] > 9 up to 62	nonuniform, STL	43.0
Fig. 3.28	15.1 ± 0.95, [3.8, 48.9] 16.7 ± 2.6, [2.2, 60]	> 20, [0.04, 17] > 10 up to 49 > 5.5 up to 69.5	nonuniform, STL	15.94
Ref.	**Couplers in Inhomogeneous Media**			
[46]	3.4 ± 1.1, [3.1, 10.6]*	> 10 up to 10.6*	uniform, 3 layers	32.6
[47]	3.7 ± 1, [3.1, 10.6]	> 15, [3.1, 10.6]	slot coupled MSL, 3 layers	8.4 •

Table 3.2: Comprehensive Comparison of Reported and Commercially Available Couplers

	av. coup. ± ripple $\overline{C} \pm r$ [dB], $f \in [f_1, f_2]$ [GHz]	min. directivity D_{min} [dB], $f \in [f_1, f_2]$ [GHz]	realization type	size or length [mm]
[48]	3 ± 0.9, [3, 10]	≥ 5, [3, 10] ≥ 15, [3, 7.3]	slot coupled coplanar waveguide, 3 layers	20.0
[49–51]	4.5 ± 1.5, [2, 20]	> 12 up to 20	cMSL, tandem config., wiggly-line	14.2
[52]	11 ± 1, [2, 18]	> 9 up to 12.4 3 at 18	rectangularly-wiggled edges	n/a
[53]	20 − 5.0, [0.8, 3.6]*	> 30 up to 3.6*	uniform, cMSL, wiggly-line	20.0
[54]	20 − 10, [0.5, 3.25]*	> 15 up to 2.3* 0 at 3*	uniform, cMSL	19.0
[26]	15 − 3, [1, 2]*	> 40 up to 4*	uniform, cMSL, interdigital cap.	n/a
[55]	3.0 − 3.0, [2.2, 3.8]	2.5 at 3	nonuniform, cMSL, electromagn. bandgap	51.3 •
[56]	19 ± 1.5, [1, 8]	> 10, [1, 8]	asym., uniform, wiggly-line	$\approx \frac{1}{2}\frac{3\lambda}{4}$
[29]	11 − 5, [0.5, 1.5]	≤ 8, [0.5, 1.5] 50 at 0.9	uniform, cMSL, shunt inductances	53.4
[57]	15.2 − 5.8, [1, 5]◇	≤ 5 up to 5◇	uniform, cMSL with floating conductor	9.5
[58]	20 ± 6, [0.3, 3]◇	> 30, [0.3, 2.5]◇	uniform, cMSL, wiggly-line	25.0
[59]	11 − 4, [0.5, 2.5]	≥ 20 up to 2.5	uniform, cMSL with planar comp. elements	n/a
[30]	16 ± 0.5, [4, 30]◇	> 18 up to 17◇ 0 at 28◇	asym., nonuniform, cMSL with res. matching element	n/a
[60]	21.5 ± 0.5, [20, 30]	> 15, [20, 30] 40 at 24.8	suspended STL combined with MSL	20.3
[28]	17 ± 3, [10, 40]	≥ 10, [10, 38]	meandered edge coupled transm. lines, 0.18 µm CMOS	0.240
[27]	4.5 ± 1.2, [30, 69.3]	≥ 4.3 up to 69.3	uniform, cMSL, lumped cap. comp., 100 µm GaAs	0.44
Fig. 3.37	16.6 ± 2.4, [2, 50]	> 10, [0.01, 33.8] 8, [0.01, 48.4]	nonuniform, cMSL, wiggly-line	8.0

• estimated from device illustration, * valid for TRL calibrated planar reference planes,
° commercial product, datasheet information, ◇ data from EM simulation.

3.3 Codirectional Couplers

As it is with backward wave directional couplers of the preceding section, codirectional or forward wave couplers are reciprocal four port devices with double symmetry.

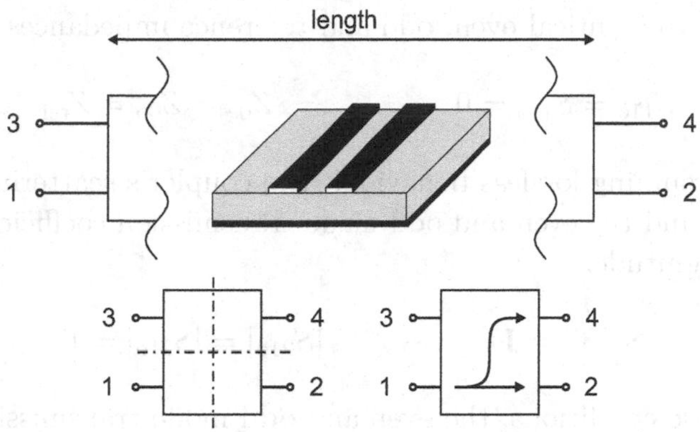

Figure 3.40: Block diagrams of codirectional couplers.

Hence, there are only four independent scattering parameters S_{11}, S_{21}, S_{31}, and S_{41}.

$$\underbrace{\left[\begin{array}{cc|cc} S_{11} & S_{12} & S_{13} & S_{14} \\ S_{21} & S_{22} & S_{23} & S_{24} \\ \hline S_{31} & S_{32} & S_{33} & S_{34} \\ S_{41} & S_{42} & S_{43} & S_{44} \end{array}\right]}_{\mathbf{S}} = \left[\begin{array}{cc|cc} S_{11} & S_{21} & S_{31} & S_{41} \\ S_{21} & S_{11} & S_{41} & S_{31} \\ \hline S_{31} & S_{41} & S_{11} & S_{21} \\ S_{41} & S_{31} & S_{21} & S_{11} \end{array}\right] = \left[\begin{array}{cc|cc} \bigcirc & \blacksquare & 0 & \square \\ \blacksquare & \bigcirc & \square & 0 \\ \hline 0 & \square & \bigcirc & \blacksquare \\ \square & 0 & \blacksquare & \bigcirc \end{array}\right] \tag{3.83}$$

If the realization is based on coupled symmetric transmission lines, the modal and nodal scattering parameters of Eq. (3.20) and Eq. (3.39) are valid. Assuming matched conditions $S_{11} = S_{22} = S_{33} = S_{44} = 0$ leads to Eq. (3.84).

$$S_{11\mathrm{e}} = -S_{11\mathrm{o}} \tag{3.84}$$

Isolation requires $S_{13} = S_{24} = S_{31} = S_{42} = 0$ and results in Eq. (3.85).

$$S_{11\text{e}} = S_{11\text{o}} \tag{3.85}$$

Fulfilling both, is possible only if the even and odd mode reflection coefficients equal zero. Evaluating Eq. (3.20), it turns out, this is equivalent to identical even, odd and reference impedances.

$$S_{11\text{e}} = S_{11\text{o}} = 0 \qquad \rightarrow \qquad Z_{0\text{e}} = Z_{0\text{o}} = Z_{\text{ref}} \tag{3.86}$$

Further assuming lossless behaviour, the coupler's scattering matrix is unitary and the even and odd mode transmission coefficients equal one in magnitude.

$$\mathbf{S} \cdot \mathbf{S}^{\text{H}} = \mathbf{I} \qquad \rightarrow \qquad |S_{21\text{e}}| = |S_{21\text{o}}| = 1 \tag{3.87}$$

Under these conditions, the even and odd mode transmission coefficients can be expressed as simple exponential functions with imaginary propagation coefficients $\gamma_\text{e} = j\beta_\text{e}$, $\gamma_\text{o} = j\beta_\text{o}$ and coupler length ℓ_0.

$$S_{21\text{e}} = e^{-\gamma_\text{e}\ell_0} \qquad S_{21\text{o}} = e^{-\gamma_\text{o}\ell_0} \tag{3.88}$$

Consequently, the coupled arm response $S_{14} = S_{23} = S_{32} = S_{41}$ is given by Eq. (3.89).

$$\begin{aligned}
S_{41} &= \frac{1}{2}\left(S_{21\text{e}} - S_{21\text{o}}\right) = \frac{1}{2}\left(e^{-\gamma_\text{e}\ell_0} - e^{-\gamma_\text{o}\ell_0}\right) \\
|S_{41}| &= \sin\left(\frac{1}{2}\left(\beta_\text{e} - \beta_\text{o}\right)\ell_0\right) = \sin\left(\left(v_\text{e}^{-1} - v_\text{o}^{-1}\right)\pi f \ell_0\right) \\
|S_{41}| &= \sin\left(\left(\sqrt{\epsilon_{\text{reff e}}} - \sqrt{\epsilon_{\text{reff o}}}\right)\frac{\pi}{c} f \ell_0\right)
\end{aligned} \tag{3.89}$$

According to Eq. (3.89), coupling occurs only if the propagation coefficients, phase velocities or effective permittivity values of both modes differ $\beta_\text{e} - \beta_\text{o} \neq 0$, $v_\text{e} - v_\text{o} \neq 0$, $\epsilon_{\text{reff e}} - \epsilon_{\text{reff o}} \neq 0$, which is the case for non transversal electromagnetic (TEM) waves. The enourmous design effort to equalize phase velocities for backward wave directional couplers in the preceding section is therefore obsolete, which makes codirectional coupler design much easier. In case of almost equalized modes, high operating frequency f and/or large coupler length ℓ_0 are required. The latter allows for complete power transfer to the coupled port $|S_{41}| = 1$, which is useful for diplexer operation, but also leads to a minimum operation frequency due to parasitic backward wave coupling $S_{31} \neq 0$ at low frequencies.

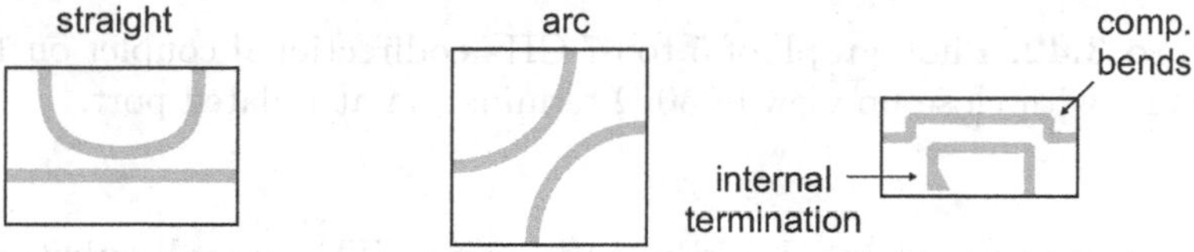

Figure 3.41: Different codirectional architectures. Straight (left), arc shaped (middle) and tight, shielded (right).

Fig. 3.41 shows different top views of microstrip line (MSL) codirectional coupler realizations. Both strip widths w_1, w_2 are fixed by Eq. (3.86). Eq. (3.86) further restricts the strip spacing s to sufficiently large values or loose coupling. Coupler length ℓ_0 and strip spacing s are the only remaining design parameters. Unshielded (open boundary) assemblies, following the uniform / straight and arc shaped architectures of Fig. 3.41 are preferred at lower operating frequencies (e.g. up to $f < 67$ GHz). If small substrate width and shielding are inevitable, the main design problems are realization of compensated bends and internal impedance terminations on the small substrate size. In the following, an unshielded coupler on 10 mil alumina from 5 to 67 GHz and two couplers on 5 mil alumina are proposed. The latter are used for power monitoring within the source modules of chapter 6 from 20 to 40 GHz and 60 to 110 GHz.

10 mil Al_2O_3 Codirectional Coupler for 5 to 67 GHz Fig. 3.42 shows a photograph of the 5 to 67 GHz codirectional coupler, realized

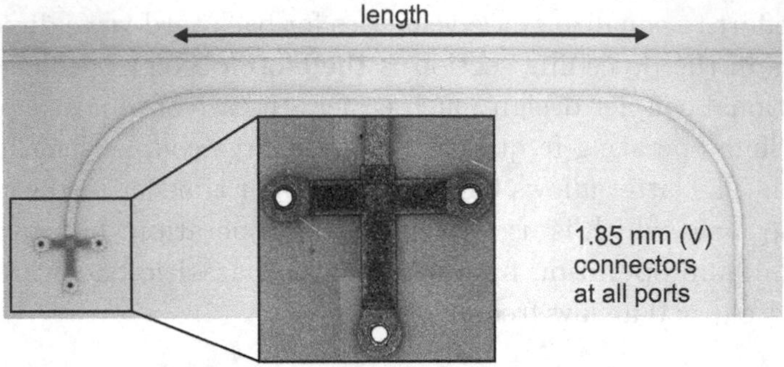

Figure 3.42: Photograph of 5 to 67 GHz codirectional coupler on 10 mil alumina with close up view of 50 Ω termination at isolated port.

on 10 mil alumina by thin-film processing. The overall substrate dimensions are (31.8 × 15.9) mm^2. Strip widths are $w_1 = w_2 = 220$ µm with strip spacing of $s = 1030$ µm and a coupler length $\ell_0 = 12$ mm. The internal 50 Ω termination at the isolated port is discussed in chapter 3.4 (Fig. 3.57, Fig. 3.58). The coupled arm response is between 25 dB and 15 dB from 5 to 67 GHz. Maximum insertion loss of the direct path is about 2 dB at 67 GHz (Fig. 3.43). Fig. 3.44 illustrates measured and simulated directivity D and reflection coefficients S_{11}, S_{44} versus frequency. Directivity values are greater than the obligatory 10 dB at the focused frequency range. Deviations between results from measurement and 3D EM fullwave simulations are caused by imperfections of the 50 Ω termination and the 1.85 mm (V) coaxial connectors at all three ports. Simulations do only cover the transition from MSL to the first air filled coaxial waveguide, but not the entire coaxial connector. Fig. 3.45 and Fig. 3.46 provide simulation results of the same codirectional coupler without internal termination and transitions to coaxial waveguide. Whereas the coupled arm and direct path responses are almost the same, predicted directivity values are much higher.

The author achieved similar results on organic RO3010 laminates using standard PCB processing. Slightly decreased directivity values have to be accepted.

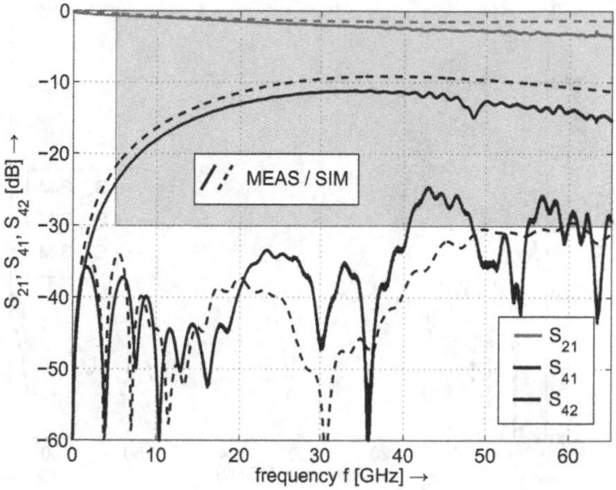

Figure 3.43: Measured (solid) and simulated (dashed) transmission coefficients to direct port (red), coupled port (blue) and isolated port (black) of 5 to 67 GHz codirectional coupler on 10 mil alumina.

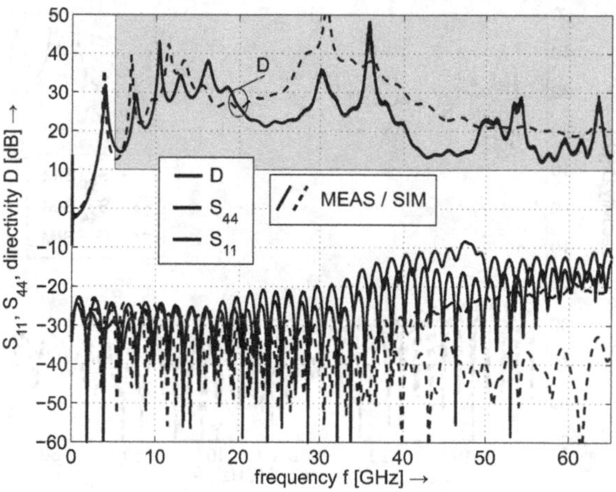

Figure 3.44: Measured (solid) and simulated (dashed) directivity and reflection coefficients of 5 to 67 GHz codirectional coupler on 10 mil alumina.

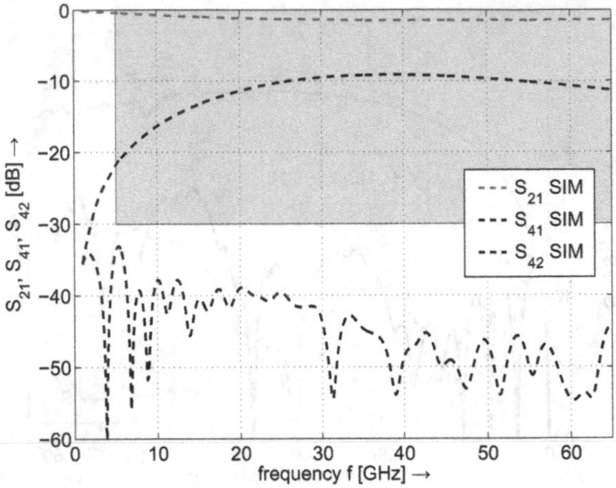

Figure 3.45: Simulated transmission coefficients to direct port (red), coupled port (blue) and isolated port (black) of four port 5 to 67 GHz codirectional coupler on 10 mil alumina without internal 50 Ω termination.

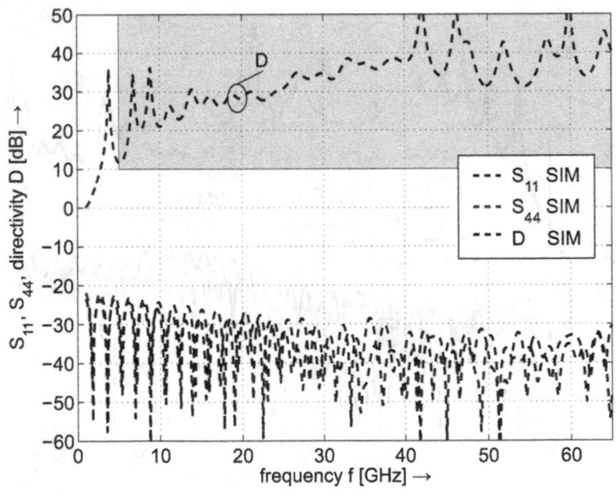

Figure 3.46: Simulated directivity and reflection coefficients of four port 5 to 67 GHz codirectional coupler on 10 mil alumina without internal 50 Ω termination.

5 mil Al_2O_3 Codirectional Coupler for 20 to 40 GHz and 60 to 110 GHz The codirectional couplers of Fig. 3.47 and Fig. 3.50 establish output power level monitoring within the source modules of chapter 6 at 20 to 40 GHz and 60 to 110 GHz, respectively. Both use the same 50 Ω termination with only three sheet resistors and two plated through via holes (Fig. 3.59 of chapter 3.4). The lower frequency coupler operates without shielding, but is manufactured together with the shielded high frequency coupler, which restricts the substrate width to a maximum of 1000/1100 µm. The couplers ports are equipped with transitions to short sections of grounded coplanar waveguide (gCPW), which are designed for usage with three double bondwires or GSG on-wafer probes. The available test equipment did not allow for three port on-wafer scattering parameter measurements. Proper functionality is ensured, by performed measurements of the entire source modules of chapter 6. The main problem is the design of compensated bends to integrate the gCPW sections and impedance terminations on the small substrate area. The results from 3D EM simulation, Fig. (3.48, 3.49) and Fig. (3.51, 3.52), include 3D models of the bondwire transitions at all ports. Simulation results predict directivity values in the order of 15 dB and 20 dB at the focussed frequency ranges.

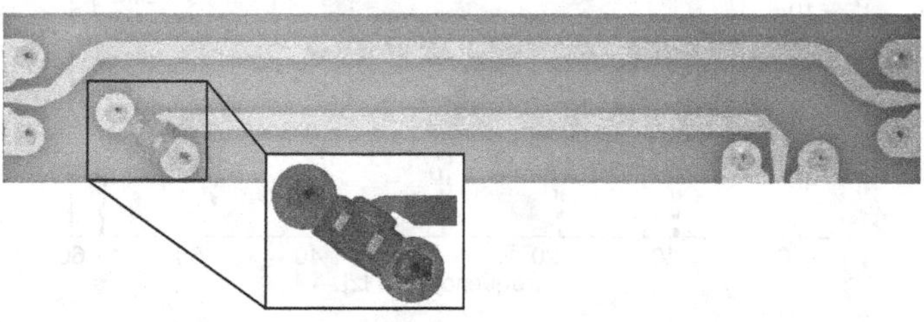

Figure 3.47: Photograph of 20 to 40 GHz codirectional coupler on 5 mil alumina with close up view of 50 Ω termination at isolated port. Substrate dimensions 6000 × 1100 µm^2, strip widths $w_1 = w_2 = 110$ µm, strip spacing $s = 350$ µm.

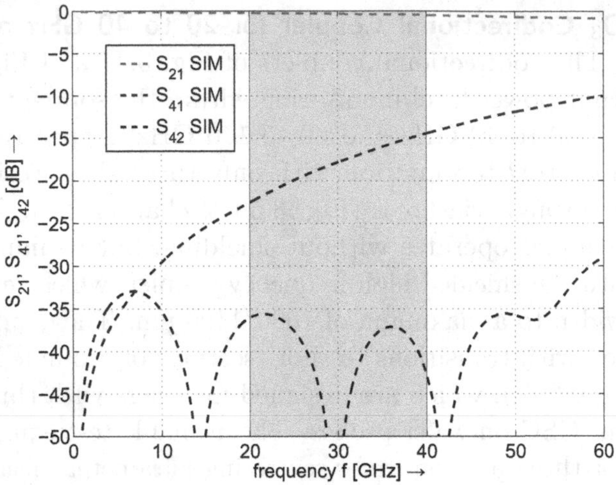

Figure 3.48: Simulated transmission coefficients to direct port (red), coupled port (blue) and isolated port (black) of 20 to 40 GHz codirectional coupler on 5 mil alumina.

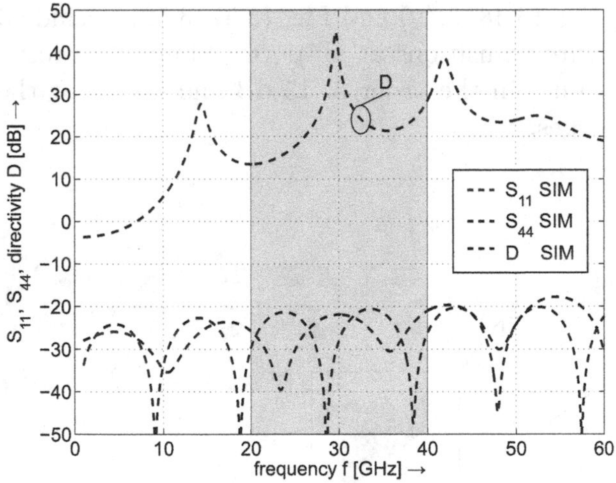

Figure 3.49: Simulated directivity and reflection coefficients of 20 to 40 GHz codirectional coupler on 5 mil alumina.

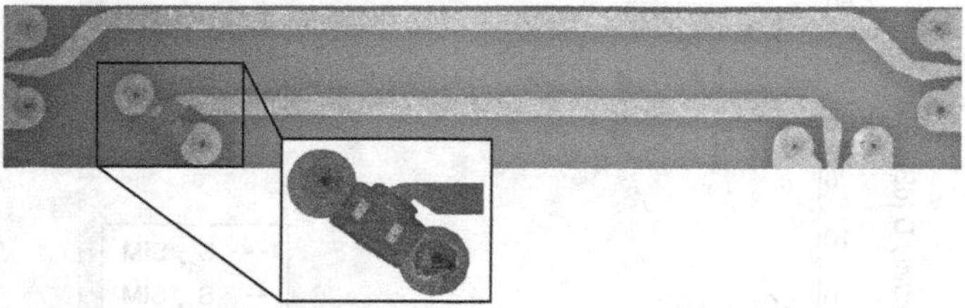

Figure 3.50: Photograph of 60 to 110 GHz codirectional coupler on 5 mil alumina with close up view of 50 Ω termination at isolated port. Substrate dimensions (6000 × 1000) µm^2, strip widths $w_1 = w_2 = 110$ µm, strip spacing $s = 415$ µm.

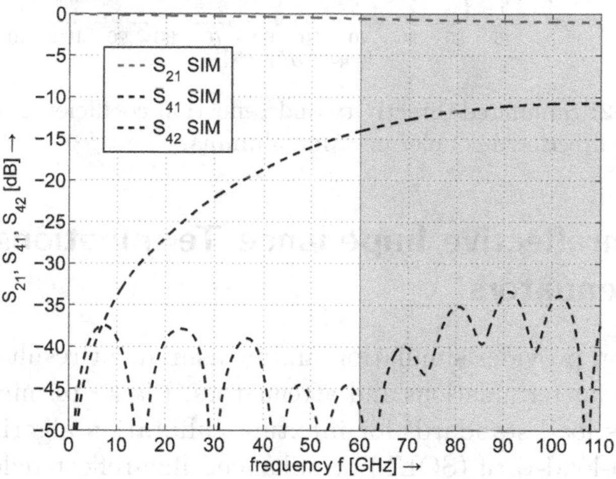

Figure 3.51: Simulated transmission coefficients to direct port (red), coupled port (blue) and isolated port (black) of 60 to 110 GHz codirectional coupler on 5 mil alumina.

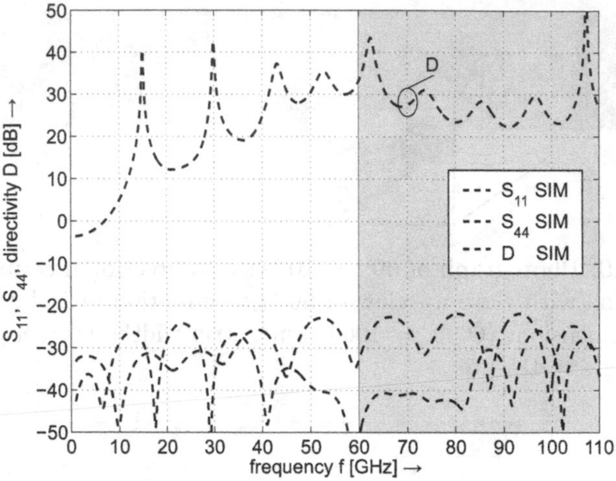

Figure 3.52: Simulated directivity and reflection coefficients of 60 to 110 GHz codirectional coupler on 5 mil alumina.

3.4 Nonreflective Impedance Terminations and Attenuators

This section provides simulation and measurement results of nonreflective planar terminations and attenuators. Planar terminations are required as load standards for infixture calibration algorithms, like short-open-load-thru (SOLT) or enhanced line-reflect-reflect-match (eLRRM) [68]. We distinguish terminations realized as lossy waveguides, which are several wavelengths long, and lumped realizations. Due to size restrictions in all focused applications, the latter are discussed exclusively. As outlined in the preceding sections, directional couplers make use of nonreflective terminations. These couplers are essential parts of scattering parameter testsets for homodyne (six-port) [41, 69] and heterodyne vector network analyzer (VNA) architectures [I1].

Within integrated front end modules, establishing defined power levels at certain ports is necessary in many cases. This is established

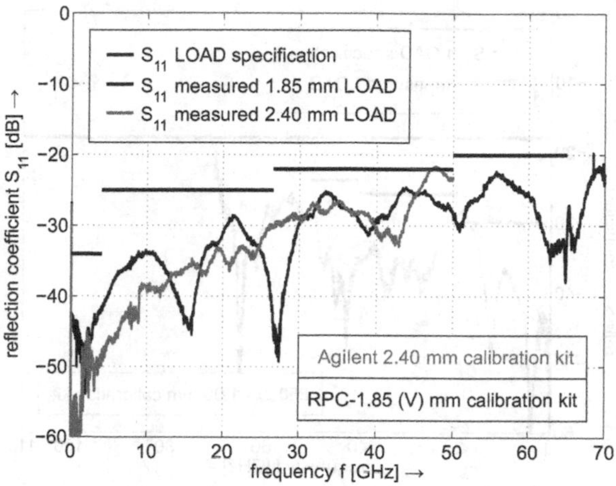

Figure 3.53: Measured reflection coefficient versus frequency of load standards from an Agilent 2.40 mm (red) and Rosenberger 1.85 mm (blue) calibration kit, together with datasheet specifications (black).

by fixed planar attenuators. Lowering the signal's power level is required before mixers or amplifiers to guarantee sufficient linear operation. Another application is enhancing the return loss value of inherently poor matched devices, like the power detectors of chapter 5. Step attenuators consisting of mechanically or electronically switched attenuators constitute a further application.

Nonreflective Coaxial Waveguide Terminations In the focused frequency range of DC to 110 GHz, coaxial terminations are mechanically sophisticated constructions but achieve the best performance compared to planar realizations. Therefore, datasheet specifications (black) of 2.40 mm, 1.85 mm [70] and 1.00 mm [71] load standards are shown together with measurement results (blue, red) in Fig. 3.53 and Fig. 3.54. From DC up to 70 GHz, return loss values greater than 20 dB are guaranteed, whereas the minimum return loss value up to 110 GHz is only 18 dB.

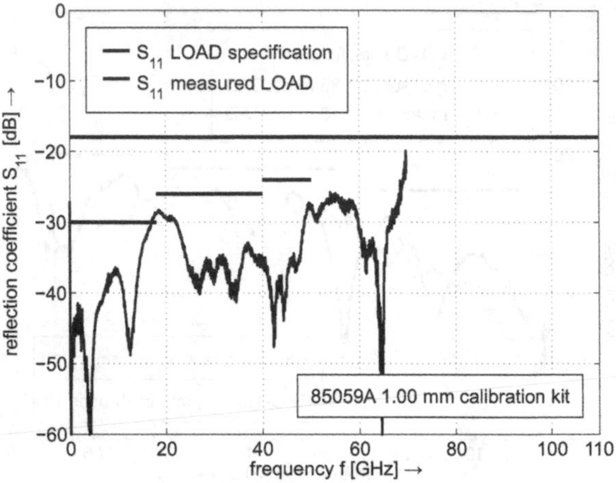

Figure 3.54: Measured reflection coefficient versus frequency of Hewlett Packard 1.00 mm load standard, together with datasheet specifications (black).

Nonreflective Planar Terminations Within the scope of this work, only lumped realizations of planar terminations are discussed. These consist of 40Ni-60Cr thin-film sheet resistors [72]. If DC or low frequency capability is not required, open circuited sheet resistors with several geometries are used (e.g. dot terminations [73]). To allow for DC to 110 GHz operating frequency range, plated through via holes, short circuiting the sheet resistors, have to be used. Fig. 3.55 shows realizations with one (left), two (middle) and three (right) plated through via holes. Which solution shows best high frequency performance mainly depends on the parasitics of the plated through via holes (Fig. 3.56). In general, the realization with shortest overall dimensions performs best. Symmetric architectures reduce the amount of power leaking through the termination without experiencing the resistive effect. The sheet's width should not extend the width of the planar waveguide, hence substrate permittivity also takes influence on the sheet dimensions. Fig. 3.56 contains the real and imaginary parts of the complex via impedance versus frequency on 5 mil alumina.

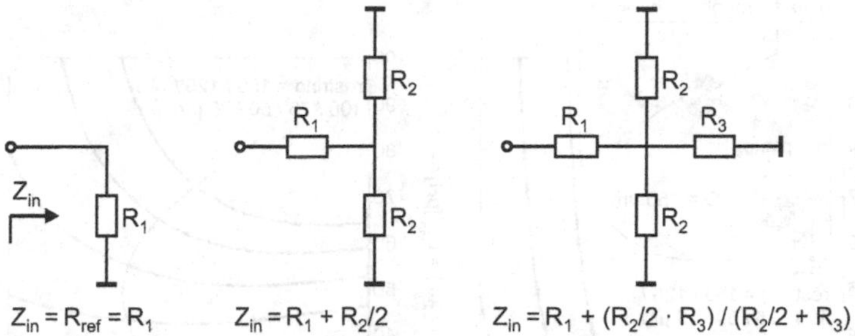

Figure 3.55: Equivalent circuits of lumped resistor terminations.

With a fixed via diameter of $\varnothing = 150$ µm, the influence of several different pad restrings on Z_{via} is shown. The sharp rise of $\text{Re}\{Z_{\text{via}}\}$ and $\text{Im}\{Z_{\text{via}}\}$ indicates the resonant behaviour of the via barrel inductance and capacitance. The maximum operating frequency is below this resonant frequency. Another way to realize terminations is cascading the circuits of Fig. 3.55 to planar attenuators, which are discussed in the next paragraph. In this context, attenuators with different input and output impedances show the greatest potential for short overall dimensions.

Fig. 3.57 shows the measured performance of a planar 50 Ω termination, realized with four sheet resistances and three plated through via holes on 10 mil alumina. This termination is used for the wiggly-line directional coupler of section 3.2.3 and codirectional coupler of section 3.3. With a utilized sheet resistance of $R_\square = 40\ \Omega$, return loss values stay below 20 dB up to 55 GHz. Eq. (3.90, 3.91) summarize the sheet dimensions. The NiCr sheets appear black in photographs, but are covered by green photopatternable solder resist (Fig. 3.58)

$$R_1 = \frac{\ell_{\text{R}_1}}{w_{\text{R}_1}} \cdot R_\square = \frac{90\ \mu\text{m}}{210\ \mu\text{m}} \cdot R_\square = 0.43 \cdot R_\square \qquad R_\square = 40\ \Omega \quad (3.90)$$

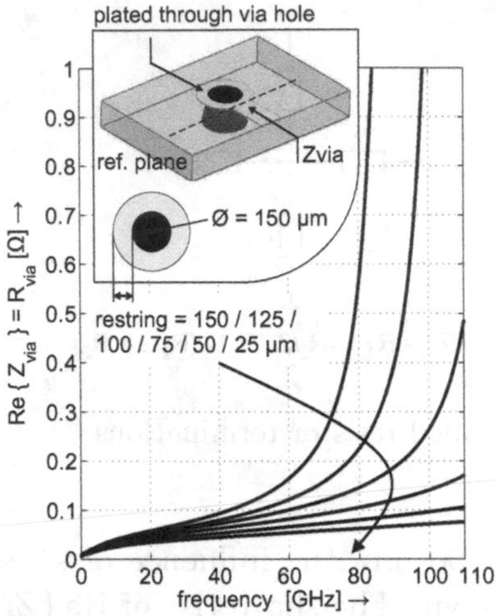

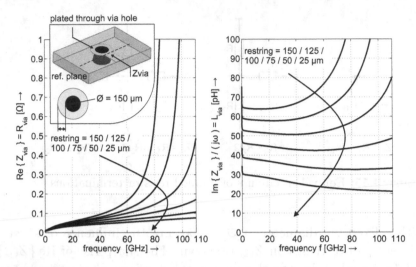

Figure 3.56: Frequency dependent equivalent impedance of plated through via hole with ⌀ = 150 µm on 5 mil alumina from 3D EM simulation.

$$\begin{aligned} R_2 &= \frac{\ell_{\mathrm{R}_2}}{w_{\mathrm{R}_2}} \cdot R_\square = \frac{434\ \mu\mathrm{m}}{210\ \mu\mathrm{m}} \cdot R_\square = 2.06 \cdot R_\square \qquad R_\square = 40\ \Omega \\ R_3 &= \frac{\ell_{\mathrm{R}_3}}{w_{\mathrm{R}_3}} \cdot R_\square = \frac{798\ \mu\mathrm{m}}{210\ \mu\mathrm{m}} \cdot R_\square = 3.80 \cdot R_\square \end{aligned} \tag{3.91}$$

The 50 Ω termination of Fig. 3.59 is realized with three sheet resistors and only two plated through via holes on 5 mil alumina. It achieves return loss values greater than 14 dB up to 110 GHz. In conjunction with a compensated 45° bend, this termination is used with the codirectional couplers of section 3.3, which provide level control within the source modules of chapter 6.

Although it requires more space, even better results are achieved with the termination of Fig. 3.60. Up to 80 GHz return values are greater than 25 dB and up to 110 GHz they are at least 15 dB, outperforming the coaxial terminations of Fig. 3.53 and Fig. 3.54.

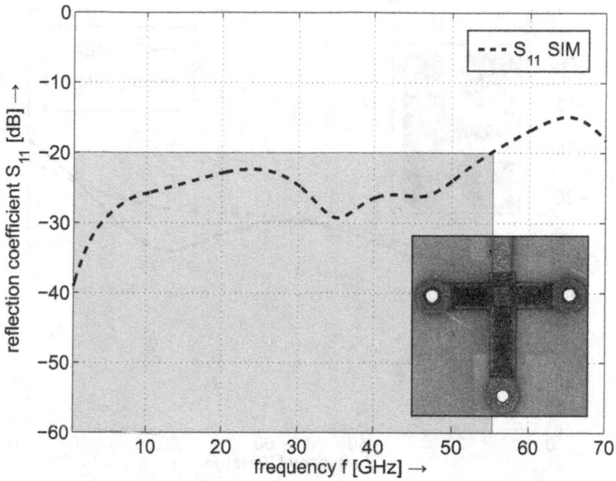

Figure 3.57: Measured reflection coefficient versus frequency of a manufactured planar 50 Ω termination, realized with four sheet resistances and three plated through via holes on 10 mil alumina.

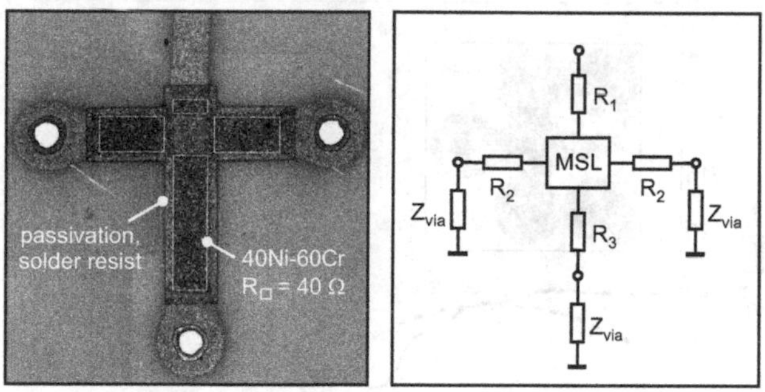

Figure 3.58: Photograph (left) and equivalent circuit (right) of the manufactured planar 50 Ω termination, realized with four sheet resistances and three plated through via holes on 10 mil alumina.

The 1.00 mm coaxial loads guarantee greater than 18 dB return loss up to 110 GHz, but require lossy, liquid waveguides with about 70 mm overall length.

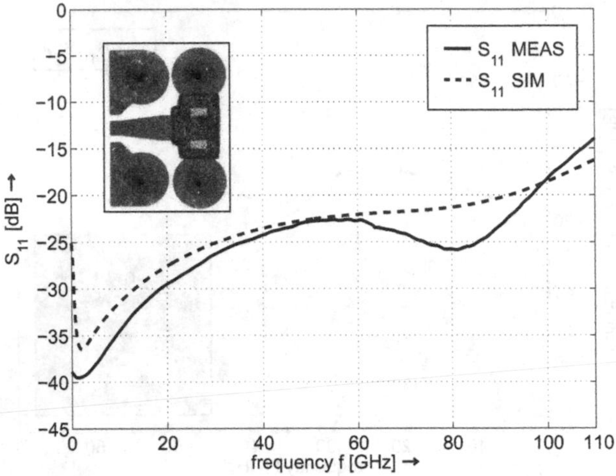

Figure 3.59: Measured reflection coefficient versus frequency of a manufactured planar 50 Ω termination, realized with three sheet resistances and two plated through via holes on 5 mil alumina.

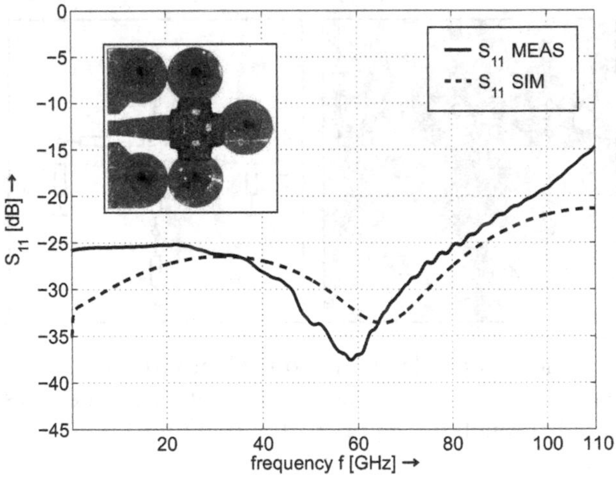

Figure 3.60: Measured reflection coefficient versus frequency of a manufactured planar 50 Ω termination, realized with four sheet resistances and three plated through via holes on 5 mil alumina.

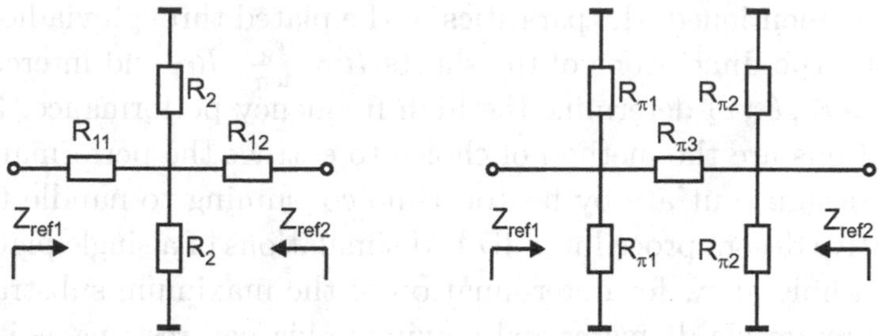

Figure 3.61: T (left) and Pi (right) fixed attenuator equivalent circuits.

Planar Attenuators Fig. 3.61 illustrates lumped attenuator circuits in T (left) and Pi (right) network realizations. The latter requires twice as much plated through via holes, which is in contrast to good high frequency performance. Hence, only T attenuator designs are presented. Eq. (3.92) relates the magnitude of the transmission coefficient $|S_{21}|$ to the required resistance values R_{11}, R_2, R_{12}.

$$\begin{aligned} R_{11} &= Z_{\text{ref1}}\left(\frac{|S_{21}|^2+1}{|S_{21}|^2-1}\right) - 2\sqrt{Z_{\text{ref1}}Z_{\text{ref2}}}\left(\frac{|S_{21}|}{|S_{21}|^2-1}\right) \\ R_{12} &= Z_{\text{ref2}}\left(\frac{|S_{21}|^2+1}{|S_{21}|^2-1}\right) - 2\sqrt{Z_{\text{ref1}}Z_{\text{ref2}}}\left(\frac{|S_{21}|^2+1}{|S_{21}|^2-1}\right) \\ R_2 &= \sqrt{Z_{\text{ref1}}Z_{\text{ref2}}}\left(\frac{|S_{21}|}{|S_{21}|^2-1}\right) \end{aligned} \tag{3.92}$$

If $Z_{\text{ref1}} = Z_{\text{ref2}} = Z_{\text{ref}}$, which is assumed in the following, Eq. (3.92) simplifies to Eq. (3.93).

$$\begin{aligned} R_1 = R_{11} = R_{12} &= Z_{\text{ref}}\left(\frac{|S_{21}|-1}{|S_{21}|+1}\right) \\ R_2 &= Z_{\text{ref}}\left(\frac{|S_{21}|}{|S_{21}|^2-1}\right) \end{aligned} \tag{3.93}$$

As already mentioned, the parasitics of the plated through via holes $\ell_{\text{via}}, Z_{\text{via}}$ and the dimensions of the sheets $R = \frac{\ell_{\text{R}}}{w_{\text{R}}} \cdot R_{\square}$ and interconnection lines ℓ_0, ℓ_1, ℓ_2 determine the high frequency performance. 3D EM simulations are the method of choice to extract the performance of a given design, but are by far too time consuming to handle the entire iterative design procedure. 3D EM simulations of a single plated through via hole allow for determination of the maximum substrate height, maximum via diameter and maximum via pad restring as it is shown in Fig. 3.56. Together with 2D EM simulations of the planar waveguide, characteristic impedance Z_{c} and propagation coefficient γ (Fig. 3.63), this results in a parametric, semi-analytical equivalent model according to Fig. 3.62.

The frequency dependent, complex via impedance Z_{via}, feeding transmission line with length ℓ_{via}, the shunt resistor R_2 and transmission line with length ℓ_2 build the chain matrix $\mathbf{C}'_{\text{shunt}}$, as shown in Eq. (3.94, 3.95). $\mathbf{C}_{\text{shunt}}$ considers both shunt arms of Fig. 3.62.

$$\begin{aligned}
Z_{\text{via}} &= R_{\text{via}} + j\omega L_{\text{via}} \\
Z'_{\text{shunt}} &= R_2 + \frac{Z_{\text{via}} \cosh(\gamma \cdot \ell_{\text{via}}) + Z_{\text{c}} \cdot \sinh(\gamma \cdot \ell_{\text{via}})}{(Z_{\text{via}}/Z_{\text{c}}) \sinh(\gamma \cdot \ell_{\text{via}}) + \cosh(\gamma \cdot \ell_{\text{via}})} \\
Z_{\text{shunt}} &= \frac{Z'_{\text{shunt}} \cosh(\gamma \cdot \ell_2) + Z_{\text{c}} \cdot \sinh(\gamma \cdot \ell_2)}{(Z'_{\text{shunt}}/Z_{\text{c}}) \sinh(\gamma \cdot \ell_2) + \cosh(\gamma \cdot \ell_2)}
\end{aligned} \tag{3.94}$$

$$\begin{aligned}
Y'_{\text{shunt}} &= Z^{-1}_{\text{shunt}} \\
Y_{\text{shunt}} &= 2Y'_{\text{shunt}} \\
\mathbf{C}_{\text{shunt}} &= \begin{bmatrix} 1 & 0 \\ Y_{\text{shunt}} Z_{\text{c}} & 1 \end{bmatrix}
\end{aligned} \tag{3.95}$$

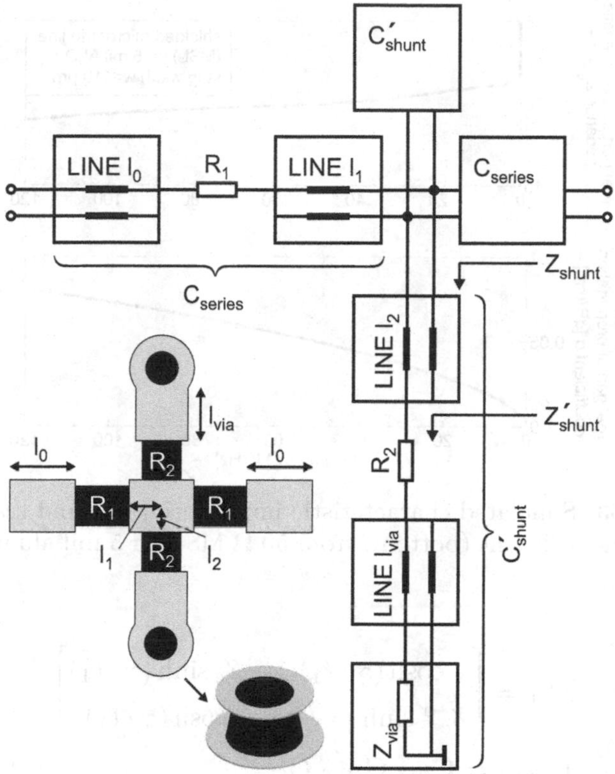

Figure 3.62: Single parts of a planar T attenuator equivalent model for circuit analysis.

In full analogy, a single series arm of the T attenuator is characterized by the chain matrix $\mathbf{C}_{\text{series}}$ (Eq. (3.94, 3.95)).

$$\mathbf{C}_{\ell 0} = \begin{bmatrix} \cosh(\gamma \cdot \ell_0) & Z_\text{c} \sinh(\gamma \cdot \ell_0) \\ Z_\text{c}^{-1} \sinh(\gamma \cdot \ell_0) & \cosh(\gamma \cdot \ell_0) \end{bmatrix}$$
$$\mathbf{C}_{R1} = \begin{bmatrix} 1 & (R_1/Z_\text{c}) \\ 0 & 1 \end{bmatrix} \tag{3.96}$$

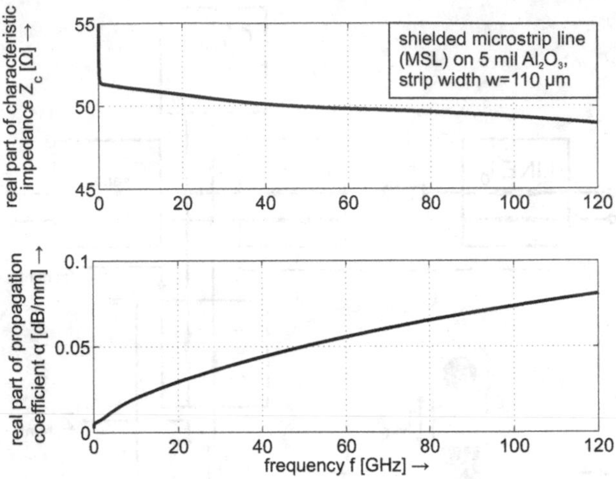

Figure 3.63: Simulated characteristic impedance (top) and real part of propagation coefficient (bottom) from 50 Ω MSL on 5 mil alumina versus frequency.

$$\mathbf{C}_{\ell 1} = \begin{bmatrix} \cosh(\gamma \cdot \ell_1) & Z_\mathrm{c} \sinh(\gamma \cdot \ell_1) \\ Z_\mathrm{c}^{-1} \sinh(\gamma \cdot \ell_1) & \cosh(\gamma \cdot \ell_1) \end{bmatrix} \tag{3.97}$$

$$\mathbf{C}_\mathrm{series} = \mathbf{C}_{\ell 0} \cdot \mathbf{C}_{R1} \cdot \mathbf{C}_{\ell 1}$$

The chain matrix **C** of the complete T attenuator is then given by the matrix multiplication of Eq. (3.99), which is easily converted to a scattering[17] matrix **S**.

[17]The **C** parameters relate voltage and current components from port 2 $[v_2, i_2]^\mathrm{T}$ with port 1 $[v_1, i_1]^\mathrm{T}$. These voltages and currents are not normalized to the reference impedance, which is the only difference to the common **ABCD** parameters. Eq. (3.98) includes conversion equations from **C** to scattering parameters **S**.

$$\Delta = (C_{11}C_{22} - C_{12}C_{21}) \qquad D = (C_{11} + C_{12} + C_{21} + C_{22})$$

$$\mathbf{S} = \begin{bmatrix} (C_{11} + C_{12} - C_{21} - C_{22})/D & (2\Delta)/D \\ 2/D & (-C_{11} + C_{12} - C_{21} + C_{22})/D \end{bmatrix} \tag{3.98}$$

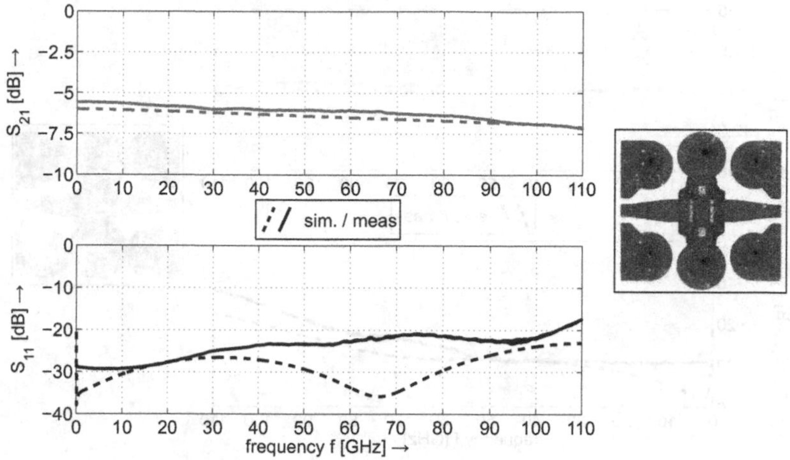

Figure 3.64: Measured transmission coefficient (top) and reflection coefficient (bottom) versus frequency of 6 dB planar attenuator, realized as T structures on 5 mil alumina.

$$\begin{aligned} \mathbf{C} &= \mathbf{C}_{\text{series}} \cdot \mathbf{C}_{\text{shunt}} \cdot \mathbf{C}_{\text{series}} \\ \mathbf{C} &\rightarrow \mathbf{S} \end{aligned} \tag{3.99}$$

Such a semi-analytic synthesis approach does not allow for exact prediction of the attenuators high frequency performance, but accuracy is sufficient to check feasibility of various different sheet and conductor geometries. This approach is valid for the design of planar terminations from the preceding paragraph, as well.

Following this synthesis procedure, the 6 dB, 10 dB and 15 dB T attenuators from Fig. 3.64, Fig. 3.65 and Fig. 3.66 have been designed. 3D EM simulation results and on-wafer measurements are in reasonable agreement. Deviations are due to the transition from grounded coplanar waveguide (gCPW) to microstrip line (MSL) and the passivation layer, which are not part of the simulation model.

Table 3.3 summarizes the dimensions of the manufactured attenuators and Fig. 3.67 compares the attenuators transmission coefficients

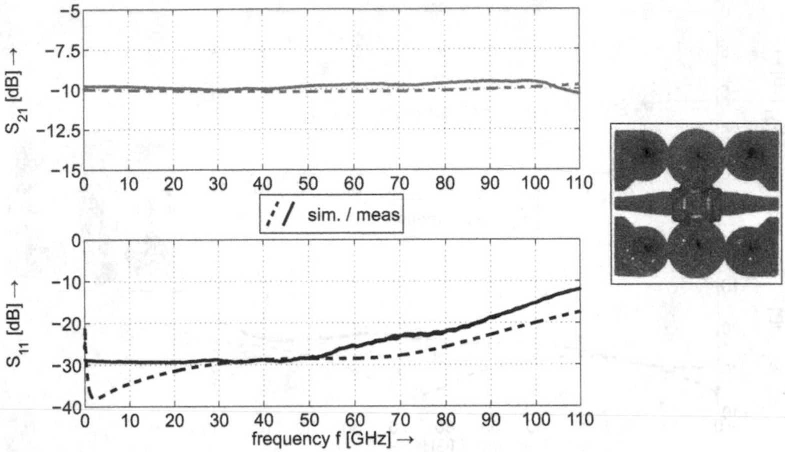

Figure 3.65: Measured transmission coefficient (top) and reflection coefficient (bottom) versus frequency of 10 dB planar attenuator, realized as T structures on 5 mil alumina.

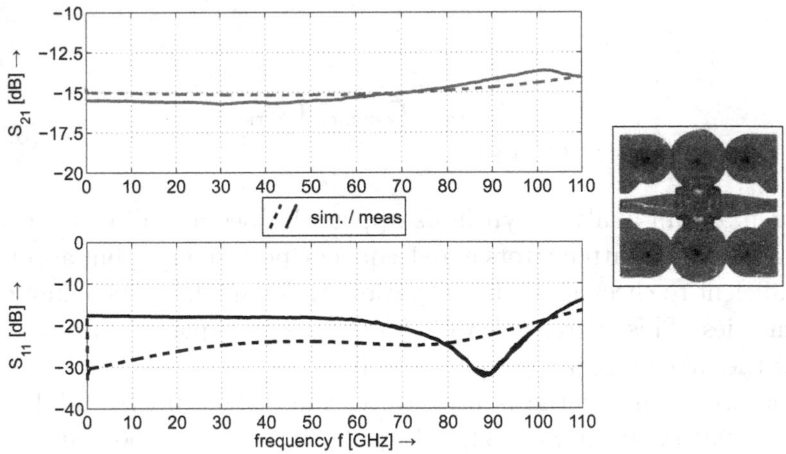

Figure 3.66: Measured transmission coefficient (top) and reflection coefficient (bottom) versus frequency of 15 dB planar attenuator, realized as T structures on 5 mil alumina.

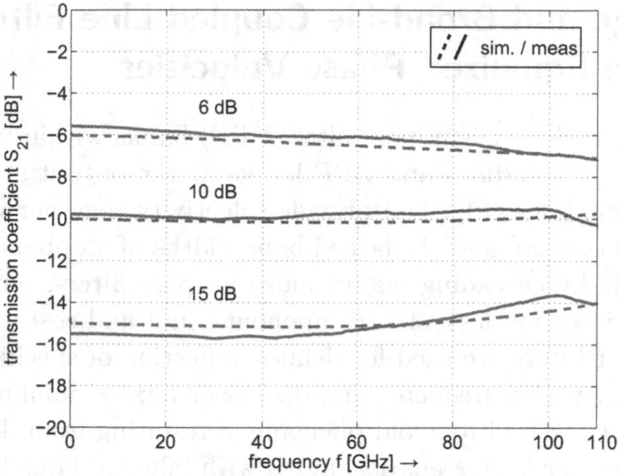

Figure 3.67: Measured transmission coefficient versus frequency of 6 dB, 10 dB and 15 dB attenuators, realized as T structures on 5 mil alumina.

Table 3.3: Dimensions of manufactured 6, 10 and 15 dB attenuators

attenuation [dB]	6	10	15
ℓ_0 [μm]	20	100	100
R_1 [Ω] / ℓ_{R_1} [μm] / w_{R_1} [μm]	16.6/20.3/122	26/15.6/60	34.8/20.9/60
ℓ_1 [μm]	45	50	53
R_2 [Ω] / ℓ_{R_2} [μm] / w_{R_2} [μm]	133.8/53.5/40	70.2/35.1/50	36.7/20.2/55
ℓ_2 [μm]	86	55	55
ℓ_{via} [μm]	39	9	15
via pad diameter [μm]	250	300	300

40Ni-60Cr sheet resistance $R_\square = 100\ \Omega$, via diameter at top and bottom layer $\varnothing_{TOP} = 150$ μm, $\varnothing_{BOT} = 200$ μm.

from simulation and measurement. The presented attenuators outperform commercially available GaAs attenuator MMICs like [74] from Hittite Microwave Corp. or [75] from GigOptix Inc.

3.5 Edge and Broadside Coupled Line Filters with Equalized Phase Velocities

Switched preselector bandpass filter (BPF) banks within broadband downconverter modules and BPF banks at the output stage of upconverter modules (Fig. 1.4) provide selectivity together with large bandwidth of stopband. Enlarged bandwidths of stopbands of BPF are possible by cascading one or more lowpass filters, but leads to additional insertion loss, size, component count and cost.

Whenever filters are used for defined reflection of specific spectral content, e.g. within frequency multipliers or mixers, minimum reflection phase is desired to avoid phenomena resulting from destructive interference. For both scenarios, filters with inherent large bandwidth of stopband are the method of choice.

From section 3.2 it is known, that coupled waveguides with equalized phase velocities are necessary to build high directivity backward wave directional couplers.

Within edge or broadside coupled halfwave BPF (Fig. 3.71 and Fig. 3.69), such coupled waveguides are utilized as impedance inverters J. The parasitic second passband of such BPF strongly depends on the ideality of the impedance inverters. Inverters with equalized phase velocities $v_\mathrm{e} = v_\mathrm{o}$ achieve up to 40 dB rejection around twice the center frequency $2 \cdot f_0$. It is shown, that applying wiggly-line technique to the first and last inverter only, allows for about 20 dB rejection at $2 \cdot f_0$ while keeping the design procedure unchanged. In the following, the analytical design procedure [12] and simulation results of a stripline realization are presented. Then, simulation and measurement results of a microstrip line filter realization, with and without applied wiggly-line technique, are shown.

3.5.1 Synthesis Procedure

Edge or broadside coupled halfwave BPF consist of N parallel halfwave ($\lambda/2$) resonators, coupled by $(N+1)$ impedance inverters J_k (Fig. 3.68). The $\lambda/2$ resonators overlap at half of their lengths $\lambda/4$ with each other,

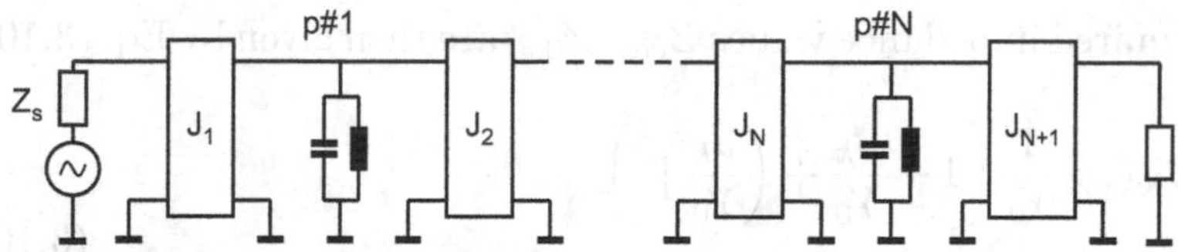

Figure 3.68: Bandpass filter consisting of N parallel resonators and $(N+1)$ impedance inverters

building $(N+1)$ coupled waveguide sections (Fig. 3.71 and Fig. 3.69). These are characterized by their lengths ℓ_k and impedances of their even and odd modes Z_{0ek}, Z_{0ok} [12, 76]. Fractional bandwidth FBW is defined as the bandwidth of passband $f_2 - f_1$ divided by the geometric average of upper f_2 and lower f_1 three dB corner frequencies.

$$\mathrm{FBW} = \frac{f_2 - f_1}{f_0} \qquad f_0 = \sqrt{f_1 f_2} \tag{3.100}$$

The inverter's lengths are given by the geometric average of the even and odd modes quarterwave lengths, which differ only if both mode's phase velocities $v_{\mathrm{e/o}} = \lambda_{\mathrm{e/o}} \cdot f_0$ are not equalized.

$$\ell_k = \lambda_{\mathrm{f_0}}/4 = \frac{1}{4}\sqrt{\lambda_{\mathrm{e}}\lambda_{\mathrm{o}}} = \frac{\lambda_0}{4\sqrt{\sqrt{\epsilon_{\mathrm{re}}\epsilon_{\mathrm{ro}}}}} \qquad k = 1, \ldots, N+1 \tag{3.101}$$

The inverter values, normalized to the reference admittance $Y_0 = Y_{\mathrm{ref}} = g_0 Y_{\mathrm{s}} = g_{N+1} Y_\ell$, are derived from the **g** parameters of the lowpass prototype filter (compare appendix).

$$\begin{aligned} \frac{J_1}{Y_0} &= \sqrt{\frac{\pi \mathrm{FBW}}{2 g_1 g_2}} \qquad \frac{J_{N+1}}{Y_0} = \sqrt{\frac{\pi \mathrm{FBW}}{2 g_{N+1} g_{N+2}}} \\ \frac{J_k}{Y_0} &= \frac{\pi \mathrm{FBW}}{2\sqrt{g_k g_{k+1}}} \qquad k = 2, \ldots, N \end{aligned} \tag{3.102}$$

The required impedance values Z_{0ek}, Z_{0ok} are then given by Eq. (3.103).

$$
\begin{aligned}
Z_{0ek} &= \frac{1}{Y_0}\left(1 + \frac{J_k}{Y_0} + \left(\frac{J_k}{Y_0}\right)^2\right) \\
Z_{0ok} &= \frac{1}{Y_0}\left(1 - \frac{J_k}{Y_0} + \left(\frac{J_k}{Y_0}\right)^2\right) \qquad k = 1, \ldots, N+1
\end{aligned}
\tag{3.103}
$$

In full analogy to the design of backward wave directional couplers, the effective permittivities in Eq. (3.101) and the waveguide dimensions, which correspond to the impedances in Eq. (3.103), have to be extracted from 2D EM simulation results. The open end effect is considered by scaling the lengths ℓ_k. Analytical modelling of each coupled waveguide section is possible using the procedure described in section 3.1.

3.5.2 Experimental Results

Fig. 3.69 illustrates a 3D EM simulation model of an edge coupled halfwave BPF, realized on multilayer hydrocarbon RO4350/4450 laminates ($\epsilon_r = 3.48/3.54 \pm 0.05$). The stackup ensures almost identical phase velocities of the even and odd mode on the utilized stripline (STL) waveguides. The chosen dimensions of the 7th order BPF with fractional bandwidth FBW = 43 % and center frequency $f_c = 24.45$ GHz are summarized in Table 3.5. Low cost hydrocarbon is normally used in conjunction with standard PCB processing (panel or pattern plating). The latter does not allow for realization of the necessary minimum strip widths and spacings[18]. The discussion is therefore based on 3D EM simulation results, which are shown in Fig. 3.70. The filter achieves an excellent second passband rejection of about 40 dB, which is sufficient for many applications without cascading additional lowpass filters. Within highly integrated front end modules, it is often preferred to keep the same waveguide type,

[18] The stackup of Fig. 3.69 would allow for realization of broadside coupled halfwave BPF to overcome problems with the required minimum strip spacings, but problems with minimum strip widths remain.

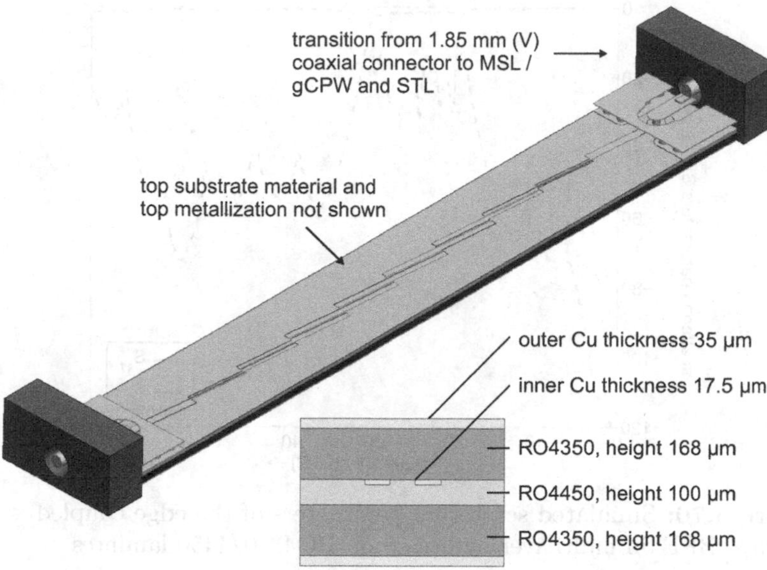

Figure 3.69: 3D EM simulation model of edge coupled halfwave BPF, realized on multilayer hydrocarbon RO4350/4450 laminates. The illustrated stackup ensures almost identical phase velocities of the even and odd mode on the utilized stripline (STL) waveguides.

e.g. microstrip line, and substrate material throughout the entire design. Not least, because waveguide transitions add insertion loss and complicate the assembly process.

Fig. 3.71 shows two edge coupled halfwave BPF on 10 mil alumina with close up view of the first / last impedance inverters. Specifications are similar to the filter realization on hydrocarbon laminates (Table 3.5). The upper part of Fig. 3.71 depicts an initial design according to Eq. (3.100 to 3.103). The bottom part shows a second variant with wiggly-line technique applied to the first / last impedance inverters J_1, J_8. These inverters are the main contributor to the parasitic second passband. Hence, equalization of these even and odd mode phase velocities shows most effect. In the sense of subsection 3.2.3 (compare Fig. 3.30), the wiggle depth d, tight coupling

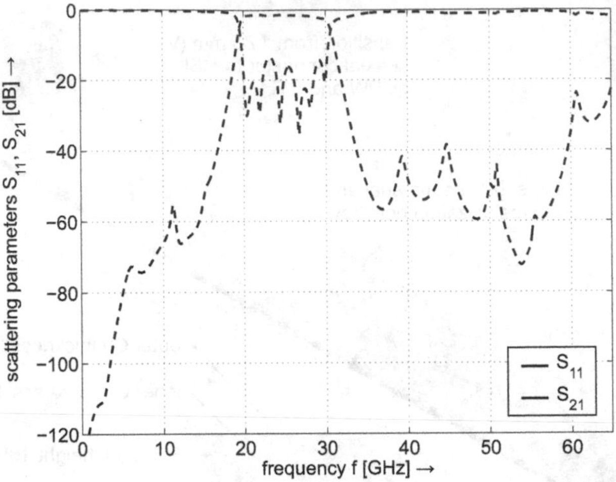

Figure 3.70: Simulated scattering parameters of the edge coupled halfwave BPF on multilayer hydrocarbon RO4350/4450 lamintes.

Table 3.4: 7th order halfwave resonator BPF (hydrocarbon four layer stackup)

impedance inverters J_i	**even and odd mode impedances $Z_{0e}/Z_{0o}[\Omega]$,**	**strip widths / spacings [μm]**	**lengths [mm]**
1 and 8	131/43	39/35	1.507
2 and 7	99/38	96/32	1.507
3 and 6	79/39	136/44	1.507
4 and 5	77/40	143/47	1.507

filter order $N = 7$, fractional bandwidth FBW $= 43$ %,
center frequency $f_0 = \sqrt{19.8 \cdot 30.2}$ GHz $= 24.45$ GHz.

spacing s_{t}, loose coupling spacing s_ℓ and conductor width w are given by Eq. (3.104).

$$d = 57 \text{ ţm} \qquad s_{\mathrm{t}} = 34 \text{ ţm} \qquad s_\ell = 48 \text{ ţm} \qquad w = 45 \text{ ţm} \qquad (3.104)$$

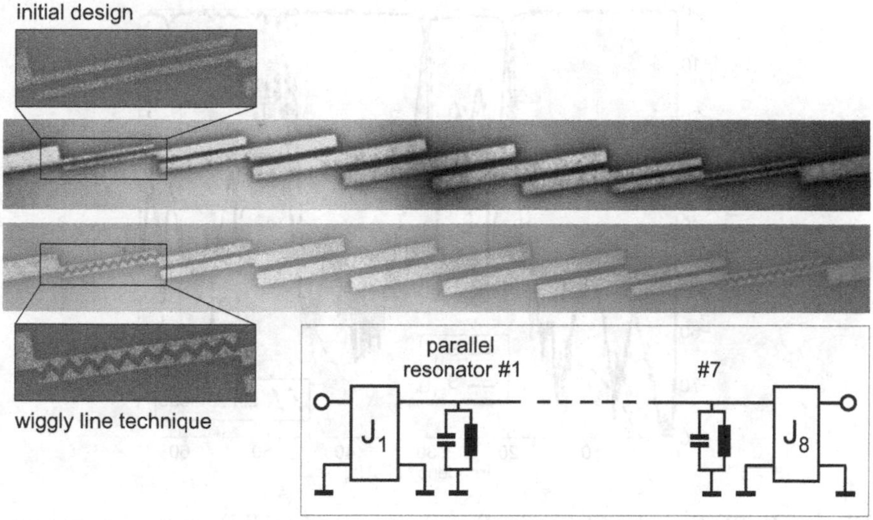

Figure 3.71: Photographs of edge coupled halfwave BPF, realized on 10 mil alumina, together with close up view of the first/last impedance inverter with and without applied wiggly-line technique.

Simulation (dashed) and measurement (solid) results of both BPF are shown in Fig. 3.72 and Fig. 3.73. The applied wiggly-line technique results in additional 12 dB rejection of the parasitic passband yielding an overall rejection of 22 dB. This result cannot compete with stripline designs, that benefit from an inherent phase equalization. Anyway, a difference of 20 dB between the power level of the desired signal and the spurious signal is a typical specification for many applications. There are reported equalization techniques, e.g. defected ground structures (DGS) [77], which allow for better rejection. These are complicated to apply and do not generally lead to a successful result if FBW, f_0 or the substrate material are changed. The presented method relies on the simple procedure of Eq. (3.100 to 3.103).

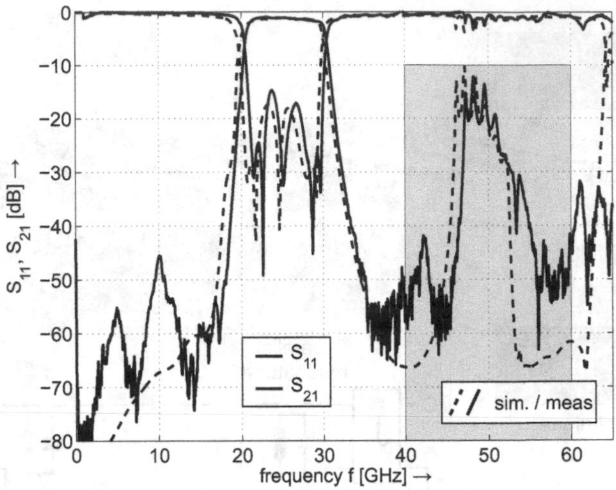

Figure 3.72: Simulated (dashed) and measured (solid) scattering parameters of edge coupled halfwave BPF, realized on 10 mil alumina, without applied wiggly-line technique.

Table 3.5: 7th order halfwave resonator BPF (10 mil Al_2O_3)

impedance inverters J_i	even and odd mode impedances $Z_{0e}/Z_{0o}[\Omega]$,	strip widths / spacings [µm]	lengths [mm]
1 and 8	119/43	45/48	1.186
2 and 7	90/37	105/55	1.141
3 and 6	75/36	153/81,	1.114
4 and 5	73/35	160/88	1.110

filter order $N = 7$, fractional bandwidth FBW $= 43$ %,
center frequency $f_0 = \sqrt{19.8 \cdot 30.2}$ GHz $= 24.45$ GHz.

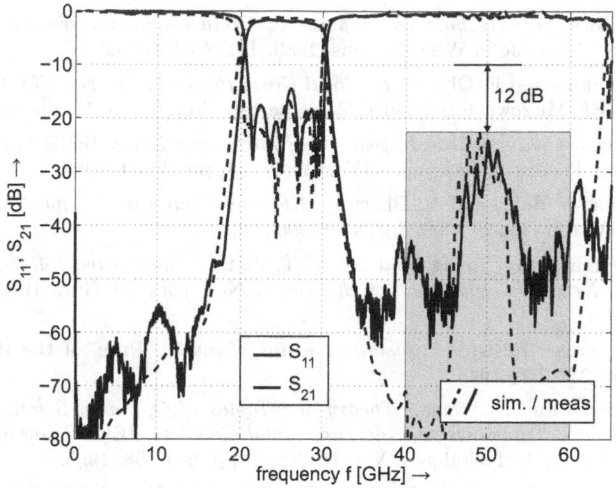

Figure 3.73: Simulated (dashed) and measured (solid) scattering parameters of edge coupled halfwave BPF, realized on 10 mil alumina, with applied wiggly-line technique.

References

[1] R. B. Marks and D. F. Williams. *A General Waveguide Circuit Theory.* Journal of Research of the National Institute of Standards and Technology, Vol. 97, No. 5, 1992.

[2] D. F. Williams, L. A. Hayden, and R. B. Marks. *A Complete Multimode Equivalent-Circuit Theory for Electrical Design.* Journal of Research of the National Institute of Standards and Technology, Vol. 102, No. 4, 1997.

[3] V. K. Tripathi. *Asymmetric Coupled Transmission Lines in an Inhomogeneous Medium.* IEEE Transactions on Microwave Theory & Techniques, Vol. 23, No. 9, pp. 734-739, 1975.

[4] V. K. Tripathi. *Equivalent Circuits and Characteristics of Inhomogeneous Nonsymmetrical Coupled-Line Two-Port Circuits.* IEEE Transactions on Microwave Theory & Techniques, Vol. 25, No. 2, pp. 140-142, 1977.

[5] V. K. Tripathi. *Properties and Applications of Asymmetric Coupled Line Structures in an Inhomogeneous Media.* Proc. of the 5th European Microwave Conference, pp. 278-282, 1975.

[6] D. G. Swanson and W. J. R. Hoefer. *Microwave Circuit Modeling Using Electromagnetic Field Simulation.* Boston: Artech House, 2003. ISBN 1580533086.

[7] A. D. Jimenez. *2-D Electromagnetic Simulation of Passive Microstrip Circuits.* New York: CRC Press, 2008. ISBN 9781420087055.

[8] T. Itoh. *Numerical Techniques for Microwave and Millimeter-Wave Passive Structures.* New York: John Wiley & Sons, 1989. ISBN 0471625639.

[9] D. F. Williams and F. Olyslager. *Modal Cross Power in Quasi-TEM Transmission Lines.* IEEE Microwave & Guided Wave Letters, Vol. 6, No. 11, pp. 413-415, 1996.

[10] J. R. Brews. *Characteristic Impedance of Microstrip Lines.* IEEE Transactions on Microwave Theory & Techniques, Vol. 35, No. 1, pp. 30-34, 1987.

[11] R. Mongia, I. Bahl, and P. Bhartia. *RF and Microwave Coupled-Line Circuits.* Boston: Artech House, 1999. ISBN 0890068305.

[12] G. L. Matthaei, L. Young, and E. M. T. Jones. *Microwave Filters, Impedance Matching Networks, and Coupling Structures.* New York: McGraw-Hill, 1964. ISBN 0890060991.

[13] P. I. Richards. *Resistor-Transmission-Line Circuits.* Proc. of the IRE, Vol. 36, No. 2, pp. 217-220, 1948.

[14] E. G. Cristal and L. Young. *Theory and Tables of Optimum Symmetrical TEM-Mode Coupled-Transmission-Line Directional Couplers.* IEEE Transactions on Microwave Theory & Techniques, Vol. 13, No. 5, pp. 544-558, 1965.

[15] C. P. Tresselt. *Design and Computed Theoretical Performance of Three Classes of Equal-Ripple Nonuniform Line Couplers.* IEEE Transactions on Microwave Theory & Techniques, Vol. 17, No. 4, pp. 218-230, 1966.

[16] D. W. Kammler. *The Design of Discrete N-Section and Continuously Tapered Symmetrical Microwave TEM Directional Couplers.* IEEE Transactions on Microwave Theory & Techniques, Vol. 17, No. 8, pp. 577-590, 1969.

[17] S. Uysal and J. Watkins. *Novel Microstrip Multifunction Directional Couplers and Filters for Microwave and Millimeter-Wave Applications.* IEEE Transactions on Microwave Theory & Techniques, Vol. 39, No. 6, pp. 977-985, 1991.

[18] C. B. Sharpe. *An Alternative Derivation of Orlov's Synthesis Formula for Non-Uniform Lines.* Proc. of the IEE - Part C: Monographs, Vol. 109, No. 15, pp. 226-229, 1962.

[19] S. L. Orlov. *Concerning the Theory of Nonuniform Transmission Lines.* Journal of Technical Physics of the USSR, Vol. 26, p. 2361, 1956.

[20] A. Bergquist. *Wave Propagation on Nonuniform Transmission Lines.* IEEE Transactions on Microwave Theory & Techniques, Vol. 20, No. 8, pp. 557-558, 1972.

[21] S. C. Dutta Roy. *Comments on Wave Propagation on Nonuniform Transmission Lines.* IEEE Transactions on Microwave Theory & Techniques, Vol. 22, No. 2, pp. 149-150, 1974.

[22] B. G. Kazansky. *Outline of a Theroy of Non-Uniform Transmission Lines.* Proc. of the IEE - Part C: Monographs, Vol. 105, No. 7, pp. 126-138, 1958.

[23] L. R. Walker and N. Wax. *Non-Uniform Transmission Lines and Reflection Coefficients.* Journal of Applied Physics, Vol. 17, No. 12, pp. 1043-1045, 1946.

[24] E. N. Protonotarios and O. Wing. *Delay and Rise Time of Arbitrarily Tapered RC Transmission Lines.* IEEE Int. Convention Record, Part 7, pp. 1-6, 1965.

[25] E. N. Protonotarios and O. Wing. *Analysis and Intrinsic Properties of the General Nonuniform Transmission Line.* IEEE Transactions on Microwave Theory & Techniques, Vol. 15, No. 3, pp. 142-150, 1967.

[26] J. Mueller, M. N. Pham, and A. F. Jacob. *Directional Coupler Compensation With Optimally Positioned Capacitances.* IEEE Transactions on Microwave Theory & Techniques, Vol. 59, No. 11, pp. 2824-2832, 2011.

[27] K. Nishikawa, M. Kawashima, T. Seki, and K. Hiraga. *Broadband and Compact 3-dB MMIC Directional Coupler with Lumped Element.* IEEE MTT-S Int. Microwave Symposium Digest, p. 1, 2010.

[28] M.-J. Chiang, H.-S. Wu, and C.-K. C. Tzuang. *Artificial-Synthesized Edge-Coupled Transmission Lines for Compact CMOS Directional Coupler Designs.* IEEE Transactions on Microwave Theory & Techniques, Vol. 57, No. 12, pp. 3410-3417, 2009.

[29] S. Lee and Y. Lee. *An Inductor-Loaded Microstrip Directional Coupler for Directivity Enhancement.* IEEE Microwave & Wireless Components Letters, Vol. 19, No. 6, pp. 362-364, 2009.

[30] E. V. Andronov, G. G. Goshin, O. Y. Morozov, A. V. Semenov, and A. V. Fateyey. *Ultra Wideband Directional Coupler with Matching Resistive Elements.* 20th Int. Crimean Conference on Microwave and Telecommunication Technology (CriMiCo), pp. 639-640, 2010.

[31] B. M. Oliver. *Directional Electromagnetic Couplers.* Proc. of the IRE, Vol. 42, pp. 1686-1692, 1954.

[32] M. Sterns, R. Rehner, D. Schneiderbanger, S. Martius, and L.-P. Schmidt. *Broadband, Highly Integrated Receiver Frontend Up To 67 GHz.* Proc. 41th European Microwave Conference, pp. 1185-1188, 2011.

[33] Marki Microwave Inc. *C10-0450 Directional Coupler.* www.markimicrowave.com.

[34] Marki Microwave Inc. *C13-0150 Directional Coupler.* www.markimicrowave.com.

[35] KRYTAR. *501040010K Directional Coupler.* www.krytar.com.

[36] KRYTAR. *501040010K Directional Coupler, scattering parameters provided by Mike Romero.* www.krytar.com.

[37] KRYTAR. *101065013 Directional Coupler.* www.krytar.com.

[38] Electromagnetic Technologies Industries Inc. *C-165-13 Directional Coupler.* www.etiworld.com.

[39] C. M. Potter and G. Hjipieris. *Improvements to Broadband TEM Coupler Design.* IEE Colloquium on Measurements and Modelling of Microwave Devices and Circuits, pp. 1/1-1/6, 1989.

[40] C. M. Potter and G. Hjipieris. *Improvements in Ultra-Broadband TEM Coupler Design.* IEE Proceedings of Microwaves, Antennas and Propagation, Vol. 139, No. 2, pp. 171-178, 1992.

[41] C. M. Potter, G. Hjipieris, and N. J. Fanthom. *A Novel 250 MHz - 26.5 GHz Reflection Analyzer.* Proc. 23rd Eur. Microw. Conf., pp. 302-304, 1993.

[42] J. P. Shelton. *Impedances of Offset Parallel-Coupled Strip Transmission Lines.* IEEE Transactions on Microwave Theory & Techniques, Vol. 14, No. 1, p. 7, 1966.

[43] J. P. Shelton. *Impedances of Offset Parallel-Coupled Strip Transmission Lines (Corrections).* IEEE Transactions on Microwave Theory & Techniques, Vol. 17, No. 5, pp. 249, 1966.

[44] A. Stark and A. F. Jacob. *A Broadband Vertical Transition for Millimeter-Wave Applications.* Proc. 38th European Microwave Conference, p. 476, 2008.

[45] A. Podell. *A High Directivity Microstrip Coupler Technique.* Int. Microwave Symposium G-MTT, Vol. 17, No. 8, pp. 33-36, 1970.

[46] M. Leib, D. Mack, F. Thurow, and W. Menzel. *Design of a Multilayer Ultra-Wideband Directional Coupler.* Proc. of the 3rd German Microwave Conference (GeMiC), pp. 1-4, 2008.

[47] M. Nedil and T. A. Denidni. *A New Ultra Wideband Directional Coupler based on a Combination Between CB-CPW and Microstrip Technologies.* IEEE MTT-S Int. Microwave Symposium Digest, pp. 1219-1222, 2008.

[48] L. Abdelghani, T. A. Denidni, and M. Nedil. *Design of a Broadband Multilayer Coupler for UWB Beamforming Applications.* Proc. of the 41st European Microwave Conference (EuMC), pp. 810-813, 2011.

[49] S. Uysal and A. H. Aghvami. *Synthesis and Design of Wideband Symmetrical Nonuniform Directional Couplers for MIC Applications.* IEEE MTT-S Int. Microwave Symposium Digest, Vol. 2, pp. 587-590, 1988.

[50] S. Uysal and A. H. Aghvami. *Synthesis, Design, and Construction of Ultra-Wide-Band Nonuniform Quadrature Directional Couplers in Inhomogeneous Media.* IEEE Transactions on Microwave Theory & Techniques, Vol. 37, No. 6, pp. 969-976, 1989.

[51] S. Uysal and A. H. Aghvami. *Improved Wideband -3 dB Nonuniform Directional Coupler.* Electronics Letters, Vol. 25, No. 8, pp. 541-542, 1989.

[52] J.-L. Chen, S.-F. Chang, Y.-H. Jeng, and C.-Y. Lin. *Wiggly Technique for Broadband Non-Uniform Line Couplers.* Electronics Letters, Vol. 39, No. 20, pp. 1451-1453, 2003.

[53] J. Mueller and A. F. Jacob. *Advanced Characterization and Design of Compensated High Directivity Quadrature Coupler.* IEEE MTT-S Int. Microwave Symposium Digest, pp. 724-727, 2010.

[54] J. Mueller, C. Friesicke, and A. F. Jacob. *Stepped Impedance Microstrip Couplers with Improved Directivity.* IEEE MTT-S Int. Microwave Symposium Digest, pp. 621-624, 2009.

[55] M. Chudzik, I. Arnedo, A. Lujambio, I. Arregui, F. Teberio, M. A. G. Laso, and T. Lopetegi. *Microstrip Coupled-Line Directional Coupler with Enhanced Coupling based on EBG Concept.* Electronics Letters, Vol. 47, No. 23, pp. 1284-1286, 2011.

[56] D. Brady. *The Design, Fabrication and Measurement of Microstrip Filter and Coupler Circuits.* High Frequency Electronics, July, pp. 22-30, 2002.

[57] T. Jayanthy and S. Maheswari. *Design of High Directivity Microstrip Coupler with Single Floating Conductor on Coupled Edges.* 10th Annual Wireless and Microwave Technology Conference (WAMICON), pp. 1-4, 2009.

[58] A. A. Shauerman, A. V. Borisov, M. S. Zharikov, A. K. Shauerman, and F. S. Kroshin. *Development and Investigation of Microstrip Directional Coupler with Phase Velocity Compensation based on Sawtooth Configuration of Coupled Lines.* XII Int. Conference and Seminar of Young Specialists on Micro/Nanotechnologies and Electron Devices (EDM), Sec. IV, pp. 191-194, 2011.

[59] S. Gruszczynski and K. Wincza. *Generalized Methods for the Design of Quasi-Ideal Symmetric and Asymmetric Coupled-Line Sections and Directional Couplers.* IEEE Transactions on Microwave Theory & Techniques, Vol. 59, No. 7, pp. 1709-1718, 2011.

[60] S. Lin, M. Eron, S. Turner, and J. Sepulveda. *Development of Wideband Low-Loss Directional Coupler with Suspended Stripline and Microstrip Line.* Electronics Letters, Vol. 47, No. 25, pp. 1377-1379, 2011.

[61] E. Hammerstad and O. Jensen. *Accurate Models for Microstrip Computer-Aided Design.* IEEE MTT-S Int. Microwave Symposium Digest, pp. 407-409, 1980.

[62] R. H. Jansen and M. Kirschning. *Arguments and an Accurate Model for the Power-Current Formulation of Microstrip Characteristic Impedance.* Arch. Elek. Uebertragung (AEU), Vol. 37, pp. 108-112, 1983.

[63] R. H. Jansen and M. Kirschning. *Accurate Wide-Range Design Equations for the Frequency-Dependent Characteristic of Parallel Coupled Microstrip Lines.* IEEE Transactions on Microwave Theory & Techniques, Vol. 32, pp. 83-90, 1984.

[64] M. Kirschning and R. H. Jansen. *Corrections to Accurate Wide-Range Design Equations for the Frequency-Dependent Characteristic of Parallel Coupled Microstrip Lines.* IEEE Transactions on Microwave Theory & Techniques, Vol. 33, No. 3, page 288, 1985.

[65] S. D. Shamasundara and K. C. Gupta. *Sensitivity Analysis of Coupled Microstrip Directional Couplers.* IEEE Transactions on Microwave Theory & Techniques, Vol. 26, No. 10, pp. 788-794, 1978.

[66] S. D. Shamasundara and K. C. Gupta. *Correction to Sensitivity Analysis of Coupled Microstrip Directional Couplers.* IEEE Transactions on Microwave Theory & Techniques, Vol. 27, No. 2, p. 208, 1979.

[67] P. Salem, C. Wu, and M.C.E Yagoub. *Non-uniform Tapered Ultra Wideband Directional Coupler Design and Modern Ultra Wideband Balun Integration.* Proc. Asia Pacific Microwave Conference, pp. 803-806, 2006.

[68] L. Hayden. *An Enhanced Line-Reflect-Reflect-Match Calibration.* Proc. of the 67th ARFTG Conference, pp. 143-149, 2006.

[69] G. Vinci, F. Barbon, B. Laemmle, R. Weigel, and A. Koelpin. *Promise of a Better Position.* IEEE Microwave Magazine, Vol. 1, No. 7, pp. 41-49, 2012.

[70] Rosenberger Hochfrequenztechnik GmbH & Co. KG. *RPC-1.85 mm Calibration Kit.* www.rosenberger.com.

[71] Agilent Technologies Inc. *85059A 1.0 mm Calibration Kit.* www.home.agilent.com.

[72] L. I. Maissel and R. Glang. *Handbook of Thin Film Technology.* New York: McGraw-Hill, 1970. ISBN 0070397422.

[73] B. Oldfield. *Connector and Termination Construction above 50 GHz.* Applied Microwave & Wireless Magazine, Vol. 13, pp. 56-66, 2001.

[74] GigOptix Inc. *EWA65xxZZ, DC-105 GHz GaAs MMIC Attenuator Series.* www.gigoptix.com.

[75] Hittite Microwave Corp. *HMC650 to HMC658, DC-50 GHz GaAs MMIC Attenuator Series.* www.hittite.com.

[76] J.-S. Hong and M. J. Lancaster. *Microstrip Filters for RF / Microwave Applications.* New York: John Wiley & Sons, 2001. ISBN 0471388777.

[77] R. Rehner, D. Schneiderbanger, M. Sterns, S. Martius, and L.-P. Schmidt. *Novel Coupled Microstrip Wideband Filters with Spurious Response Suppression.* Proc. of the 10th European Microwave Conference, pp. 858-861, 2007.

[78] D. M. Kerns and R. W. Beatty. *Basic Theory of Waveguide Junctions and Introductory Microwave Network Analysis.* Oxford: Pergamon Press, 1967. ISBN 0080120644.

[79] N. Fache, F. Olyslager, and D. De Zutter. *Electromagnetic and Circuit Modeling of Multiconductor Transmission Lines.* Oxford: Clarendon Press, 1993. ISBN 0198562500.

[80] R. E. Collin. *Field Theory of Guided Waves.* New York: McGraw-Hill, 1960. ISBN 0070118027.

[81] W. Klein. *Grundlagen der Theorie elektrischer Schaltungen, Teil 1 Mehrtortheorie.* Berlin: Akademie-Verlag, 1967.

[82] R. E. Collin. *Foundations for Microwave Engineering.* New York: McGraw-Hill, 1966. ISBN 0780360311.

[83] S. Ramo, T. R. Whinnery, and T. Van Duzer. *Fields and Waves in Communication Electronics.* New York: John Wiley & Sons, 1965. ISBN 0471585513.

[84] D. C. Youla. *On Scattering Matrices Normalized to Complex Port Numbers.* Proc. of the IRE, Vol. 49, p. 122, 1961.

[85] K. Kurokawa. *Power Waves and the Scattering Matrix.* IEEE Transactions on Microwave Theory & Techniques, Vol. 13, No. 2, pp. 194-202, 1965.

[86] H. Howe. *Stripline Circuit Design.* Boston: Artech House, 1974. ISBN 0890060207.

[87] J. A. G. Malherbe. *Microwave Transmission Line Couplers.* Boston: Artech House, 1988. ISBN 0890063001.

[88] B. Bhat and S. Koul. *Stripline-Like Transmission Lines for Microwave Integrated Circuits.* New Delhi: New Age International, 1989. ISBN 8122400523.

[89] G. Schaller. *Untersuchungen an Leitungsrichtkopplern insbesondere in Mikrostreifenleitungstechnik.* Ph.D. Thesis, Friedrich-Alexander University of Erlangen-Nuremberg, Germany, 1976.

[90] R. W. Klopfenstein. *A Transmission Line Taper of Improved Design.* Proc. of the IRE, Vol. 44, No. 1, pp. 31-35, 1956.

[91] D. Kajfez and J. O. Prewitt. *Correction to A Transmission Line Taper of Improved Design.* IEEE Transactions on Microwave Theory & Techniques, Vol. 21, No. 5, page 364, 1973.

[92] S. Uysal. *Nonuniform Line Microstrip Directional Couplers and Filters.* Boston: Artech House, 1993. ISBN 0890066833.

[93] D. E. Bockelman. *The Theory, Measurement, and Applications of Mode Specific Scattering Parameters with Multiple Modes of Propagation.* Ph.D. Thesis, University of Florida, USA, 1997.

4 Triple Balanced Mixers

Dedicated designs of single balanced mixers (SBM) often enable outstanding system performance, whereas double balanced mixers (DBM) constitute the industry's workhorse mixer for various fields of application. This chapter presents the author's investigations on triple balanced mixers (TBM), which are the only mixers providing overlapping RF, LO and IF frequency ranges together with large IF bandwidth and enhanced RF large signal handling capability.

The first section highlights major differences and similarities between frequency multipliers and mixers. This is followed by a section about analytical and numerical methods for mixer analysis. Different methods are listed together with its strengths and weaknesses. With the results of single tone large signal analysis (harmonic balance) of section 2.3 as a starting point, subsection 4.2.1 derives large signal / small signal (LSSS) mixer analysis based on conversion matrices. The minimum achievable conversion loss at different embedding impedances and also optimum embedding impedances are calculated for a simplified case, which includes nonzero diode series resistances. Section 4.3 compares different mixer configurations (SBM, DBM and TBM) from a small signal analysis point of view. It further includes a comparison of the mixers' spurious tone behaviour, following a method from Henderson [1].

Subsection 4.4 presents a planar TBM realization on SiO_2 using commercial silicon crossed quad diodes. The TBM operates from 1 to 45 / 50 GHz at RF / LO and achieves at least 20 GHz IF bandwidth. Simulation and measurement results of the component are presented. Measurement results of the TBM within an automatic test system front end module are included.

4.1 General Comments on Commutative Mixers

From a simplified equivalent circuit point of view, frequency multipliers are mixers without RF and IF signals but a strong LO signal with its harmonics. This alleged similarity is used throughout this work (e.g. section 2.1). Anyway, there are some severe differences from a circuit designers point of view.

Both, mixers and multipliers use balancing methods, symmetric / antisymmetric configurations of identical diodes, to avoid undesired harmonics / spurious tones and therefore achieve inherent filtering properties with the diode arrangement. In case of frequency multipliers, real filter circuits are placed at the input and output additionally. The fact, that real diodes perform different than ideal switches (section 2.2) does influence all multiplier parameters, like

- the required minimum input power level,
- achievable minimum conversion loss,
- optimum embedding impedances,

but the working principle is not violated. As long as these nonidealities are identical for all diodes, filter effects from balancing still work. Frequency multipliers are harmonic generators with processing of the undesired harmonics defined by passive circuit environment.

This is different for commutative diode mixers (polarity switching mixers), which are considered exclusively in this work. These are intended to perform linear frequency translation from RF to IF port and vice versa, without distorting the original RF / IF signal and without adding spectral components. For such mixers the RF / IF signals need to be small signals and the local oscillator signal is switching between two states. Hence commutative switching is required but diode nonideality is not [2, 3]. This linear behaviour is optimized over a certain bandwidth at an almost constant local oscillator power level $P_{\mathrm{LO\,opt}}$. The mixer's spurious tone behaviour stays approximately the same for $P_{\mathrm{LO}} = P_{\mathrm{LO\,opt}} \pm 3$ dBm. Whereas the output power level of frequency multipliers at the desired output harmonic is generally increasing with an increasing input power level[1].

Interaction with preceding and cascaded circuits constitutes another difference between frequency multipliers and mixers. All involved harmonics within a broadband frequency multiplier are confined inside the multiplier circuit, except for the fundamental signal at the input and the desired harmonic at the output. In contrast, the ports of broadband diode mixers (multiple octave bandwidth) are designed to operate with resistive loads. Hence, the generated spurious tones leave the mixer's ports. This assumption often leads to performance degradation if the mixers are connected to reactive loads [4] within front end modules. Back reflected harmonics and spurs lead to self-mixing phenomena.

4.2 Analytical and Numerical Methods for Mixer Analysis

Fig. 4.1 gives an overview about some analytical and numerical methods for analysis of frequency mixers. The list is without claim to completeness.

Solving the harmonic balance Eq. (2.78) of section 2.3 has first been done by Hicks, Khan [5] and Kerr [6] with the relaxation methods. These are limited to simple problems and have become obsolete. Commercial solvers utilize Newton-Raphson method with either LU decomposition or generalized minimal residual method (GMRES) as mentioned in section 2.3. The following subsection applies large signal / small signal (LSSS) analysis to study mixer circuits. LSSS analysis is an extension of the single tone large signal analysis from section 2.3. The derivations make clear, analytical calculation of mixer performance requires severe simplifications. Conversion loss

[1]This is only partly true, due to fixed input matching circuits. With increasing input drive level the diode's input impedance decreases and therefore differs from the fixed source impedance causing reflections (saturation effect).

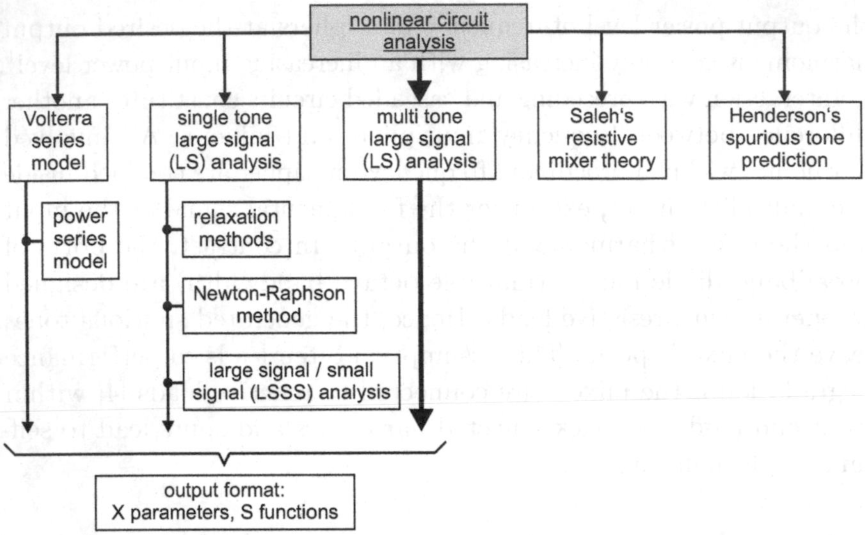

Figure 4.1: Overview about analytical and numerical methods for frequency mixer analysis.

and optimum small signal RF / IF impedances are calculated and compared for different mixer architectures.

In section 4.3, the spurious tone behaviour of the same mixer architectures are compared qualitatively. This follows the procedure from Henderson [1]. It assumes identical diodes and considers the cancellation (destructive interference) of spurious tones at the ports of balanced mixers.

If the nonlinear circuit is assumed to be weakly nonlinear, the Volterra[2] series is applicable, which is a generalized case of the power (Taylor) series model. Weakly nonlinear means the nonlinear transfer function is single valued and a series with a limited number of terms (< 5) is sufficient for description. Volterra modelling [7, 8] allows for prediction of intermodulation behaviour in case of multiple small signal excitations without DC component.

[2]Vito Volterra (1860–1940), Italian mathematician.

The analytical and numerical analysis of Saleh [9] allows for extraction of a mixer's required optimum conductance waveform for minimum conversion loss, without answering the question of how to practically realize the waveform. Several conductance waveforms and embedding impedance configurations are assumed and the related conversion parameters are derived as it is done in the next subsection.

The most versatile method to solve nonlinear circuit problems is multi tone large signal analysis, which is utilized by several commercial circuit simulators like Agilent ADS, Ansys Designer, Cadence SpectreRF and AWR Microwave Office. All signal stimuli with non-commensurate frequencies are assumed to be large signals. Generated DC component, harmonics and mixing frequencies are considered. The implementation is similar to the single tone large signal analysis but requires a generalized time to frequency mapping [8]. Simple Fourier transform as it is with single tone harmonic balance method is not sufficient.

Agilent's X parameters [10] or NMDG's[3] S-functions [11] are adequate file formats to cover the spectral output power level of mixers as a function of frequency, input power level and source and load impedance. These are based on polyharmonic distortion models [12, 13] and include information about phase relation of all involved signals. Modern nonlinear vector network analyzers allow for measurements of X parameters but are limited to a single tone large signal at the time of writing.

4.2.1 Large Signal / Small Signal (LSSS) Analysis

In general, the applied LO, RF and IF signals (including DC components) are large signals and their influence on the solution of the nonlinear problem has to be considered. In this general case, transient methods or multi tone harmonic balance are adequate algorithms. In the following, the local oscillator (LO) signal is considered as the only large signal. Hence, the method from chapter 2.3 allows for

[3]Since October 2012, Network Measurement and Description Group (NMDG) belongs to National Instruments (NI).

determination of large signal currents $i_{\mathrm{LS}}(t)$ and voltages $v_{\mathrm{LS}}(t)$, including DC quantities $i_{\mathrm{DC}}, v_{\mathrm{DC}}$, at all circuit branches. The functional dependency of $i_{\mathrm{LS}}(t)$ and $v_{\mathrm{LS}}(t)$ is called y_{LS} in Eq. (4.1).

$$i_{\mathrm{LS}}(t) = y_{\mathrm{LS}}\left(v_{\mathrm{LS}}(t)\right) \tag{4.1}$$

Taylor series approximation describes how additional small signals (RF and / or IF signals) $v_{\mathrm{SS}}(t), i_{\mathrm{SS}}(t)$ contribute to the overall solution (Eq. (4.2) and Eq. (4.3)).

$$\begin{aligned} i_{\mathrm{LS}}(t) &= y_{\mathrm{LS}}\left(v_{\mathrm{LS}}(t)\right) \\ i_{\mathrm{LS}}(t) + i_{\mathrm{SS}}(t) &= y_{\mathrm{LS}}\left(v_{\mathrm{LS}}(t) + v_{\mathrm{SS}}(t)\right) \end{aligned} \tag{4.2}$$

$$\begin{aligned} i_{\mathrm{LS}} + i_{\mathrm{SS}} = y_{\mathrm{LS}}\left(v_{\mathrm{LS}}\right) &+ \frac{\mathrm{d}y_{\mathrm{LS}}\left(v_{\mathrm{LS}}\right)}{\mathrm{d}v_{\mathrm{LS}}} \cdot v_{\mathrm{SS}} \\ &+ \frac{1}{2}\frac{\mathrm{d}^2 y_{\mathrm{LS}}\left(v_{\mathrm{LS}}\right)}{\mathrm{d}v_{\mathrm{LS}}^2} \cdot v_{\mathrm{SS}}^2 \\ &+ \frac{1}{6}\frac{\mathrm{d}^3 y_{\mathrm{LS}}\left(v_{\mathrm{LS}}\right)}{\mathrm{d}v_{\mathrm{LS}}^3} \cdot v_{\mathrm{SS}}^3 + \dots \end{aligned} \tag{4.3}$$

Truncating the Taylor series after the linear (second) term and separating the small signal part of the equation leads to Eq. (4.4), with the small signal admittance waveform $y_{\mathrm{SS}}(t)$. The latter builds the basis of the large signal / small signal (LSSS) analysis using conversion matrices, which is discussed in the following.

$$\begin{aligned} i_{\mathrm{SS}} &= \frac{\mathrm{d}y_{\mathrm{LS}}\left(v_{\mathrm{LS}}\right)}{\mathrm{d}v_{\mathrm{LS}}} \cdot v_{\mathrm{SS}} \\ \frac{i_{\mathrm{SS}}(t)}{v_{\mathrm{SS}}(t)} &= y_{\mathrm{SS}}(t) = \frac{\mathrm{d}y_{\mathrm{LS}}\left(v_{\mathrm{LS}}\right)}{\mathrm{d}v_{\mathrm{LS}}} \end{aligned} \tag{4.4}$$

All harmonic and mixing frequencies are expressed by Eq. (4.5) and hence, identified by two integers $\langle \mathrm{RF/IF}, \mathrm{LO} \rangle = \langle n, m \rangle$.

$$\omega_{nm} = 2\pi f_{nm} = n\,\omega_{\mathrm{RF/IF}} + m\,\omega_{\mathrm{LO}} \qquad n, m \in \pm\mathbb{N}_0 \tag{4.5}$$

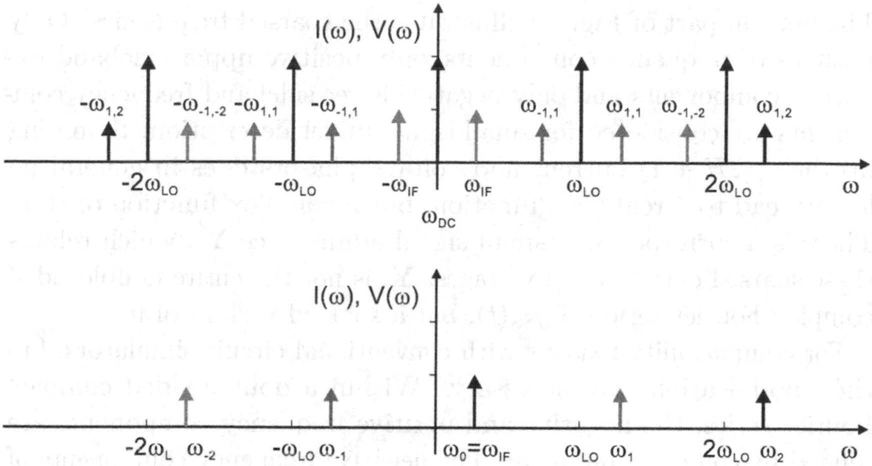

Figure 4.2: Spectral components at mixer ports if a single large signal is applied only (top) and spectral components for small signal mixer analysis (botttom) based on conversion matrices.

In the present LSSS analysis case, it is $n = 1$ and $m \in \{1, \ldots, H\}$ with the maximum considered large signal harmonic H. In the upper part of Fig. 4.2, only the first two LO harmonics $m = 2$ and the corresponding upper and lower sidebands $n = 1$ are shown. Each spectral component appears with a positive and negative frequency, as usual for double sided complex Fourier series. This yields a DC component, two IF components and $2{\cdot}3H$ LO harmonics and sidebands. For small signal description only $(2 \cdot 2H + 2)$ frequencies, IF and sidebands, are of importance. $y_{\mathrm{SS}}(t)$ from Eq. (4.4) is a periodic function of time and the corresponding double sided complex Fourier series connects these $(2 \cdot 2H + 2)$ voltages and currents.

A more compact description, which is fully compatible to conventional AC circuit theory and SPICE-like circuit simulators, is based on the sparsed frequencies of Eq. (4.6) [8, 9].

$$\omega_k = 2\pi f_k = \omega_{\mathrm{IF}} + k\,\omega_{\mathrm{LO}} \qquad k \in \pm\mathbb{N}_0 \tag{4.6}$$

The bottom part of Fig. 4.2 illustrates the sparsed frequencies. Only positive IF frequency components, only positive upper sideband frequency components and only negative lower sideband frequency components are considered for small signal mixer description. Summing up these $(2H+1)$ current and voltage phasors does in general no longer lead to a real time function, but a complex function of time. There is a corresponding small signal admittance $\mathbf{Y}^{'}$, which relates these sparsed currents and voltages. $\mathbf{Y}^{'}$ is not the entire double sided complex Fourier series of $y_{\mathrm{SS}}(t)$, but a sparsed variant of it.

For compatibility reasons with conventional circuit simulators, further modifications are necessary. Within a double sided complex Fourier series, the negative and positive frequency components are related by complex conjugate. The negative frequency components of lower sidebands appear at negative values of k, but are assumed as phasors of positive frequencies, and are therefore written as complex conjugate (Eq. (4.7)).

$$V_{-k}^{\star}e^{-j\omega_{-k}t} = V_{-k}^{\star}e^{-j|\omega_{-k}|t} = V_{-k}^{\star}e^{j|\omega_{IF}+k\omega_{LO}|t} \tag{4.7}$$

It is important to note, that $V_{-k}^{\star}$ and V_k belong, contrary to the double sided complex Fourier series, to different spectral components.

The large signal analysis from section 2.3 considered N nonlinear diode models, connected to a linear circuit part and stimulus. Based on the sparsed frequency notation, we derive small signal conversion matrices of the diode's conductance and junction capacitance.

Complex current and voltage time waveforms $i^{'}(t), v^{'}(t)$ with their frequency domain representations at an arbitrary circuit branch are given by Eq. (4.8).

$$
\begin{aligned}
i^{'}(t) &= \sum_{k=-\infty}^{\infty} I_k \exp(j\omega_k t) \approx \sum_{k=-H}^{H} I_k \exp(j\omega_k t) \\
\mathbf{I}^{'} &= [I_{-H}, \ldots, I_0, \ldots, I_H] \\
v^{'}(t) &= \sum_{k=-\infty}^{\infty} V_k \exp(j\omega_k t) \approx \sum_{k=-H}^{H} V_k \exp(j\omega_k t) \\
\mathbf{V}^{'} &= [V_{-H}, \ldots, V_0, \ldots, V_H]
\end{aligned}
\tag{4.8}
$$

The time varying conductance waveform $g^{'}(t)$ of a diode is shown in Eq. (4.9).

$$
g^{'}(t) = \sum_{k=-\infty}^{\infty} G_k \exp(jk\omega_{\mathrm{LO}} t) \approx \sum_{k=-H}^{H} G_k \exp(jk\omega_{\mathrm{LO}} t) \tag{4.9}
$$

The $(2H+1)$ current and voltage components of Eq. (4.8) are related by Ohm's law.

$$
i^{'}(t) = g^{'}(t) v^{'}(t) \tag{4.10}
$$

Inserting Eq. (4.8) and Eq. (4.9) into Eq. (4.10) leads to Eq. (4.11).

$$
\begin{aligned}
i^{'}(t) &= \sum_{a=-H}^{H} I_a \exp(j\omega_a t) = \sum_{b=-H}^{H} \sum_{c=-H}^{H} G_b V_c \exp(j\omega_{b+c} t) \\
\mathbf{I}^{'} &= \mathbf{G}^{'}\mathbf{V}^{'} = \mathbf{I}_{\mathrm{G}}
\end{aligned}
\tag{4.11}
$$

$\mathbf{G}'$ is the small signal conversion matrix of the diode's conductance. A matrix form is shown in Eq. (4.12).

$$\begin{bmatrix} I^{\star}_{-H} \\ I^{\star}_{-H+1} \\ I^{\star}_{-H+2} \\ \cdots \\ I^{\star}_{-1} \\ I_0 \\ I_1 \\ \cdots \\ I_{H-2} \\ I_{H-1} \\ I_H \end{bmatrix} = \begin{bmatrix} G_0 & G_{-1} & G_{-2} & \cdots & G_{-2H} \\ G_1 & G_0 & G_{-1} & \cdots & G_{-2H+1} \\ G_2 & G_1 & G_0 & \cdots & G_{-2H+2} \\ \cdots & \cdots & \cdots & \cdots & \cdots \\ G_{H-1} & G_{H-2} & G_{H-3} & \cdots & G_{-H-1} \\ G_H & G_{H-1} & G_{H-2} & \cdots & G_{-H} \\ G_{H+1} & G_H & G_{H-1} & \cdots & G_{-H+1} \\ \cdots & \cdots & \cdots & \cdots & \cdots \\ G_{2H-2} & G_{2H-1} & G_{2H-2} & \cdots & G_{2H-1} \\ G_{2H-1} & G_{2H-2} & G_{2H-3} & \cdots & G_{2H} \\ G_{2H} & G_{2H-1} & G_{2H-2} & \cdots & G_0 \end{bmatrix} \begin{bmatrix} V^{\star}_{-H} \\ V^{\star}_{-H+1} \\ V^{\star}_{-H+2} \\ \cdots \\ V^{\star}_{-1} \\ V_0 \\ V_1 \\ \cdots \\ V_{H-2} \\ V_{H-1} \\ V_H \end{bmatrix} \tag{4.12}$$

In full analogy to the derivation of $\mathbf{G}'$, the diode's time varying capacitor charge $q'(t)$, capacitance $c'(t)$, displacement current $i'_{\mathrm{C}}(t)$, voltage $v'(t)$ waveforms and their frequency domain representations are given in Eq. (4.13, 4.14, 4.15, 4.16).

$$\begin{aligned} q'(t) &= c'(t)v'(t) \\ q'(t) &= \sum_{k=-\infty}^{\infty} Q_k \exp\left(jk\omega_{\mathrm{LO}}t\right) \approx \sum_{k=-H}^{H} Q_k \exp\left(jk\omega_{\mathrm{LO}}t\right) \\ c'(t) &= \sum_{k=-\infty}^{\infty} C_k \exp\left(jk\omega_{\mathrm{LO}}t\right) \approx \sum_{k=-H}^{H} C_k \exp\left(jk\omega_{\mathrm{LO}}t\right) \end{aligned} \tag{4.13}$$

$$i'_{\mathrm{C}}(t) = \frac{\mathrm{d}}{\mathrm{d}t} q'(t) \tag{4.14}$$

$$q'(t) = \sum_{a=-\infty}^{\infty} Q_a \exp(j\omega_a t) = \sum_{b=-\infty}^{\infty} \sum_{c=-\infty}^{\infty} C_b V_c \exp(j\omega_{b+c} t) \quad (4.15)$$

$$i'_{\mathrm{C}}(t) = \sum_{a=-H}^{H} I_a \exp(j\omega_a t) = \sum_{b=-H}^{H} \sum_{c=-H}^{H} j\omega_{b+c} C_b V_c \exp(j\omega_{b+c} t)$$

$$\mathbf{I}'_{\mathrm{C}} = j\boldsymbol{\Omega}\mathbf{C}'\mathbf{V}' = \mathbf{I}_{\mathrm{C}} \quad (4.16)$$

Eq. (4.17) shows the small signal conversion matrix $j\boldsymbol{\Omega}\mathbf{C}'$ of the diode's junction capacitances.

$$\begin{bmatrix} I^\star_{-H} \\ I^\star_{-H+1} \\ I^\star_{-H+2} \\ \cdots \\ I^\star_{-1} \\ I_0 \\ I_1 \\ \cdots \\ I_{H-2} \\ I_{H-1} \\ I_H \end{bmatrix} = \begin{bmatrix} \omega_{-H}C_0 & \omega_{-H}C_{-1} & \dots & \omega_{-H}C_{-2H} \\ \omega_{-H+1}C_1 & \omega_{-H+1}C_0 & \dots & \omega_{-H+1}C_{-2H+1} \\ \omega_{-H+2}C_2 & \omega_{-H+2}C_1 & \dots & \omega_{-H+2}C_{-2H+2} \\ \cdots & \cdots & \cdots & \cdots \\ \omega_{-1}C_{H-1} & \omega_{-1}C_{H-2} & \dots & \omega_{-1}C_{-H-1} \\ \omega_0 C_H & \omega_0 C_{H-1} & \dots & \omega_0 C_{-H} \\ \omega_1 C_{H+1} & \omega_1 C_H & \dots & \omega_1 C_{-H+1} \\ \cdots & \cdots & \cdots & \cdots \\ \omega_{H-2}C_{2H-2} & \omega_{H-2}C_{2H-1} & \dots & \omega_{H-2}C_{2H-1} \\ \omega_{H-1}C_{2H-1} & \omega_{H-1}C_{2H-2} & \dots & \omega_{H-1}C_{2H} \\ \omega_H C_{2H} & \omega_H C_{2H-1} & \dots & \omega_H C_0 \end{bmatrix} \begin{bmatrix} V^\star_{-H} \\ V^\star_{-H+1} \\ V^\star_{-H+2} \\ \cdots \\ V^\star_{-1} \\ V_0 \\ V_1 \\ \cdots \\ V_{H-2} \\ V_{H-1} \\ V_H \end{bmatrix} \quad (4.17)$$

Conversion matrices of time invariant elements are simple diagonal matrices with the corresponding admittances or impedances at each small signal mixing frequency ω_k.

$$\mathbf{Z} = \mathrm{diag}\left[Z^\star(\omega_{-H}), \dots, Z(\omega_0), \dots, Z(\omega_H)\right] \quad (4.18)$$

Eq. (4.11), Eq. (4.16) and Eq. (4.18) together with conventional multiport circuit analysis allow for small signal description of frequency mixer circuits. The necessary steps for derivation of the conversion matrix are summarized in the following.

1. Build conversion matrices of all N diodes and construct $\mathbf{Z}_\mathrm{d}$ according to Eq. (4.19), which also includes the diode series resistances (gray box in Fig. 4.3).

$$\begin{aligned}
\mathbf{Y}_\mathrm{j} &= \mathbf{G}_\mathrm{j} + j\boldsymbol{\Omega}\mathbf{C}_\mathrm{j} \\
\mathbf{Z}_\mathrm{j} &= \mathbf{Y}_\mathrm{j}^{-1} = (\mathbf{G}_\mathrm{j} + j\boldsymbol{\Omega}\mathbf{C}_\mathrm{j})^{-1} \\
\mathbf{Z}_\mathrm{d} &= \mathbf{R}_\mathrm{j} + \mathbf{Z}_\mathrm{j} = \mathbf{R}_\mathrm{j} + (\mathbf{G}_\mathrm{j} + j\Omega\mathbf{C}_\mathrm{j})^{-1}
\end{aligned} \tag{4.19}$$

2. Combine all other linear time invariant circuit elements to a small signal impedance or admittance conversion matrix. The resulting diagonal matrices consist of the embedding network $\mathbf{Z}_\mathrm{e}$, $\mathbf{Y}_\mathrm{e}$.

3. Add embedding impedances, $\mathbf{Z}_\mathrm{d} + \mathbf{Z}_\mathrm{e}$, but leave out the embedding impedances of the interesting ports (e.g upper sideband RF, IF, lower sideband RF = image).

4. Build the admittance matrix $(\mathbf{Z}_\mathrm{d} + \mathbf{Z}_\mathrm{e})^{-1}$ and short circuit all ports except for the interesting ones.

5. Finally, a conversion matrix relating the interesting ports, while incorporating the influence of all embedding impedances, is derived. Fig. 4.3 illustrates a 3×3 conversion matrix, covering the mixer's conversion properties between upper sideband RF (RFUS) port, lower sideband RF (RFLS = image) port and IF port.

The outlined procedure is implemented in several commercial circuit simulators and ideally suited to design or analyze frequency mixers with rather low RF and IF power levels.

To proceed with general calculations of conversion loss, more simplifications are necessary. The components of the conversion matrix $\hat{\mathbf{Y}}$ from Fig. 4.3 are complex and depend on all embedding impedances. Embedding impedances at frequencies other than $\omega_k, k \in \{-1, 0, 1\}$ are neglected (short idlers) and all source impedances Z_RFUS, Z_IF, Z_RFLS are assumed to be real. If the diodes' junction capacitances are further neglected and so called weak reciprocity $\arg(\hat{Y}_{12}) = \arg(\hat{Y}_{21}) + \arg(\hat{Y}_{13})$ ([14], page 124) is valid, choosing an appropriate time zero, and adding

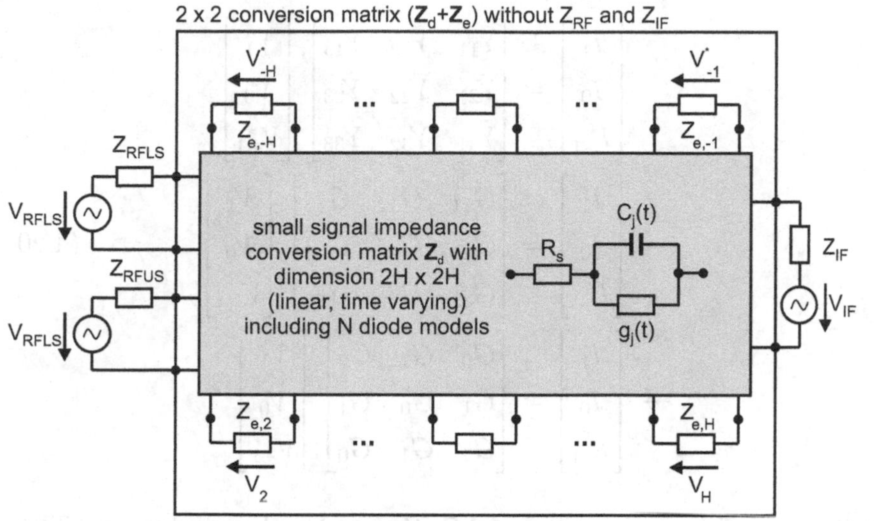

Figure 4.3: Derivation of a mixer's small signal conversion matrix, relating the upper and lower RF sidebands and IF port only but incorporating the influence of all embedding impedances.

a lossless transmission line and susceptance to one of the mixer's ports allows for making the conversion matrix components real and positive. In Eq. (4.20), positive frequency indices and therefore upper sidebands are written before the negative indices or lower sidebands. Although unfavourable for sum representations, this convention is maintained in the following. With these significant limitations, the real and positive components of the conversion matrix equal the double sided Fourier coefficients of the time varying diode conductance waveform (Eq. (4.9)). Taking into account $G^{\star}_{-k} = G_{-k} = G_k$, only three different matrix elements exist in Eq. (4.20) and the conversion matrix is a submatrix of $\mathbf{G}'$ from Eq. (4.12).

$$\begin{bmatrix} I_1 \\ I_0 \\ I_{-1}^{\star} \end{bmatrix} = \begin{bmatrix} \tilde{Y}_{11} & \tilde{Y}_{12} & \tilde{Y}_{13} \\ \tilde{Y}_{21} & \tilde{Y}_{22} & \tilde{Y}_{23} \\ \tilde{Y}_{31} & \tilde{Y}_{32} & \tilde{Y}_{33} \end{bmatrix} \begin{bmatrix} V_1 \\ V_0 \\ V_{-1}^{\star} \end{bmatrix}$$
$$\rightarrow \begin{bmatrix} I_1 \\ I_0 \\ I_{-1} \end{bmatrix} = \begin{bmatrix} G_{11} & G_{12} & G_{13} \\ G_{21} & G_{22} & G_{23} \\ G_{31} & G_{32} & G_{33} \end{bmatrix} \begin{bmatrix} V_1 \\ V_0 \\ V_{-1} \end{bmatrix} \tag{4.20}$$
$$\rightarrow \begin{bmatrix} I_1 \\ I_0 \\ I_{-1} \end{bmatrix} = \begin{bmatrix} G_0 & G_1 & G_2 \\ G_1 & G_0 & G_1 \\ G_2 & G_1 & G_0 \end{bmatrix} \begin{bmatrix} V_1 \\ V_0 \\ V_{-1} \end{bmatrix}$$

Minimum Conversion Loss and Optimum Load Impedances Then the usual matrix manipulations can be used to derive conversion loss and study the effects of different load impedances. At first a lower sideband (image) load admittance G_{RFLS} is connected to the three port conductance matrix of Eq. (4.20).

$$I_{-1} = -Y_{-1}V_{-1} = -G_{\mathrm{RFLS}}V_{-1} \tag{4.21}$$

This yields a two port conversion matrix, whereas its components are functions of G_{RFLS} (Eq. (4.23)).

$$\begin{bmatrix} I_1 \\ I_0 \end{bmatrix} = \begin{bmatrix} Y_{11} & Y_{12} \\ Y_{21} & Y_{22} \end{bmatrix} \begin{bmatrix} V_1 \\ V_0 \end{bmatrix} \tag{4.22}$$

$$
\begin{aligned}
Y_{11} &= G_{11} - \frac{G_{13}G_{13}}{G_{11}+G_{\mathrm{RFLS}}} = G_0 - \frac{G_2^2}{G_0+G_{\mathrm{RFLS}}} \\
Y_{12} &= G_{12} - \frac{G_{13}G_{12}}{G_{11}+G_{\mathrm{RFLS}}} = G_1 - \frac{G_2G_1}{G_0+G_{\mathrm{RFLS}}} \\
Y_{21} &= G_{21} - \frac{G_{13}G_{21}}{G_{11}+G_{\mathrm{RFLS}}} = G_1 - \frac{G_2G_1}{G_0+G_{\mathrm{RFLS}}} \\
Y_{22} &= G_{22} - \frac{G_{21}G_{12}}{G_{11}+G_{\mathrm{RFLS}}} = G_0 - \frac{G_1^2}{G_0+G_{\mathrm{RFLS}}}
\end{aligned}
\tag{4.23}
$$

In Eq. (4.24), the transducer power gain[4] $\mathcal{G}_{\mathrm{T}}$, which is the reciprocal of conversion loss L, is applied to the conversion matrix in Eq. (4.22).

$$
\mathcal{G}_{\mathrm{T}} = \frac{4Y_{21}^2 G_{\mathrm{RFUS}} G_{\mathrm{IF}}}{\left[(Y_{11}+G_{\mathrm{IF}})(Y_{22}+G_{\mathrm{RFUS}}) - Y_{12}Y_{21}\right]^2}
\tag{4.24}
$$

This allows for calculating the conversion loss with respect to different load impedance scenarios. Minimization of $L = \mathcal{G}_{\mathrm{T}}^{-1}$ is applied for three special cases.

- L_1 image short circuit $G_{\mathrm{RFLS}} \to \infty$
- L_2 broadband case $G_{\mathrm{RFUS}} = G_{\mathrm{RFLS}}$
- L_3 image open circuit $G_{\mathrm{RFLS}} = 0\ \Omega^{-1}$

[4]Transducer power gain $\mathcal{G}_{\mathrm{T}}$ (dt. Betriebsleistungsverstärkung) is defined as the power delivered to an arbitrary load Y_{L}, divided by power available from the source with source admittance Y_{S}.

$$
\mathcal{G}_{\mathrm{T}} = \frac{4\left|Y_{21}\right|^2 \mathrm{Re}\left\{Y_{\mathrm{L}}\right\} \mathrm{Re}\left\{Y_{\mathrm{S}}\right\}}{\left|(Y_{11}+Y_{\mathrm{S}})(Y_{22}+Y_{\mathrm{L}}) - Y_{12}Y_{21}\right|^2}
$$

Expressions in concise form are summarized in Eq. (4.25) [14, 15].

$$\begin{aligned}
L_1 &= \left(\frac{G_{12}}{G_{21}}\right)\frac{1+\sqrt{1-\epsilon_1}}{1-\sqrt{1-\epsilon_1}} \\
L_3 &= \left(\frac{G_{12}}{G_{21}}\right)\frac{1+\sqrt{1-\epsilon_3}}{1-\sqrt{1-\epsilon_3}} \\
L_2 &= 2\left(\frac{G_{12}}{G_{21}}\right)\frac{1+\sqrt{1-\kappa_2}}{1-\sqrt{1-\kappa_2}}
\end{aligned} \tag{4.25}$$

$$\begin{aligned}
\epsilon_1 &= \frac{G_{12}G_{21}}{G_{11}G_{22}} = \left(\frac{G_1}{G_0}\right)^2 \qquad \theta = \frac{G_{13}}{G_{11}} = \frac{G_2}{G_0} \\
\epsilon_3 &= \frac{\epsilon_1}{1-\epsilon_1}\frac{1-\theta}{1+\theta} \qquad \kappa_2 = \epsilon_1 + \epsilon_3 - \epsilon_1\epsilon_3
\end{aligned} \tag{4.26}$$

Beside minimization of Eq. (4.24), L_1, L_2, and L_3 can also be determined by application of the maximum available power gain[5] $\mathcal{G}_\mathrm{M}$ formulas to the conversion matrix in Eq. (4.22), e.g. $L_1 = \mathcal{G}_\mathrm{M}^{-1}$.

$$L_1 = \mathcal{G}_\mathrm{M}^{-1} = \frac{1+\sqrt{1-\left(\frac{G_1}{G_0}\right)^2}}{1-\sqrt{1-\left(\frac{G_1}{G_0}\right)^2}} \tag{4.28}$$

[5] Maximum available power gain (MAG) $\mathcal{G}_\mathrm{M}$ (dt. maximal verfügbare Leistungsverstärkung) is defined as power gain with input and output ports connected to conjugately matched optimum source admittance $Y_{\mathrm{S\,opt}}$ and optimum load admittance $Y_{\mathrm{L\,opt}}$.

$$\begin{aligned}
\mathcal{G}_\mathrm{M} &= \left|\frac{Y_{21}}{Y_{12}}\right|\left(K-\sqrt{K^2-1}\right) \\
K &= \frac{2\,\mathrm{Re}\{Y_{11}\}\,\mathrm{Re}\{Y_{22}\} - \mathrm{Re}\{Y_{12}Y_{21}\}}{|Y_{12}Y_{21}|} \\
Y_{\mathrm{S\,opt}} &= \frac{Y_{12}Y_{21} + |Y_{12}Y_{21}|\left(K+\sqrt{K^2-1}\right)}{2\,\mathrm{Re}\{Y_{22}\}} \\
Y_{\mathrm{L\,opt}} &= \frac{Y_{12}Y_{21} + |Y_{12}Y_{21}|\left(K+\sqrt{K^2-1}\right)}{2\,\mathrm{Re}\{Y_{11}\}}
\end{aligned} \tag{4.27}$$

$$L_2 = -2 \frac{\left(1+\sqrt{\frac{G_0^2+G_2G_0-2G_1^2}{G_0(G_0+G_2)}}\right)}{\left(-1+\sqrt{\frac{G_0^2+G_2G_0-2G_1^2}{G_0(G_0+G_2)}}\right)} \tag{4.29}$$

$$L_3 = \frac{\left(-1-\sqrt{\frac{(G_0^2+G_2G_0-2G_1^2)G_0}{(G_0^2-G_1^2)(G_0+G_2)}}\right)}{\left(-1+\sqrt{\frac{(G_0^2+G_2G_0-2G_1^2)G_0}{(G_0^2-G_1^2)(G_0+G_2)}}\right)} \tag{4.30}$$

The calculations, shown in the following, are based on the diode parameters of the UMS DBES105a, which are saturation current $I_S = 35$ fA, ideality factor $\eta = 1.2$ and series resistance $R_s = 4.4\ \Omega$. Analytical and numerical calculations of the Fourier components is explained in a separate paragraph in the end of this subsection (page 285 et seqq.).

Fig. 4.4 illustrates the minimum conversion loss values L_1, L_2, L_3 versus local oscillator peak voltage $\hat{v}_{LO}$. It is $L_3 < L_1 < L_2$. The current flow in case of a short circuit at the image frequency, implies additional loss compared to an open circuit image, if the diode's series resistance is nonzero. L_1 and L_3 both approach 0 dB conversion loss, if the pump voltage (LO) is sufficiently high. In the broadband case, power splits equally between the upper and lower RF sideband. Hence, 3 dB constitutes the conversion loss limit if $\hat{v}_{LO} \to \infty$ (compare section 2.1).

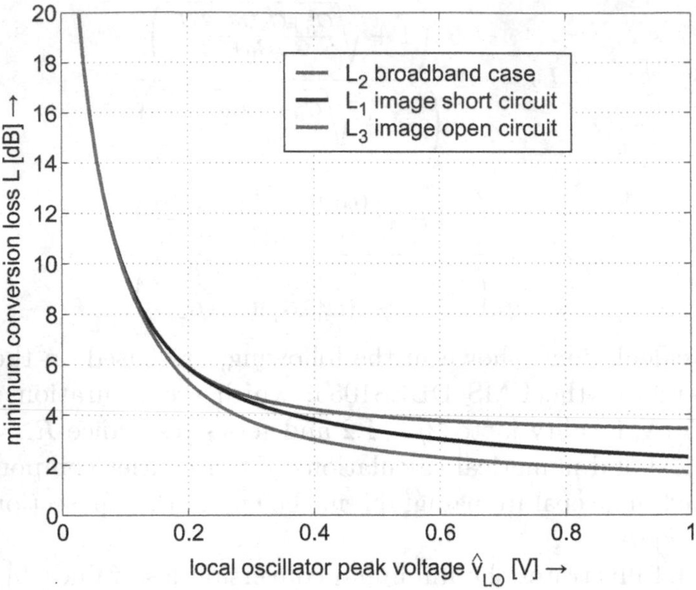

Figure 4.4: Minimum conversion loss values L versus local oscillator peak voltage $\hat{v}_{\mathrm{LO}}$ for the broadband case L_2, short circuited image L_1 and open circuited image L_3.

Calculation results with short termination $G_{\mathrm{RFLS}} \rightarrow \infty$ at the image port are shown in Fig. 4.5. The right side shows the optimum impedance values at every local oscillator peak voltage $\hat{v}_{\mathrm{LO}}$ (drive level), required to achieve the minimum conversion loss L_1. According to Eq. (4.27), these are $G_{\mathrm{IF\,opt}} = G_{\mathrm{RFUS\,opt}} = \sqrt{G_{11}^2 - G_{12}G_{21}} = \sqrt{G_0^2 - G_1^2}$. Positive DC biasing (blue) decreases the optimum impedance values, compared to zero bias operation (red). The same effect is achieved with lower barrier heights of all Schottky junctions (subsection 2.2.1), stacked diodes (section 2.4) or an integrated amplifier in the local oscillator path of the mixer. Conversion loss results, depicted on the left side of Fig. 4.5, correspond to a $G_{\mathrm{IF}}^{-1} = G_{\mathrm{RFUS}}^{-1} = 50\ \Omega$ environment. The actual conversion loss equals the minimum L_1, if the mixer's input and output impedances equal 50 Ω. This occurs at

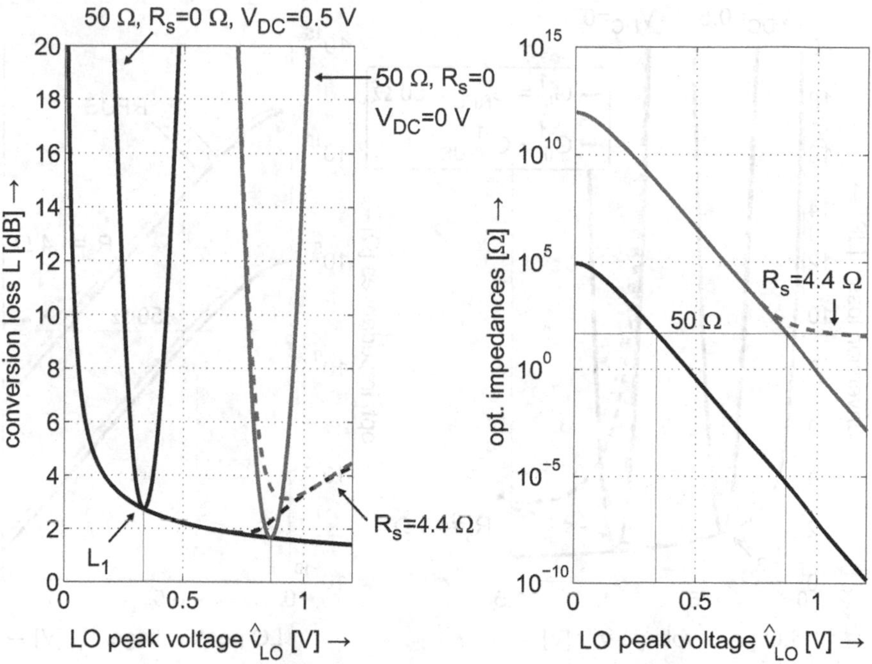

Figure 4.5: Mixer with short termination at image port. Several conversion loss values L versus local oscillator peak voltage $\hat{v}_{\mathrm{LO}}$ (left) and corresponding optimum impedance values versus $\hat{v}_{\mathrm{LO}}$ (right).

lower $\hat{v}_{\mathrm{LO}}$ values (drive levels) with applied DC bias (blue). According to these results, the mixer's conversion loss is very sensitive to $\hat{v}_{\mathrm{LO}}$, because the mixer impedances approach zero with $\hat{v} \to \infty$. This is in contrast to observation, nonzero series resistance $R_{\mathrm{s}} = 4.4\ \Omega$ (dashed) specifies a lower limit of the mixer impedances and prevents conversion loss from increase at high $\hat{v}_{\mathrm{LO}}$ values.

Fig. 4.6 includes calculation results with open termination $G_{\mathrm{RFLS}} = 0\ \Omega^{-1}$ at the image port. In analogy to the short circuited image port, conversion loss results with and without DC biasing $V_{\mathrm{DC}} = 0.5/0$ V and series resistance $R_{\mathrm{s}} = 4.4/0\ \Omega$ are shown on the left side of Fig. 4.6. In case of $G_{\mathrm{RFLS}} = 0\ \Omega^{-1}$, the optimum impedances (Eq. (4.27)) at

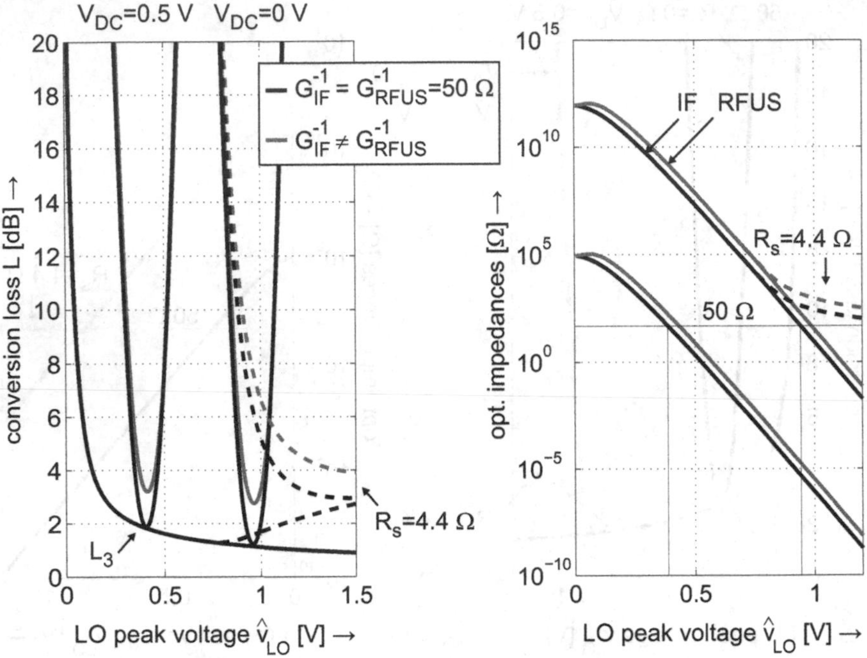

Figure 4.6: Mixer with open termination at image port. Several conversion loss values L versus local oscillator peak voltage $\hat{v}_{\mathrm{LO}}$ (left) and corresponding optimum impedance values versus $\hat{v}_{\mathrm{LO}}$ (right).

the IF and upper sideband RF ports have different values, as shown on the right side of Fig. 4.6. Conversion loss versus $\hat{v}_{\mathrm{LO}}$ is shown for $G_{\mathrm{IF}}^{-1} = G_{\mathrm{RFUS}}^{-1} = 50\ \Omega$ environment (red) and with terminations close to the optimum ($G_{\mathrm{IF}}^{-1} = 35\ \Omega$, $G_{\mathrm{RFUS}}^{-1} = 100\ \Omega$).

Conversion loss and optimum impedance values versus local oscillator peak voltage $\hat{v}_{\mathrm{LO}}$ in the broadband case are shown in Fig. 4.7. As it is the case with open termination of the lower RF sideband, the minimum conversion loss L_2 requires different terminations at IF and RFUS ports (right side of Fig. 4.7). Conversion loss results are shown for $G_{\mathrm{IF}}^{-1} = G_{\mathrm{RFUS}}^{-1} = 50\ \Omega$ environment together with the influence of positive DC biasing and series resistance.

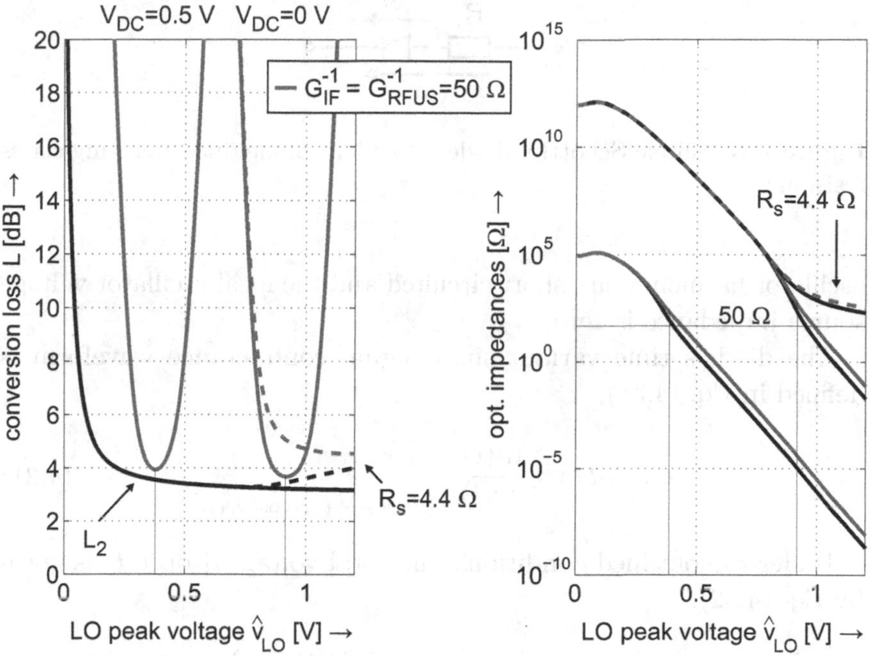

Figure 4.7: Mixer with same termination at upper and lower RF sideband $G_{\mathrm{RFUS}} = G_{\mathrm{RFLS}}$. Several conversion loss values L versus local oscillator peak voltage $\hat{v}_{\mathrm{LO}}$ (left) and corresponding optimum impedance values versus $\hat{v}_{\mathrm{LO}}$ (right).

Analytical and Numerical Calculation of Small Signal Conductance Waveform This subsection includes the derivation of the small signal conductance waveform and its frequency domain representation based on numerical integration and Bessel[6] functions.

Fig. 4.8 illustrates the voltage drop $v(t)$ across a diode's series resistance R_{s} and nonlinear junction resistance, described by the Shockley[7] equation $i(t) = I_{\mathrm{S}}(\exp(v_{\mathrm{j}}/(\eta V_{\mathrm{T}}) - 1)$. Junction capacitance $C_{\mathrm{j}}(v_{\mathrm{j}}) = C_{\mathrm{j0}}\,(1 - v_{\mathrm{j}}/\phi_{\mathrm{bi}})^{-M} = 0$ is neglected. All large signal local

[6] Friedrich Wilhelm Bessel (1784–1846), German mathematician.
[7] William Bradford Shockley Jr. (1910–1989), American physicist.

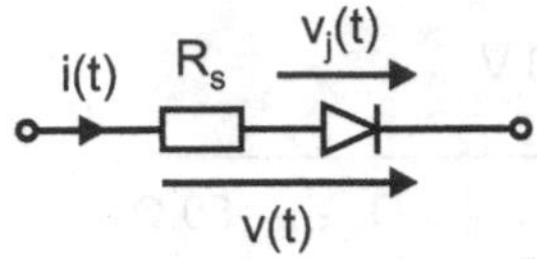

Figure 4.8: Simple Schottky diode equivalent model incorporating series resistance.

oscillator harmonics are short circuited and the local oscillator voltage source impedance is zero.

The diode's time variant small signal conductance waveform is defined in Eq. (4.31).

$$g(\omega_{\mathrm{LO}}t) := \left.\frac{\mathrm{d}i(t)}{\mathrm{d}v(t)}\right|_{v(t)=V_{\mathrm{DC}}+\hat{v}_{\mathrm{LO}}\cos(\omega_{\mathrm{LO}}t)} \tag{4.31}$$

Under the outlined conditions, the total voltage drop $v(t)$ is given by Eq. (4.32).

$$v(t) = i(t)R_{\mathrm{s}} + \eta V_{\mathrm{T}} \ln\left(\frac{i(t)}{I_{\mathrm{S}}} + 1\right) \tag{4.32}$$

Building the first current derivative of Eq. (4.32), using the chain rule and inserting the Shockley equation leads to Eq. (4.33).

$$\frac{\mathrm{d}v(t)}{\mathrm{d}i(t)} = R_{\mathrm{s}} + \eta V_{\mathrm{T}}\left(\frac{1}{i/I_{\mathrm{S}}+1}\right)\left(\frac{1}{I_{\mathrm{S}}}\right) = R_{\mathrm{s}} + \frac{\eta V_{\mathrm{T}}}{I_{\mathrm{S}}\exp\left(\frac{v}{\eta V_{\mathrm{T}}}\right) + I_{\mathrm{S}} - I_{\mathrm{S}}} \tag{4.33}$$

Hence, the conductance waveform $g(\omega_{\mathrm{LO}}t) \in \mathbb{R}$ is known (Eq. (4.34)) as a function of the DC and local oscillator stimulus.

$$g(\omega_{\mathrm{LO}}t) = \left.\frac{\mathrm{d}i(t)}{\mathrm{d}v(t)}\right|_{v(t)=V_{\mathrm{DC}}+\hat{v}_{\mathrm{LO}}\cos(\omega_{\mathrm{LO}}t)} = \left(R_{\mathrm{s}} + \frac{\eta V_{\mathrm{T}}}{I_{\mathrm{S}}\exp\left(\frac{v}{\eta V_{\mathrm{T}}}\right)}\right)^{-1} \tag{4.34}$$

After numerical integration of Eq. (4.35), the single sided Fourier components $\tilde{G}_0, \tilde{G}_\ell$ of the small signal conductance waveform are known. $\tilde{G}_0, \tilde{G}_\ell$ are related to the double sided Fourier components $G_k, k \in \{-H, \ldots, 0, \ldots, H\}$ by factor 1/2.

$$g(\omega_{\mathrm{LO}}t) = \frac{\tilde{G}_0}{2} + \sum_{\ell=1}^{\infty} \tilde{G}_\ell \cos(\ell\omega_{\mathrm{LO}}t) \approx \frac{\tilde{G}_0}{2} + \sum_{\ell=1}^{L} \tilde{G}_\ell \cos(\ell\omega_{\mathrm{LO}}t)$$

$$\tilde{G}_\ell = \frac{1}{\pi} \int_{-\pi}^{+\pi} g(\omega_{\mathrm{LO}}t) \cos(\ell\omega_{\mathrm{LO}}t) \mathrm{d}(\omega_{\mathrm{LO}}t) \quad \ell \in \{0, 1, \ldots, L, \ldots, \infty\} \tag{4.35}$$

If the diode series resistance is set to zero $R_\mathrm{s} = 0$, the Fourier integrals of Eq. (4.35) can be expressed by the modified Bessel functions I_1 of the first kind with order ℓ.

$$\tilde{G}_\ell = \frac{2I_\mathrm{S}}{\eta V_\mathrm{T}} \exp\left(\frac{V_\mathrm{DC}}{\eta V_\mathrm{T}}\right) \underbrace{\frac{1}{2\pi} \int_{-\pi}^{+\pi} \exp\left(\frac{\hat{v}_\mathrm{LO} \cos(\omega_{\mathrm{LO}}t)}{\eta V_\mathrm{T}}\right) \cos(\ell\omega_{\mathrm{LO}}t) \mathrm{d}(\omega_{\mathrm{LO}}t)}_{\text{Bessel}}$$

$$\tilde{G}_\ell = \frac{2I_\mathrm{S}}{\eta V_\mathrm{T}} \exp\left(\frac{V_\mathrm{DC}}{\eta V_\mathrm{T}}\right) I_1\left(\ell, \frac{\hat{v}_\mathrm{LO}}{\eta V_\mathrm{T}}\right) \quad \ell \in \mathbb{N}_0 \tag{4.36}$$

The conductance waveform is then given by Eq. (4.37).

$$g(\omega_{\mathrm{LO}}t) = \frac{I_\mathrm{S}}{\eta V_\mathrm{T}} \exp\left(\frac{V_\mathrm{DC}}{\eta V_\mathrm{T}}\right) I_1\left(0, \frac{\hat{v}_\mathrm{LO}}{\eta V_\mathrm{T}}\right) + \frac{2I_\mathrm{S}}{\eta V_\mathrm{T}} \exp\left(\frac{V_\mathrm{DC}}{\eta V_\mathrm{T}}\right) \left(\sum_{\ell=1}^{\infty} I_1\left(\ell, \frac{\hat{v}_\mathrm{LO}}{\eta V_\mathrm{T}}\right) \cos(\ell\omega_{\mathrm{LO}}t)\right) \tag{4.37}$$

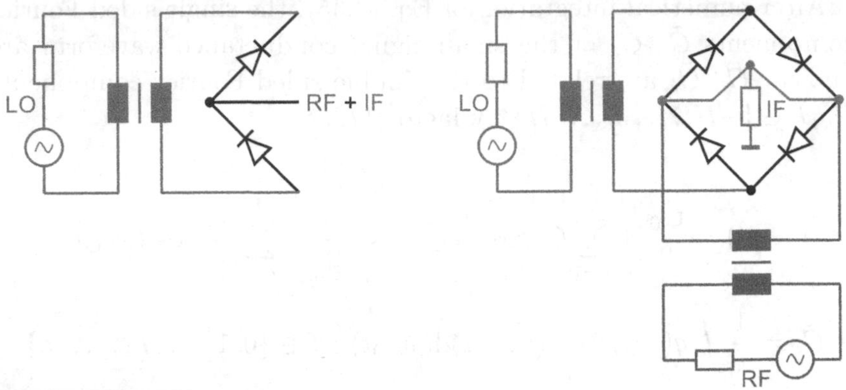

Figure 4.9: Equivalent circuits of a single balanced mixer (SBM) with balanced LO port (left) and double balanced mixer (DBM) architecture (right).

4.3 Balanced Mixer Configurations

The analysis procedure of the preceding section can handle multi diode mixers but the presented calculated results are valid for single diode mixers (SDM). Practical diode mixers make use of several diodes in balanced configurations to benefit from inherent filter properties. Fig. 4.9 shows equivalent circuits of a single balanced mixer (SBM) and a double balanced mixer (DBM) architecture. The baluns of Fig. 4.9 and Fig. 4.10 are assumed to have short circuited center tap connections in analogy to Fig. 4.11. Many planar microwave baluns show open circuit behaviour at even mode excitation, which needs to be accounted for by the circuit designer.

The single ended RF and IF signals of the SBM in Fig. 4.9 require an additional filter or diplexer circuit to be distinguished. The balanced LO signal drives the identical diodes in series connection, whereas the RF / IF circuit node constitutes a virtual short circuit providing inherent isolation between LO and RF / IF. The diodes are arranged in an antiparallel configuration with respect to the RF / IF port.

A SBM mixer with balanced LO signal is depicted (SBM LO bal), but balanced RF or IF signals are also possible (SBM RF bal). The balanced LO signal and its harmonics do not appear at RF / IF ports. The circuit is identical to a balanced frequency doubler with input signal (section 2.4) on the right (RF / IF). Hence, even order RF or IF harmonics appear at the LO port but are missing at the RF / IF port. The SBM mixer allows for overlapping LO and RF / IF frequency ranges. Due to the required filter / diplexer circuit, SBMs are designed for specific applications [16] rather than for multi purpose usage. The simple architecture, realized with discrete series tee diodes (e.g. Fig. 5.14 from chapter 2), operates up to very high frequencies and conversion loss records have been achieved.

In case of the DBM, four diodes are arranged in a quad ring. LO and RF are fed balanced with a single ended IF port. For the balanced signals, the diodes are in antiparallel configuration with two series connected diodes in each branch (stacked configuration). Hence, the virtual short circuit principle from the SBM applies mutually to the LO and RF ports. The DBM allows for overlapping LO, RF and IF frequency ranges and constitutes the industry's workhorse mixer for many applications, optimized by engineers over decades. Decoupling the single ended IF signal to the IF load is difficult. For this reason DBMs suffer from rather low IF bandwidth, which is unfavourable for front end modules with large signal bandwidth (Fig. 1.4).

This is solved by triple balanced mixers (TBM),[8] shown in Fig. 4.10, which consist of two DBMs in push pull configuration[9]. It features balanced LO, RF and IF ports. In theory, there is no IF bandwidth limitation. As the LO and RF ports are DC short circuited, there is a minimum operating LO and RF frequency. In Fig. 4.10, the LO and RF signals are balanced first and then split to the diode quads. An arrangement vice versa is also possible and has been used with monolithic TBM designs [18]. A balanced IF load is shown in Fig. 4.10.

[8]The term double DBM, abbreviated with DDB is also used in the literature.

[9]The term push pull is originated from engineers working with power amplifiers [17]. In the actual case it is understood as an antiparallel connection of nonlinear circuits that build a balanced output port (IF port of Fig. 4.10).

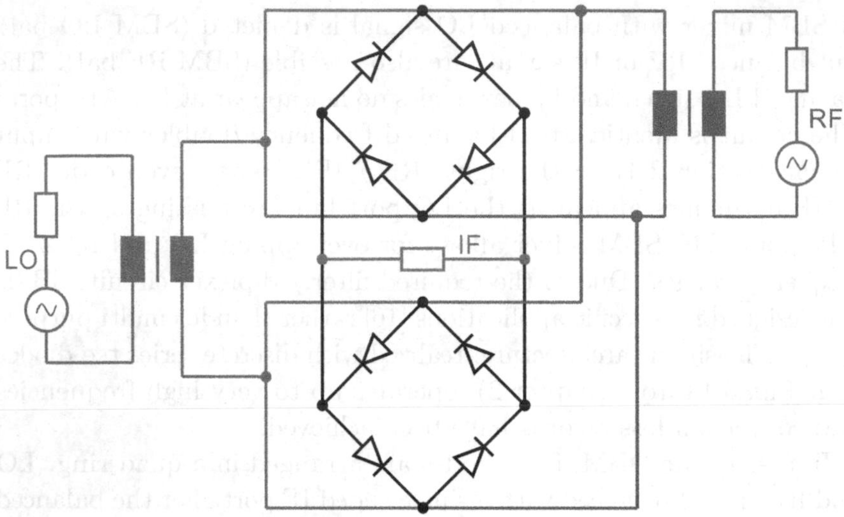

Figure 4.10: Equivalent circuit of a triple balanced mixer (TBM) architecture.

Practical TBMs are equipped with an IF balun to provide a single ended IF signal at the mixer's external port.

The TBM is the most promising candidate for front end modules of semiconductor automatic test systems, which require large IF bandwidth together with high RF drive levels. Only a few practical circuits have been reported. The author's experimental contributions to this new topic are presented in section 4.4.

LSSS Analysis of SBM, DBM, TBM In the sense of large signal / small signal (LSSS) analysis from subsection 4.2.1, which underlies severe limitations,[10] the introduced balanced mixer configurations show surprisingly little difference in their small signal behaviour.

[10]LO, RF, and IF ports are assumed to be connected to ideal bandpass filters if the mixer architecture does not provide sufficient inherent filter properties. The mixer circuit does not contain any interconnection elements, which would have to be considered by their scattering / impedance / admittance parameters. It consists of identical diodes, ideal baluns, ideal filters, ideal sources only.

Assuming short termination at the image frequency, Fig. 4.12 shows calculated conversion loss values and optimum impedances versus local oscillator peak voltage. It compares SDM, SBM, DBM and TBM at zero bias, with and without taking into account the series resistance R_s. From a small signal point of view, decreasing mixer impedance is the only difference. The minimum conversion loss L_1 is the same for all mixer types, but the required optimum impedance is lower for TBM than it is for DBM, SBM and SDM. $\hat{v}_{LO}$ in Fig. 4.12 covers the drive of a single diode only. The necessary local oscillator power level scales with the number of diodes, and is therefore higher for TBM than it is for SDM.

If the same Schottky diodes are utilized and all transformers / baluns operate with turns ratio $T = 2$ (Fig. 4.11), the relations of Eq. (4.38, 4.39) hold true. Single balanced mixers with balanced RF and LO ports are considered.

$$\begin{aligned} &\mathcal{G}_{\mathrm{M\,SDM}} = \mathcal{G}_{\mathrm{M\,SBM}} = \mathcal{G}_{\mathrm{M\,DBM}} = \mathcal{G}_{\mathrm{M\,TBM}} \\ &G^{-1}_{\mathrm{opt\,SDM}} \xrightarrow{\cdot 2} G^{-1}_{\mathrm{opt\,SBM}} \xrightarrow{\cdot 2} G^{-1}_{\mathrm{opt\,DBM}} \xrightarrow{\cdot 2} G^{-1}_{\mathrm{opt\,TBM}} \\ &P_{\mathrm{LO\,SDM}} \xrightarrow{\cdot 2} P_{\mathrm{LO\,SBM}} \xrightarrow{\cdot 2} P_{\mathrm{LO\,DBM}} \xrightarrow{\cdot 2} P_{\mathrm{LO\,TBM}} \end{aligned} \tag{4.38}$$

The increase by factor two in the last row of Eq. (4.38) guarantees the same drive power for each individual diode of the different mixers. As mentioned above, the required optimum impedance for minimum conversion loss is lower for TBM than it is for DBM, SBM, SDM and hence LO power needs to be increased by less than factor two.

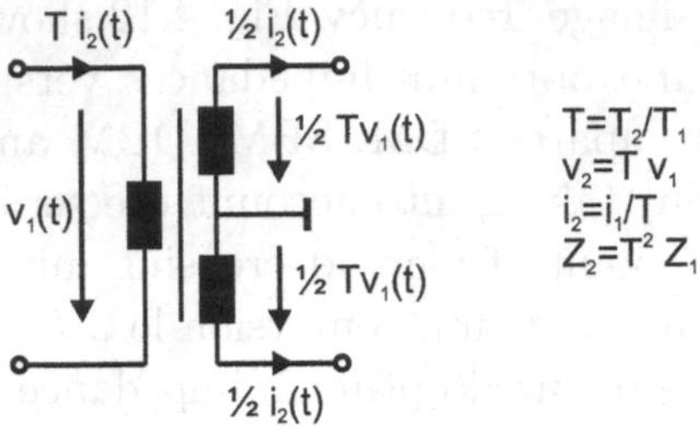

Figure 4.11: Transformer with short circuited center tap and transformer equations.

$$\begin{bmatrix} I_1 \\ I_0 \end{bmatrix} = \begin{bmatrix} I_{\mathrm{RFUS}} \\ I_{\mathrm{IF}} \end{bmatrix} = \underbrace{\begin{bmatrix} G_0 & G_1 \\ G_1 & G_0 \end{bmatrix}}_{\mathbf{G}_{\mathrm{SDM}}} \begin{bmatrix} V_1 \\ V_0 \end{bmatrix}$$

$$\begin{aligned} \mathbf{G}_{\mathrm{SBM\,LObal}} &= 2 \cdot \mathbf{G}_{\mathrm{SDM}} = 2 \cdot \begin{bmatrix} G_0 & G_1 \\ G_1 & G_0 \end{bmatrix} \\ \mathbf{G}_{\mathrm{SBM\,RFbal}} &= \begin{bmatrix} \frac{T^2}{2} G_0 & T G_1 \\ T G_1 & 2 G_0 \end{bmatrix} = \mathbf{G}_{\mathrm{SBM\,LObal}} \Big|_{T=2} \\ \mathbf{G}_{\mathrm{DBM}} &= \begin{bmatrix} T^2 G_0 & 2 T G_1 \\ 2 T G_1 & 4 G_0 \end{bmatrix} \\ \mathbf{G}_{\mathrm{TBM}} &= 2 \cdot \mathbf{G}_{\mathrm{DBM}} \end{aligned} \qquad (4.39)$$

Qualitative Spurious Tone Behaviour of SBM, DBM, TBM

Accurate prediction of a mixer's spurious tone behaviour is possible by multi tone harmonic balance simulation. Henderson reported a simple analytical method [1], which assumes identical diodes and considers the associated cancellation (destructive interference) of spurious tones at each mixer port. Following [1] does not allow for determination

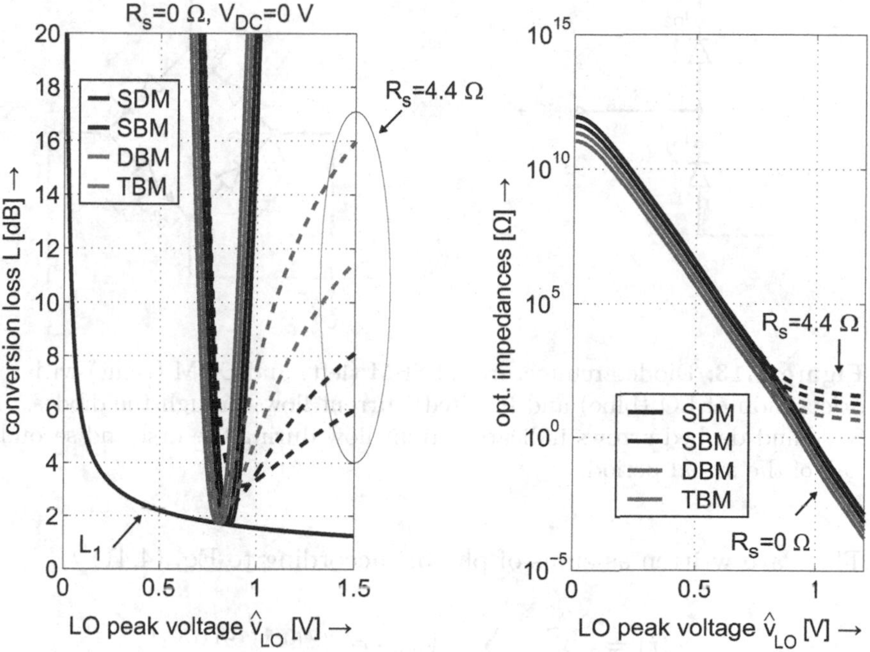

Figure 4.12: Mixer with short termination at image port. Comparison of single diode mixer (SDM), single balanced mixer (SBM), double balanced mixer (DBM) and triple balanced mixer (TBM). Conversion loss values L versus local oscillator peak voltage $\hat{v}_{\mathrm{LO}}$ (left) and corresponding optimum impedance values versus $\hat{v}_{\mathrm{LO}}$ (right).

of spurious tone magnitudes and their ratios. It does only consider the topology of the mixer and its capability to avoid certain spurs at some mixer ports.

Henderson assumes voltages v and currents i for every circuit branch with spectral components at every mixing frequency ω_{nm} (Eq. (4.40)).

$$\omega_{nm} = 2\pi f_{nm} = n\,\omega_{\mathrm{RF/IF}} + m\,\omega_{\mathrm{LO}} \qquad n, m \in \pm\mathbb{N}_0 \tag{4.40}$$

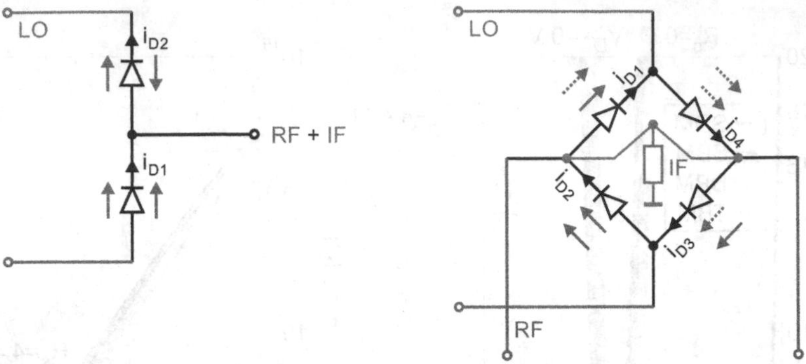

Figure 4.13: Diode architectures of SBM (left) and DBM (right) with orientation of LO (blue) and RF (red) current flow through the diodes. Solid and dashed arrows indicate current flow during the first and second half of the signal period.

These are written as sums of phasors according to Eq. (4.41)[11].

$$
\begin{aligned}
v(t) &= \sum_{n=-\infty}^{\infty} \sum_{m=-\infty}^{\infty} V_{nm} \cdot e^{j(n\,\omega_{\mathrm{RF}} + m\,\omega_{\mathrm{LO}})t} \\
i(t) &= \sum_{n=-\infty}^{\infty} \sum_{m=-\infty}^{\infty} I_{nm} \cdot e^{j(n\,\omega_{\mathrm{RF}} + m\,\omega_{\mathrm{LO}})t}
\end{aligned}
\tag{4.41}
$$

The currents through the two diodes of the single balanced mixer on the left side of Fig. 4.13 are then given by Eq. (4.42). The term $e^{jn\pi}$ accounts for the phase shift between LO and RF signals at diode D2.

[11] In [1] the current phasors are denoted as double sided Fourier series, which is incorrect as the spectral components are not only harmonics (compare explanations in subsection 4.2.1). Furthermore, the diode currents $i_{\mathrm{D}}(t)$ in [1] are written as multiplication of a conductance and voltage Fourier series, similar to the LSSS analysis approach. This is omitted in the presented equations and substituted by $I_{\mathrm{D}\,nm}$, which cannot be determined by this type of analysis.

$$i_{\mathrm{D}\,1}(t) = \sum_{n=-\infty}^{\infty} \sum_{m=-\infty}^{\infty} \underbrace{I_{nm} \cdot e^{j(n\,\omega_{\mathrm{RF}}+m\,\omega_{\mathrm{LO}})t}}_{I_{\mathrm{D}\,nm}} = \sum_{n=-\infty}^{\infty} \sum_{m=-\infty}^{\infty} I_{\mathrm{D}\,nm}$$

$$i_{\mathrm{D}\,2}(t) = \sum_{n=-\infty}^{\infty} \sum_{m=-\infty}^{\infty} I_{\mathrm{D}\,nm} \cdot e^{jn\pi} = \sum_{n=-\infty}^{\infty} \sum_{m=-\infty}^{\infty} I_{\mathrm{D}\,nm} \cdot (-1)^n \tag{4.42}$$

The IF signal current is the sum of $i_{\mathrm{D}\,1}(t)$ and $-i_{\mathrm{D}\,2}(t)$.

$$i_{\mathrm{IF}}(t) = i_{\mathrm{D}\,1}(t) - i_{\mathrm{D}\,2}(t) = \sum_{n=-\infty}^{\infty} \sum_{m=-\infty}^{\infty} I_{\mathrm{D}\,nm} \cdot [1 - (-1)^m] \tag{4.43}$$

As $i_{\mathrm{IF}}(t) = 0$ for $n = 2k,\ k \in \mathbb{N}_0$, even order spurs are suppressed at the IF port of single balanced mixers.

The diode currents in case of the double balanced mixer of the right side of Fig. 4.13 are analogously given by Eq. (4.44,4.45).

$$\begin{aligned} i_{\mathrm{D}\,1}(t) &= \sum_{n=-\infty}^{\infty} \sum_{m=-\infty}^{\infty} I_{\mathrm{D}\,nm} \cdot e^{jn\pi} \\ &= \sum_{n=-\infty}^{\infty} \sum_{m=-\infty}^{\infty} (-1)^n \cdot I_{\mathrm{D}\,nm} \\ i_{\mathrm{D}\,2}(t) &= \sum_{n=-\infty}^{\infty} \sum_{m=-\infty}^{\infty} I_{nm} \cdot e^{j(n\,\omega_{\mathrm{RF}}+m\,\omega_{\mathrm{LO}})t} \\ &= \sum_{n=-\infty}^{\infty} \sum_{m=-\infty}^{\infty} I_{\mathrm{D}\,nm} \end{aligned} \tag{4.44}$$

$$\begin{aligned}
i_{\mathrm{D}\,3}(t) &= \sum_{n=-\infty}^{\infty}\sum_{m=-\infty}^{\infty} I_{\mathrm{D}\,nm}\cdot e^{jm\pi} \\
&= \sum_{n=-\infty}^{\infty}\sum_{m=-\infty}^{\infty} (-1)^{m}\cdot I_{\mathrm{D}\,nm} \\
i_{\mathrm{D}\,4}(t) &= \sum_{n=-\infty}^{\infty}\sum_{m=-\infty}^{\infty} I_{\mathrm{D}\,nm}\cdot e^{jn\pi}\cdot e^{jm\pi} \\
&= \sum_{n=-\infty}^{\infty}\sum_{m=-\infty}^{\infty} (-1)^{n+m}\cdot I_{\mathrm{D}\,nm}
\end{aligned} \tag{4.45}$$

Applying Kirchhoff's law, the DBM's LO, RF and IF currents are known with respect to the individual diode currents (Eq. (4.46)).

$$\begin{aligned}
i_{\mathrm{DBM\,LO}}(t) &= i_{\mathrm{D}\,4} - i_{\mathrm{D}\,1} \\
&= \sum_{n=-\infty}^{\infty}\sum_{m=-\infty}^{\infty} \left[(-1)^{n+m} - (-1)^{n}\right]\cdot I_{\mathrm{D}\,nm} \\
&= \sum_{n=-\infty}^{\infty}\sum_{m=-\infty}^{\infty} (-1)^{n}\left[(-1)^{m} - 1\right]\cdot I_{\mathrm{D}\,nm} \\
i_{\mathrm{DBM\,RF}}(t) &= i_{\mathrm{D}\,1} - i_{\mathrm{D}\,2} \\
&= \sum_{n=-\infty}^{\infty}\sum_{m=-\infty}^{\infty} \left[(-1)^{n} - 1\right]\cdot I_{\mathrm{D}\,nm} \\
i_{\mathrm{DBM\,IF}}(t) &= -i_{\mathrm{D}\,1} + i_{\mathrm{D}\,2} - i_{\mathrm{D}\,3} + i_{\mathrm{D}\,4} \\
&= \sum_{n=-\infty}^{\infty}\sum_{m=-\infty}^{\infty} \left[-(-1)^{n} + 1 - (-1)^{m} + (-1)^{n+m}\right]\cdot I_{\mathrm{D}\,nm} \\
&= \sum_{n=-\infty}^{\infty}\sum_{m=-\infty}^{\infty} \left[(-1)^{n} - 1\right]\cdot\left[(-1)^{m} - 1\right]\cdot I_{\mathrm{D}\,nm}
\end{aligned} \tag{4.46}$$

Interpreting Eq. (4.46) it turns out, only $\langle\mathrm{RF},\mathrm{LO}\rangle=\langle n,m\rangle=\langle\mathrm{even},\mathrm{odd}\rangle$, $\langle\mathrm{RF},\mathrm{LO}\rangle=\langle n,m\rangle=\langle\mathrm{odd},\mathrm{even}\rangle$, $\langle\mathrm{RF},\mathrm{LO}\rangle=\langle n,m\rangle=\langle\mathrm{odd},\mathrm{odd}\rangle$ spurs appear at the LO, RF, IF port, respectively. This holds true also for the triple balanced mixer. Table 4.1 summarizes the qualitative spurious tone suppressions of SBM, DBM and TBM. Suppression of spurs due to inherent filtering is one advantage of utilizing multiple diode mixers. Another advantage is large signal handling capability. The local oscillator power level P_{LO} at each individual diode of a SDM, SBM, DBM or TBM is roughly the same (compare explanations of the last paragraph) but the applied RF / IF signal power level $P_{\mathrm{RF/IF}}$ is split between the diodes. Hence, the $P_{\mathrm{RF/IF}}/P_{\mathrm{LO}}$ ratio is smaller for TBM than it is for DBM, SBM, SDM, which shifts compression points and intercept points to higher values. A small $P_{\mathrm{RF/IF}}/P_{\mathrm{LO}}$ ratio ensures, the average diode quiescent point is determined by the local oscillator power level rather than the RF / IF power level. Mixers like TBM, which ensure linear operation up to large RF / IF power levels are ideally suited for signal generator / TX modules. Such systems (e.g. Fig. 4.34) are designed to transmit the maximum possible output power level to the output port. Lower output power levels are achieved by a cascaded step attenuator (Fig. 1.1). Hence, a high mixer input power level eliminates the need for additional signal conditioning (attenuators) before each mixing stage and thereby also benefits the system's overall noise figure performance.

Table 4.1: Comparison of Spurious Tone Behaviour of Different Mixer Architectures

	Single Balanced Mixer (SBM LO bal)
LO	no odd RF / IF harmonics no spurs $\langle n, m\rangle$ with odd RF / IF harmonics $n =$odd
RF+IF	no LO harmonics no spurs $\langle n, m\rangle$ with even RF/IF harmonics $n =$even
	Double Balanced Mixer (DBM) and Triple Balanced Mixer (TBM)
LO	only spurs $\langle n, m\rangle$ with $n =$even and $m =$odd
RF	only spurs $\langle n, m\rangle$ with $n =$odd and $m =$even
IF	only spurs $\langle n, m\rangle$ with $n =$odd and $m =$odd

4.4 Planar Crossed Quad Triple Balanced Mixer

Triple balanced mixers (TBM) offer AC coupled RF and LO ports and an IF port without an inherent IF bandwidth limitation, provide operation in overlapping RF, LO and IF frequency ranges, while achieving isolated mixer ports, a high 1 dB compression point and improved spurious tone suppression. Hence, these mixers benefit the system design of heterodyne transceivers with great signal bandwidths and high RF drive levels.

The design and construction of a hybrid broadband TBM based on commercial silicon crossed quad diodes and thin-film processed SiO_2 substrate are presented in this section. Broadband planar baluns and the use of balanced edge and broadside coupled striplines allow for covering an RF / LO frequency range of 1 to 45 GHz and at least 20 GHz IF bandwidth, if we claim for up- and downconversion loss values below 15 dB. This is achieved with local oscillator power levels $P_{\mathrm{LO}} \in [14, 18]$ dBm and an input 1 dB compression point of

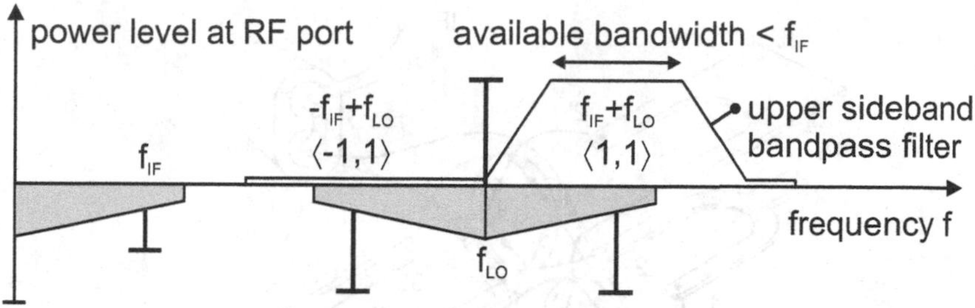

Figure 4.14: Signal power levels at the RF port versus frequency for an upconversion mixing case ⟨IF, LO⟩=⟨1, 1⟩.

12 dBm. Spectral measurement data which are in reasonable agreement with simulation results are included. The latter are based on a co-simulation procedure between 3D electromagnetic field (EM) simulations and nonlinear multi tone harmonic balance circuit simulations. The presented TBM is an improved version of the author's earlier work [I4].

Introduction and Field of Application Balanced diode mixers are used for frequency conversion in microwave spectrum analyzers, broadband up- and downconverter modules of automatic semiconductor testsystems and other heterodyne transceiver systems. The well understood double balanced mixers (DBM) with four diodes and two baluns (Fig. 4.9) can provide low conversion loss over ultrawide RF and LO bandwidths [19, 20], but are restricted to small intermediate frequency ranges from DC to some GHz.

Within the aforementioned field of applications a preferably high intermediate frequency is necessary to allow for great signal bandwidths. Fig. 4.14 illustrates an upconversion mixing case ⟨IF, LO⟩=⟨1, 1⟩ with upper sideband bandpass filter. The upconverter's signal bandwidth cannot exceed the intermediate frequency f_{IF}. The triple balanced mixer configuration with eight diodes and three baluns (Fig. 4.10) is an appropriate candidate to overcome this problem. TBMs offer a DC coupled IF port without an inherent IF bandwidth limitation. It

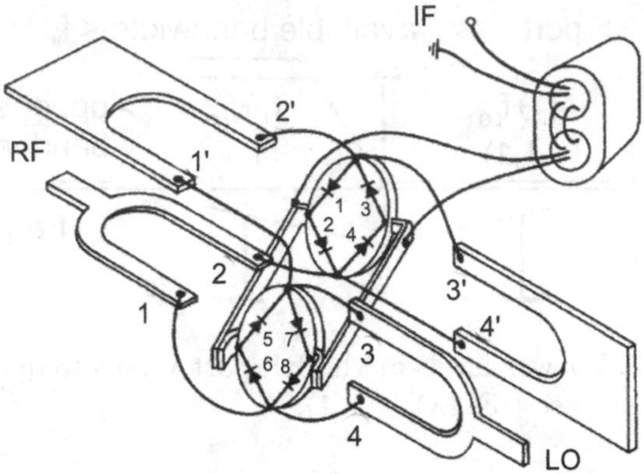

Figure 4.15: TBM realization with perpendicular orientation of RF and LO ports and two diode quad rings ([21] page 788). Courtesy of John Wiley & Sons, Inc.

is the only mixer that provides operation in overlapping RF, LO and IF frequency ranges, while achieving isolated mixer ports, a high 1 dB compression point and improved spurious tone suppression. The need for eight diodes and the corresponding higher local oscillator power level constitutes a drawback compared to DBMs.

Fig. 4.15 from [21][12] page 788, depicts a quasi-planar realization of the TBM schematic from Fig. 4.10. Only few monolithic and hybrid TBMs have been reported [18, 22–27]. To the authors knowledge the commercial products Miteq Inc. TB0250LW1 [28] and several Marki Microwave Inc. TBMs [29–32] are the only TBMs that can handle ultrawideband operation from 1 and 2 GHz up to 40 and 50 GHz. Whereas the latter are believed to be based on the realization of Fig. 4.15 with two separate organic laminate substrates in perpendicular orientation. Fig. 4.16 depicts a similar approach for low frequency TBMs from API Technologies Corp., formerly Spectrum Microwave. This realization type is unfavourable, as it cannot be used with ceramic or quartz substrates. The fragile setup has drawbacks

[12]The first edition of [21] has been published in 1990.

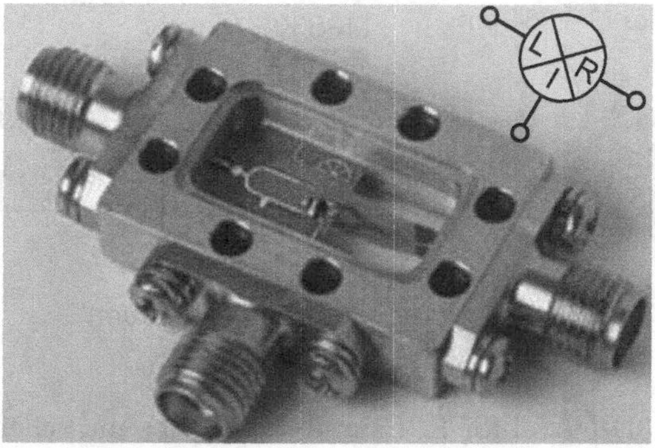

Figure 4.16: Commercial low frequency TBM, realized with two separate organic laminates for the LO / IF and RF ports, perpendicular to each other. Courtesy of API Technologies Corp., formerly Spectrum Microwave.

with respect to robustness against mechanical stress. It is further believed, that device performance will differ from assembly to assembly due to problems with reproducibility.

To overcome the aforementioned problems, the focus is on a planar TBM design using thin-film processed SiO_2 substrate with 10 mil height. Edge (eSTL) and broadside coupled striplines (bSTL) serve as balanced waveguides. In this configuration the silicon crossed quad diodes MSS30 PCR46 from Aeroflex Metelics help to simplify the planar architecture compared to the more common diode quad rings. A 3D model for EM simulation of the diodes is shown in Fig. 4.17 and photographs are depicted in Fig. 4.18. Fig. 4.19 illustrates the corresponding TBM schematic.

The design of the proposed TBM is based on a co-simulation procedure between 3D finite element (FEM) electromagnetic field (EM) simulations in the frequency domain and nonlinear harmonic balance circuit simulations. The next two paragraphs include detailed information about the planar mixer design and the applied co-simulation procedure.

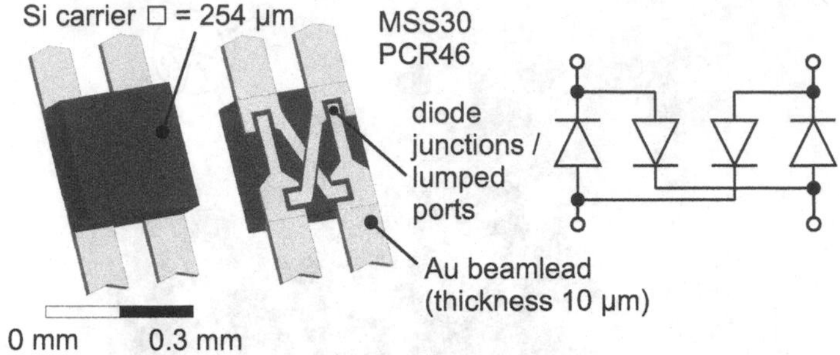

Figure 4.17: 3D model for EM simulation (left) and schematic view (right) of the utilized silicon crossed quad diodes MSS30 PCR46 from Aeroflex Metelics.

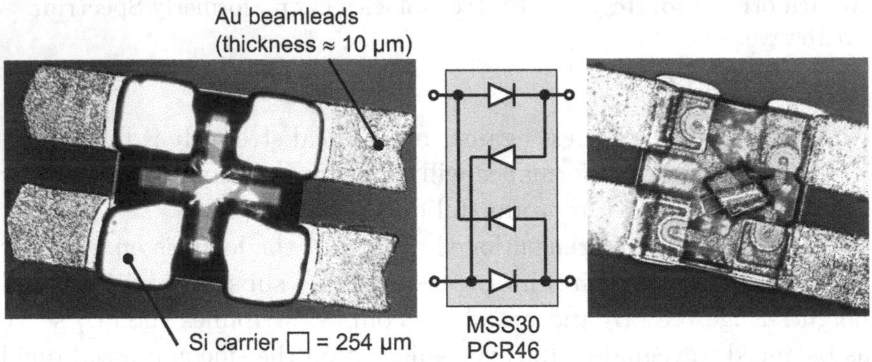

Figure 4.18: Photographs of the utilized silicon crossed quad diodes MSS30 PCR46 from Aeroflex Metelics.

Proposed Triple Balanced Mixer Design Fig. 4.20 shows a top and bottom view of the proposed mixer with its planar realization of the circuit schematic from Fig. 4.19. The IF signal enters the mixer in the coaxial mode within an UT-034-M17 semi-rigid cable. Adhesive connections are established from coaxial ground to the IF ground at the bottom layer and from the inner coaxial conductor to the microstrip line at the top layer. Tapered transitions from microstrip line (MSL) to balanced bSTL are placed at all three mixer ports. A characteristic

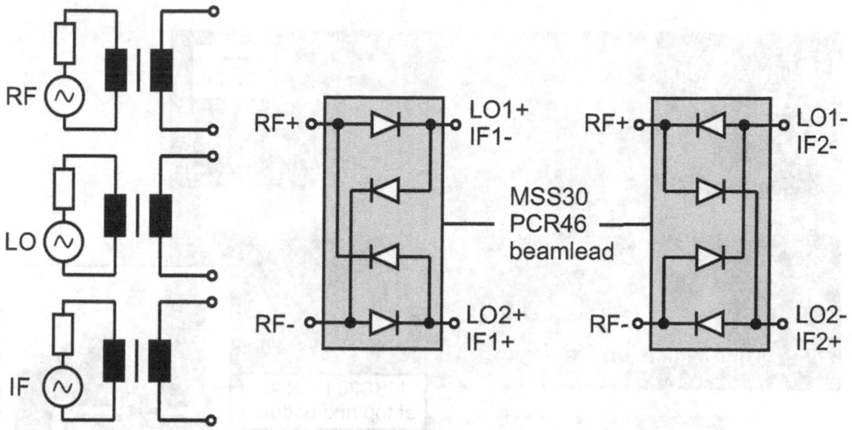

Figure 4.19: Illustration of TBM schematic with the necessary RF / LO / IF connections using the crossed quad diode configuration.

impedance contour of the taper according to the Dolph-Chebyshev[13] function [33] achieves the minimum possible reflection coefficient for a given frequency range and taper length (subsection 2.7.1). Anyway, 3D EM simulations of the simple linear tapers in Fig. 4.20 predict at least 15 dB return loss from 1 to 50 GHz. Monolithic [18, 24, 25] and multilayer [26] mixer implementations often use balun structures based on spiral inductors. The Marchand balun [34, 35] constitutes another multi-octave bandwidth alternative. However, the greatest balun bandwidth can be achieved by utilizing transitions to inherently balanced waveguides [23, 27], as it is with the proposed TBM.

The local oscillator signal in 50 Ω bSTL mode is split into two 100 Ω bSTLs by a compensated T-junction and feeds the diodes arranged in a ring structure. At the IF port two plated-through vias are used to transfer the bSTL mode into the configuration #2 of Fig. 4.21, which is denoted as double edge coupled stripline. The same kind of transition from 50 Ω bSTL mode to the four-conductor configuration of Fig. 4.21 with its three fundamental modes is also used at the

[13]The Dolph-Chebyshev taper is originated from Klopfenstein and therefore also referred to as Klopfenstein taper.

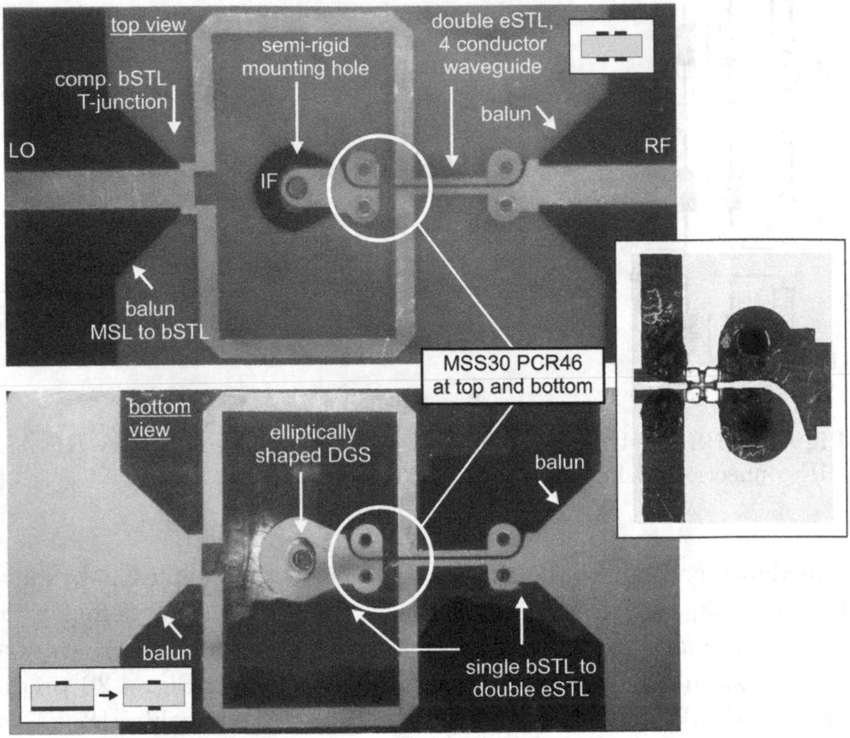

Figure 4.20: Top and bottom view of the assembled triple balanced mixer laminate substrate.

mixer's RF port. Inspecting Fig. 4.20, a DC short circuit of the LO and RF port is recognized. The IF port is DC coupled to the diodes.

Assuming ideal symmetry conditions regarding the eight diodes, the diode assembly and the applied thin-film processing, the following statements will hold true.

- The LO signal will only propagate in double bSTL mode (#1) along the four-conductor waveguide and is short circuited by two plated-through vias at the RF port, which are placed in a distance of a quarter wavelength ($\lambda_{\#1}/4$) at midband frequency of 20 GHz.

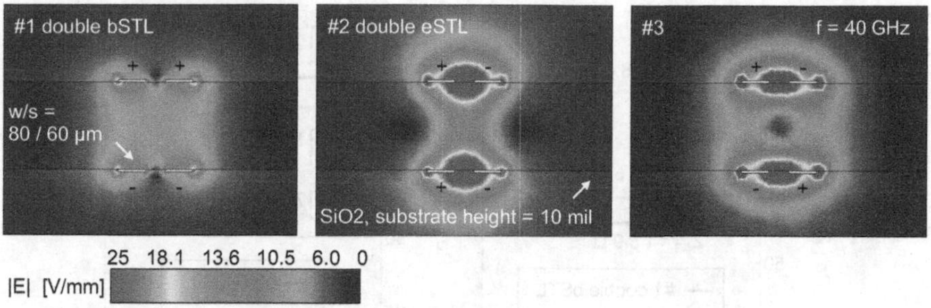

Figure 4.21: 2D electric field mode patterns at 40 GHz of the three fundamental modes within the four-conductor configuration at the mixers RF port.

- The RF signal will only propagate in double eSTL mode (#2) along the four-conductor waveguide and no propagation along the LO ring structure is possible (virtual short circuit).
- The LO ring structure constitutes a DC short circuit for the RF signal propagating in double eSTL mode (#2).

In the presence of circuit asymmetries all three mixer ports do also excite mode #3 of Fig. 4.21 in the four-conductor waveguide, prohibiting the aforementioned isolation arguments from being effective. This will predominantly influence the LO to RF isolation. As a secondary phenomenon there is also parasitic coupling between the mixer ports due to the rather compact planar design.

Fig. 4.22 includes simulated (2D EM) real parts of the characteristic impedances of all involved waveguides and their strip widths and strip spacings.

Co-Simulation Procedure To establish short simulation times the whole planar structure has to be partitioned into sensible parts that can be easily analyzed individually by 3D EM simulation. These parts are the already mentioned tapered baluns, the transition from semi-rigid coaxial mode to MSL, the compensated T-junction, the crossed quad diodes with mounting structure and the transitions

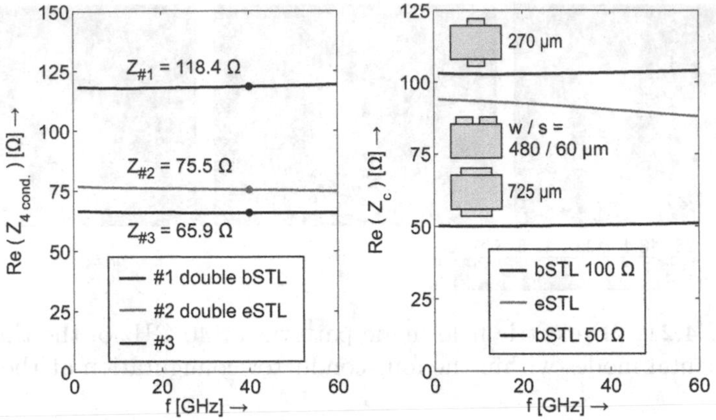

Figure 4.22: 2D EM simulated real parts of the characteristic impedances of the four-conductor waveguide, 100 Ω double bSTL, double eSTL and 50 Ω double bSTL.

from 50 Ω bSTL mode to the four-conductor configuration at the RF port. The complete mixing process up to the 4th RF and 7th LO mixing order is then covered by a nonlinear harmonic balance circuit simulation. As these individual parts of the TBM operate with balanced waveguide interfaces, the modal scattering parameters from 3D EM simulation have to be transformed to nodal parameters (conductor based description). Detailed explanations on this topic are found in section 3.1. For example the four complex scattering parameters of a MSL to balanced bSTL transition have to be converted to nine scattering parameters describing the behaviour between one unbalanced MSL port and two bSTL ports / terminals (Eq. (4.47)). The circuit simulation then allows to extract the optimum impedances and line lengths between these single mixer parts.

$$S_{\text{modal}} = \begin{pmatrix} S_{11\text{m}} & S_{12\text{m}} \\ S_{21\text{m}} & S_{22\text{m}} \end{pmatrix} \rightarrow S_{\text{nodal}} = \begin{pmatrix} S_{11\text{n}} & S_{12\text{n}} & S_{13\text{n}} \\ S_{21\text{n}} & S_{22\text{n}} & S_{23\text{n}} \\ S_{31\text{n}} & S_{32\text{n}} & S_{33\text{n}} \end{pmatrix} \quad (4.47)$$

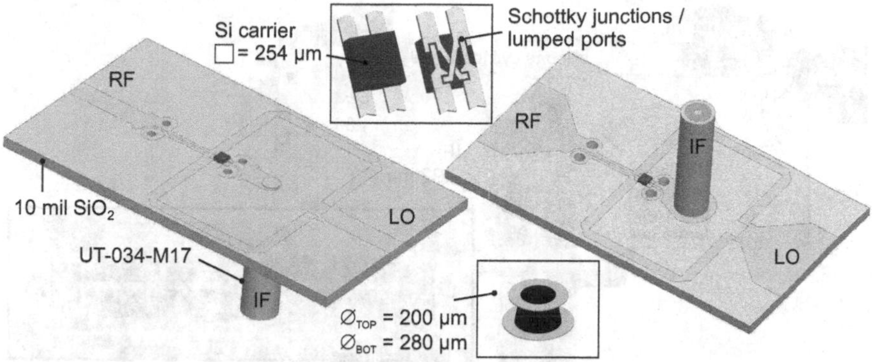

Figure 4.23: 3D model for complete EM simulation of the passive TBM mixer structure with lumped ports at the Schottky junctions.

Utilizing this design approach the diode parasitics are covered within the 3D EM simulation due to an adequate 3D diode model according to Fig. 4.17. As a consequence there is no need to model the parasitic effects occurring from the bulky diode nature within the diode junction circuit model. A 3D model for EM simulation of the entire passive TBM mixer structure is shwon in Fig. 4.23.

Measurement Results The manufactured hybrid TBM with Au plated brass housing, 1.85 mm LO and RF connectors and 2.92 mm IF connector is shown in Fig. 4.24. The IF semi-rigid coaxial cable is equipped with ferrite beads from TDK Epcos building the IF balun. The substrate does not have a soldered connection to the housing, but is pressed to it by screwed on silicone cylinders. Measurements are performed using commercial signal generators for RF and LO signal stimuli and a commercial spectrum analyzer for power detection. In addition scattering parameter measurements after SOL calibration up to 50 GHz are performed, using a commercial vector network analyzer. All the measurement data include the influence of the 1.85 mm and 2.92 mm connectors shown in Fig. 4.24. Performance of the proposed mixer in various different up- and downconversion scenarios including lower and upper sideband mixing is given in the

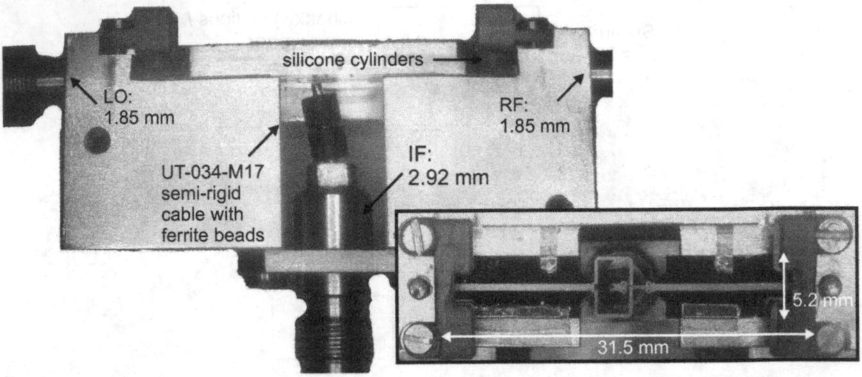

Figure 4.24: Manufactured hybrid triple balanced mixer. Substrate dimensions are (31.5 × 5.2) mm^2.

following. All these mixing cases are required if the TBM is used within front end modules for the mentioned field of applications. The proposed mixer's performance within such a system is also presented.

Although TBMs can up- and downconvert great intermediate frequency ranges, downconversion loss to $f_{\text{IF}} = 1$ GHz or 2 GHz is normally used to compare different mixers. Fig. 4.25 shows simulated and measured lower sideband downconversion loss to 1 GHz and 2 GHz versus RF frequency with local oscillator power level $P_{\text{LO}} \in [14, 18]$ dBm and $P_{\text{RF}} = -10$ dBm. To the author's knowledge the M2 mixer series from Marki Microwave Inc. achieves the lowest available conversion loss over ultrawide bandwidths. Fig. 4.25 further compares the proposed TBM with Marki Microwave Inc. M2-0250 [32] and Miteq Inc. TB0250LW1 [28]. The simulation and measurement results are in reasonable agreement, but compared to [32], the proposed TBM shows 3 to 5 dB higher conversion loss values.

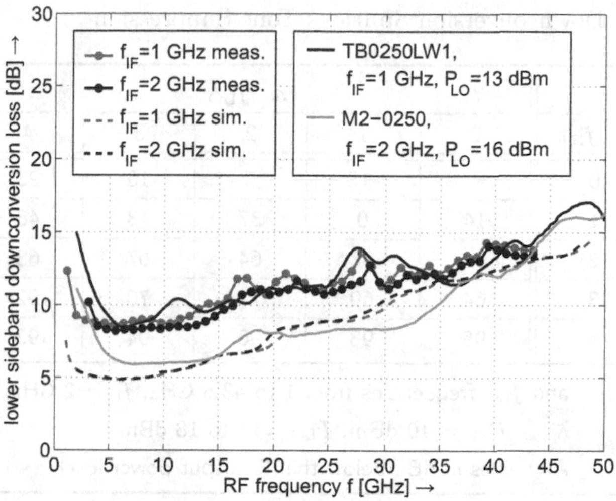

Figure 4.25: Simulated and measured lower sideband downconversion loss to f_{IF} =1 GHz and 2 GHz versus RF frequency with $P_{\mathrm{LO}} \in [14, 18]$ dBm and $P_{\mathrm{RF}} = -10$ dBm. Performance of the commercial products [28, 32] are shown for comparison. $\langle \mathrm{RF, LO} \rangle = \langle -1, 1 \rangle$.

The proposed mixer's spurious tone suppression for a downconversion scenario to f_{IF} =2 GHz is summarized in Table 4.2. The values are in dBc below the IF output power level and measured with an input power level $P_{\mathrm{RF}} = -10$ dBm. Suppression for input power levels different from $P_{\mathrm{RF}} = -10$ dBm, are calculated by scaling the dBc value with $(N-1)$, whereas N is the spur order. Spurious tone suppression for an upconversion scenario to f_{IF} =2 GHz is summarized in Table 4.3. The achieved spurious tone suppression is comparable to the best reported triple balanced mixers and especially the suppression of the most critical spurious output for system design of broadband front end modules $\langle \mathrm{RF, LO} \rangle = \langle 1, 2 \rangle$ shows at least 10 to 20 dB improvement compared to double balanced mixers.

Table 4.2: Downconversion Spurious Tone Suppression

$n \cdot f_{RF}$	$m \cdot f_{LO}$ 0	1	2	3	4
0	-	-15	5	16	23
1	14	0	37	13	48
2	51	58	64	67	63
3	64	69	77	70	92
4	95	93	90	94	92

f_{RF} and f_{LO} frequencies from 1 to 43.5 GHz, f_{IF} =2 GHz,
P_{RF} =-10 dBm, P_{LO} =14 to 18 dBm.
All values in dBc below the IF output power level.

Table 4.3: Upconversion Spurious Tone Suppression

$n \cdot f_{RF}$	$m \cdot f_{LO}$ 0	1	2	3	4
0	-	2	4	17	15
1	11	0	31	8	42
2	72	65	63	69	69
3	96	90	83	71	67
4	90	90	87	81	71

f_{RF} and f_{LO} frequencies from 1 to 43.5 GHz, f_{IF} =2 GHz,
P_{IF} =-10 dBm, P_{LO} =14 to 18 dBm.
All values in dBc below the RF output power level.

Fig. 4.26 includes measured reflection coefficients with $P_{LO}, P_{RF} \in [14, 18]$ dBm. IF and RF return loss values are greater than 5 dB up to 30 GHz at all mixer ports. Fig. 4.26 further shows isolation measurements. Except for frequencies below 5 GHz, isolation values are higher than 20 dB between all ports.

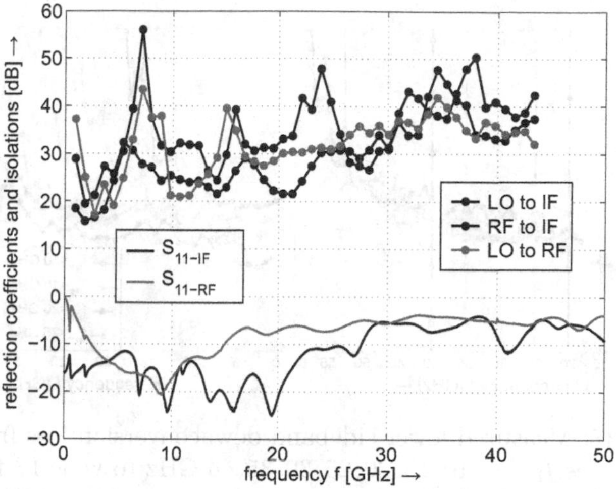

Figure 4.26: Measured IF and RF reflection coefficients and isolations (LO to IF, RF to IF, LO to RF) versus frequency with $P_{\text{LO}}, P_{\text{RF}} \in [14, 18]$ dBm.

The proposed mixer's lower sideband downconversion loss from fixed RF frequencies $f_{\text{RF}} = 10, 15, 20, 25, 30, 35, 40$ GHz to wide IF frequency range $f_{\text{IF}} = 0.1$ to 43.5 GHz is shown in Fig. 4.27. The conversion loss values stay below 15 dB as long as the local oscillator frequency does not exceed 45 GHz.

Fig. 4.28 depicts the lower sideband downconversion loss from wide RF frequency range to fixed IF frequencies $f_{\text{IF}} = 0.5, 1, 3, 5, 10, 15, 20$ GHz. Up to an LO frequency of 50 GHz, the conversion loss is below 15 dB.

The mixer's upper sideband downconversion loss from wide RF frequency range to fixed IF frequencies $f_{\text{IF}} = 0.5, 1, 3, 5, 10, 15, 20$ GHz is shown in Fig. 4.29.

The black curve in Fig. 4.30 corresponds to downconversion loss from $f_{\text{RF}} = 11$ GHz with $P_{\text{RF}} = -10$ dBm to $f_{\text{IF}} = 1$ GHz versus the local oscillator power level P_{LO}. With P_{LO} between 5 and 10 dBm, the conversion loss is below 10 dB. The red curves in Fig. 4.30 show downconversion loss (left axis) and IF output power level (right axis)

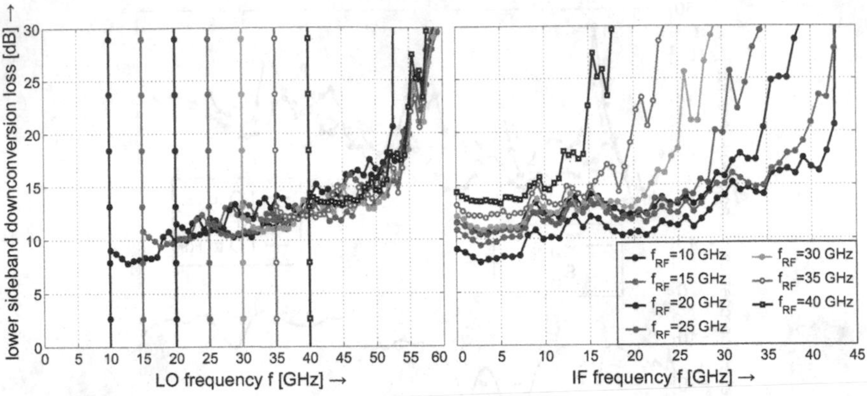

Figure 4.27: Measured lower sideband downconversion loss from fixed RF frequencies $f_{\mathrm{RF}} = 10, 15, 20, 25, 30, 35, 40$ GHz to wide IF frequency range f_{IF} =0.1 to 43.5 GHz versus LO frequency (left) and IF frequency (right). $\langle \mathrm{RF}, \mathrm{LO}\rangle = \langle -1, 1\rangle$.

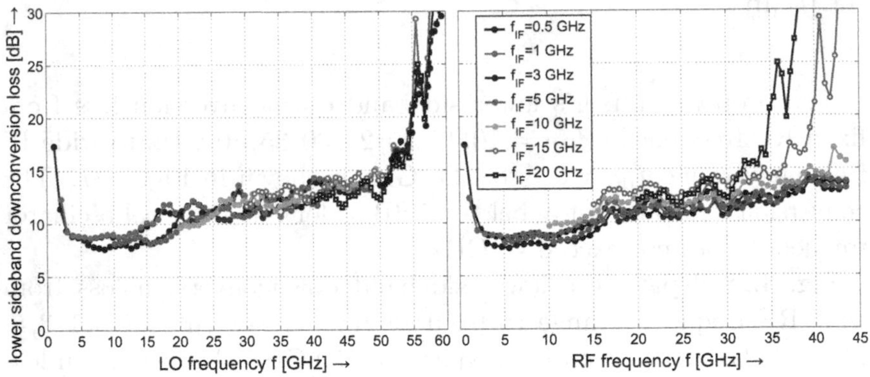

Figure 4.28: Measured lower sideband downconversion loss from wide RF frequency range to fixed IF frequencies $f_{\mathrm{IF}} = 0.5, 1, 3, 5, 10, 15, 20$ GHz versus LO frequency (left) and RF frequency (right). $\langle \mathrm{RF}, \mathrm{LO}\rangle = \langle -1, 1\rangle$.

if the RF input power level P_{RF} is swept from -20 to 18 dBm. As marked in Fig. 4.30 we measure an input 1 dB compression point of 12 dBm which is approximately 6 dB below the applied local oscillator power level.

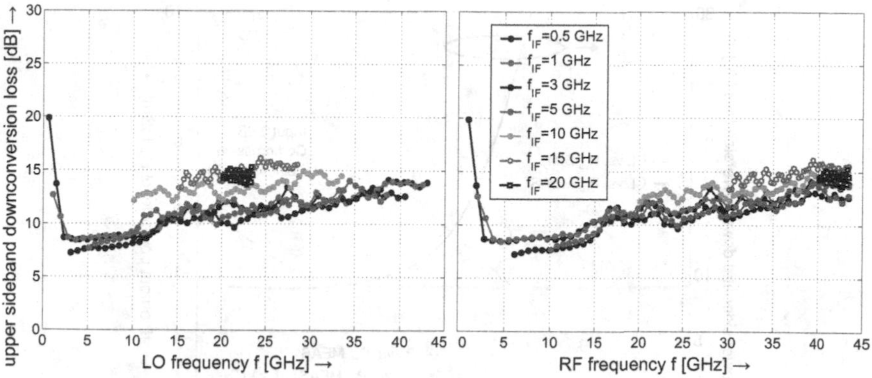

Figure 4.29: Measured upper sideband downconversion loss from wide RF frequency range to fixed IF frequencies $f_{\mathrm{IF}} = 0.5, 1, 3, 5, 10, 15$ and 20 GHz. $\langle$RF, LO$\rangle$=$\langle 1, -1\rangle$.

Fig. 4.31 illustrates the upper sideband upconversion from fixed RF frequencies $f_{\mathrm{RF}} = 1, 3, 5, 10, 15, 20, 25, 30$ GHz to wide IF frequency range. This is a necessary mixing case within receiver modules. The incoming RF signal is first upconverted to an intermediate frequency in the first mixing stage and then downconverted in the second mixing stage.

Within front end modules, the TBM's RF port is normally used for the higher frequencies (mixing to / from the maximum frequency), while the IF port handles the lower frequencies (TX / RX). TX / RX mixer stages cannot handle large output / input frequency ranges if the intermediate frequency is too low (Fig. 4.14). This is solved by applying the principle of first high IF. For instance, if detection (RX module) of signals at $f_{\mathrm{in}} \in [1, 20]$ GHz is required, the first mixing stage converts to a higher IF frequency (21 GHz ± half bandwidth), whereas the second mixing stage downconverts to baseband. Fig. 4.32 shows the mixer's lower sideband upconversion loss from wide RF frequency range to fixed IF frequencies $f_{\mathrm{out}} = f_{\mathrm{IF}} = 6, 11, 16, 21$ GHz. In the described case, it would be convincing to use the RF port for the 21 GHz signal. But if the same mixer in the same RX module does also perform downconversion from $f_{\mathrm{in}} \in [20, 40]$ GHz to 5 GHz and

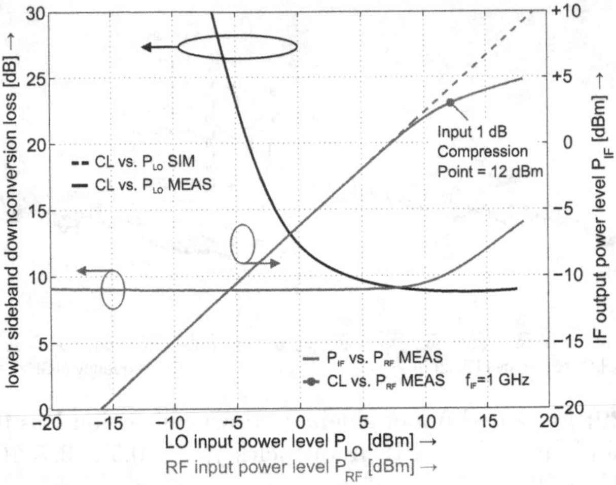

Figure 4.30: Measured lower sideband downconversion loss to f_{IF} =1 GHz versus local oscillator (black) and RF input power level (red), together with the IF output power level (red) versus RF input power level. f_{RF} =11 GHz,
P_{RF} =-10 dBm and f_{LO} =12 GHz. $\langle -1, 1\rangle$.

10 GHz ± half bandwidth, the IF port has to to cover all IF frequencies 21, 10 and 5 GHz. Choosing the right mixer port is a challenging task for the system designer. The proposed mixer's performance if the IF port is chosen as the input port is depicted in Fig. 4.33. It illustrates the lower sideband upconversion loss from wide IF frequency $f_{\mathrm{in}} = f_{\mathrm{IF}}$ range to fixed RF frequencies $f_{\mathrm{RF}} = 6, 11, 16, 21$ GHz.

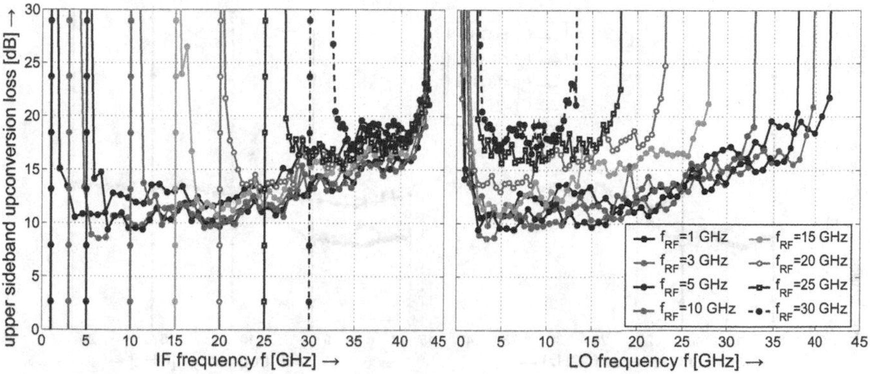

Figure 4.31: Measured upper sideband upconversion loss from fixed RF frequencies $f_{\mathrm{RF}} = 1, 3, 5, 10, 15, 20, 25, 30$ GHz to wide IF frequency range versus IF frequency (left) and LO frequency (right). $\langle\mathrm{RF}, \mathrm{LO}\rangle = \langle 1, 1\rangle$.

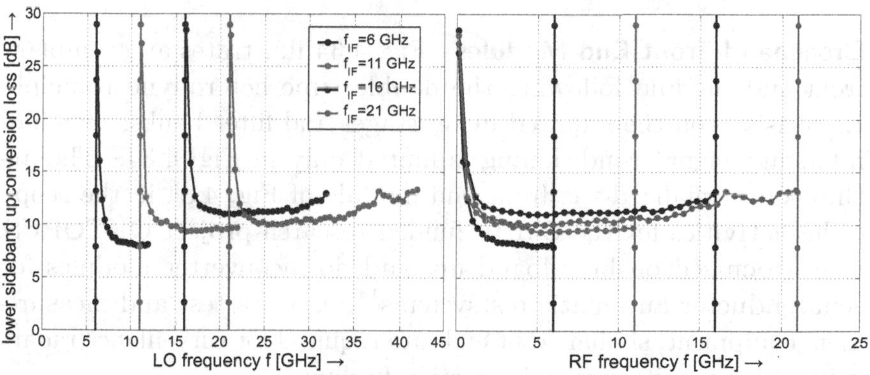

Figure 4.32: Measured lower sideband upconversion loss from wide RF frequency range to fixed IF frequencies $f_{\mathrm{IF}} = 6, 11, 16, 21$ GHz versus LO frequency (left) and RF frequency (right). $\langle\mathrm{RF}, \mathrm{LO}\rangle = \langle -1, 1\rangle$.

It is noticeable, that quite low input frequencies (100 MHz) $f_{\mathrm{in}} = f_{\mathrm{IF}}$ are upconverted with low conversion loss in Fig. 4.33, which is due to the DC coupled IF port.

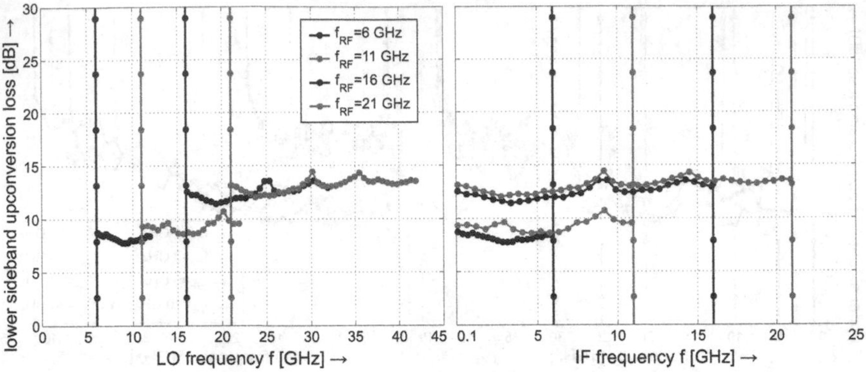

Figure 4.33: Measured lower sideband upconversion loss from wide IF frequency range to fixed RF frequencies $f_{\mathrm{RF}} = 6, 11, 16, 21$ GHz versus LO frequency (left) and IF frequency (right). $\langle$IF, LO$\rangle$=$\langle -1, 1\rangle$.

Broadband Front End Modules Fig. 4.34 illustrates an exemplary front end module following the double superheterodyne principle. Emphasis is on the required mixer stages and filter banks, therefore interstage signal conditioning is hinted only in Fig. 4.34. The author conceptualized the front end module of Fig. 4.34 in the scope of his activities for the BMBF funded research project OKTOPUS, which focussed on broadband up- and downconverter modules for semiconductor automatic testsystems[14]. Besides test and measurement equipment, similar front ends are required for surveillance radars utilized for monitoring and direction finding.

Both, transmitter (TX, top) and receiver (RX, bottom) have low baseband operating frequency (MHz) to be compatible with state of the art analog-digital converters (ADC, DAC) or arbitrary waveform generators (AWG) using direct digital synthesis (DDS). Frequency conversion from baseband to one of several intermediate frequencies (IF) and vice versa is performed best with double balanced mixers. At each intermediate frequency an individual bandpass filter (BPF), that

[14]Research project *Optimal-Konfigurierbare Test-Organisationsplattform mit Unterstützung der Synthese* (OKTOPUS), FKZ 01M3182E, funded by the German Federal Ministry of Education and Research (BMBF).

covers the system's instantaneous signal bandwidth (e.g. 1 GHz) is required (narrowband IF filter bank). In Fig. 4.34, one explicit BPF at IF frequency 5.15 GHz ±0.5 GHz is highlighted, but two more at higher frequencies[15] are necessary to cover an overall frequency range of 1 to 40 GHz with the entire up- and downconverter modules. The proposed TBM is intended to be used for the 2nd TX mixer stage and 1st mixer stage of the RX module. This could hardly be realized with DBMs, due to the large required mixer IF bandwidth. In the following, dedicated up- and downconversion cases with its associated spurious tones and bandpass filter (BPF) requirements are explained in detail. These are referred to as FE_Up_2_3 and FE_Dn_2_2 in Fig. 4.34. Measurement results of the proposed triple balanced mixer's spurs and the developed filters are included.

FE_Up_2_3 Fig. 4.35 depicts a spurious tone plot (spurs plot) for a dedicated mixing case, referred to as FE_Up_2_3 in Fig. 4.34. It shows four spurs $\langle -2, 1\rangle$, $\langle 5, -1\rangle$, $\langle 2, 0\rangle$, $\langle 3, 0\rangle$ and their frequencies at the output RF port for a mixing case with $f_{\text{in}} = f_{\text{IF}} = [4.65, 5.65]$ GHz and $f_{\text{out}} = f_{\text{RF}} = [9.7, 13.8]$ GHz. The LO frequency ranges are $f_{\text{LO}} = [15.35, 18.45]$ GHz or $f_{\text{LO}} = [14.35, 19.45]$ GHz, whether the input signal frequency is at the lower or upper edge of f_{in} (continuous wave (CW) operation only). Spurs inside the rectangle fall within the passband of the corresponding BPF (wideband filter bank) and are therefore emitted to the TX output port. Measured spectral power levels for this mixing case at the input IF and output RF port are shown in Fig. 4.36 and Fig. 4.37, respectively. Obviously, the developed input and output bandpass filters (BPF) allow for at least 45 dBc of spurious tone free dynamic range. The narrowband BPF of Fig. 4.36 and the wideband BPF of Fig. 4.37 are both designed by Gavin Ripley, Principal Production Engineer, BSC Filters Ltd. Whereas the wideband output BPF characteristics could be achieved with a single filter design approach, the narrowband BPF ($(175 \times 24 \times 12.5)$ mm^3)

[15] In fact, the number of IF frequencies constitutes a tradeoff between spectral purity and module complexity / module cost. In the actual case, at least three IF frequencies are intended ($f_{\text{IF max}} < 25$ GHz).

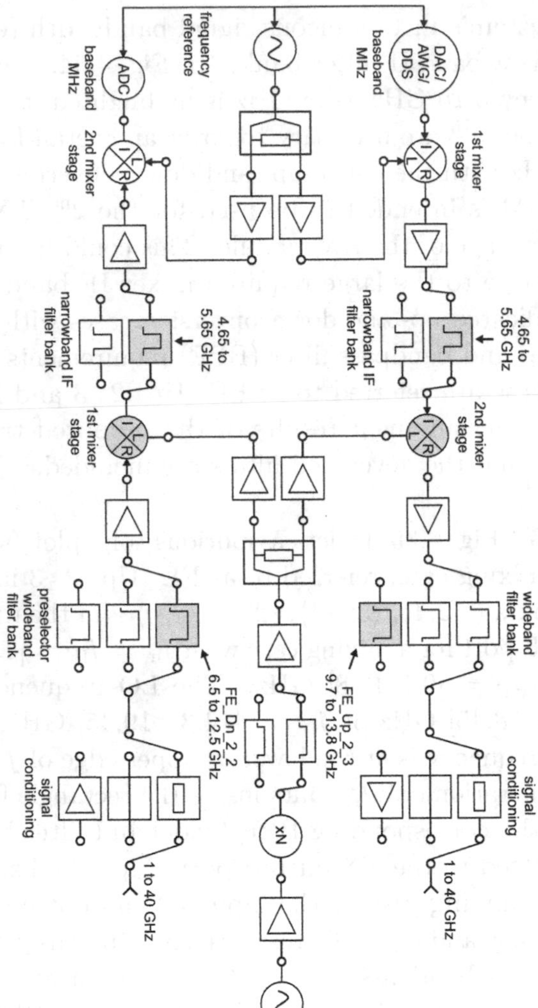

Figure 4.34: Block diagrams of an exemplary transmitter frontend module (top) and receiver frond end module (bottom).

requires a cascaded lowpass filter ($(82 \times 23 \times 13)$ mm^3) to realize stopband rejection of at least 40 dB up to 50 GHz. Fig. 4.38 shows

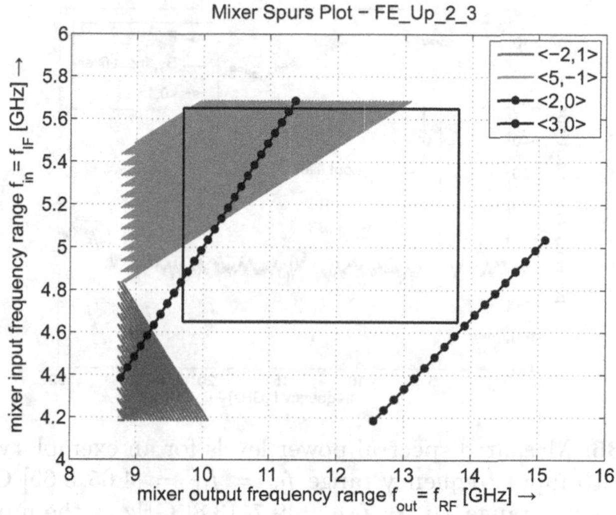

Figure 4.35: Mixer spurs plot for an exemplary upconversion case with input frequency range $f_{\text{in}} = f_{\text{IF}} = [4.65, 5.65]$ GHz, output frequency range $f_{\text{out}} = f_{\text{RF}} = [9.7, 13.8]$ GHz and four different spurs $\langle -2, 1\rangle$, $\langle 5, -1\rangle$, $\langle 2, 0\rangle$, $\langle 3, 0\rangle$. The LO frequency ranges are $f_{\text{LO}} = [15.35, 18.45]$ GHz or $f_{\text{LO}} = [14.35, 19.45]$ GHz, wether the input signal frequency is at the lower or upper edge of f_{in}. Spurs inside the rectangle fall within the passband of the corresponding BPF (wideband filter bank).

both, scattering parameters of the bare BPF and LPF (top) and the BPF cascaded with a LPF (bottom).

FE_Dn_2_2 A spurious tone plot (spurs plot) for the downconversion mixing case, referred to as FE_Dn_2_2 in Fig. 4.34, is illustrated in Fig. 4.39.

It shows four critical spurs $\langle -2, 1\rangle$, $\langle -2, 1\rangle$, $\langle -2, 2\rangle$, $\langle 3, -1\rangle$ and their frequencies at the output IF port. The input frequency range $f_{\text{in}} = f_{\text{RF}} = [6.5, 12.5]$ GHz is downconverted to the output frequency range $f_{\text{out}} = f_{\text{IF}} = [4.65, 5.65]$ GHz with local oscillator frequencies $f_{\text{LO}} = [12.15, 18.15]$ GHz. Spurs inside the rectangle fall within the passband of the corresponding BPF (narrowband IF filter bank) and

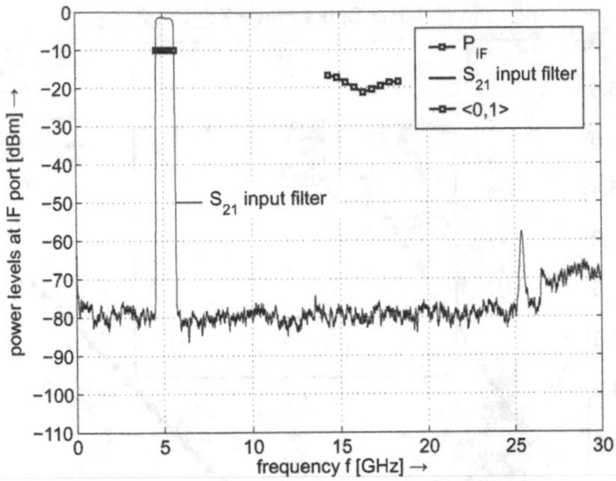

Figure 4.36: Measured spectral power levels for an exemplary upconversion case with input frequency range $f_{\text{in}} = f_{\text{IF}} = [4.65, 5.65]$ GHz and output frequency range $f_{\text{out}} = f_{\text{RF}} = [9.7, 13.8]$ GHz at the input IF port versus frequency. The desired input IF signal power levels (black) and the parasitic LO signal power levels (blue) are shown together with the transmission coefficient of the input BPF (narrowband IF filter bank).

are therefore converted to baseband. Fig. 4.40 depicts the spectral input power levels at the RF port. It shows the desired signal (black), which lies within the passband of the wideband BPF, and the LO signal (blue) leaking to the RF port. The LO frequencies, that are not filtered out are not required to cover the module's bandwidth. Fig. 4.41 includes the measured spurious tone power levels at the mixer's IF output port. The narrowband BPF is the same as in Fig. 4.38. A spurious tone free dynamic range of about 50 dBc is achieved.

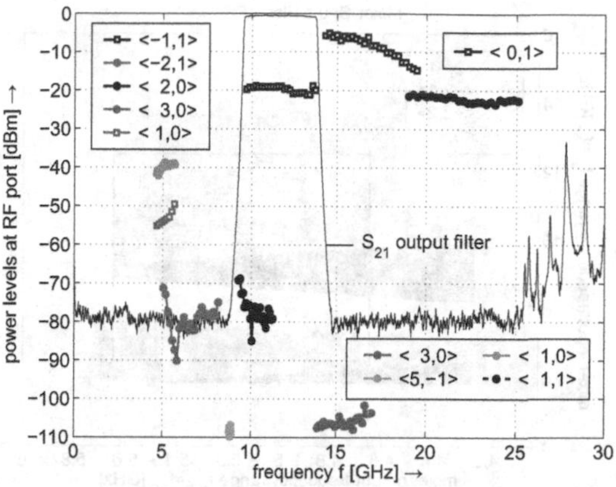

Figure 4.37: Measured spectral power levels for an exemplary upconversion case with input frequency range $f_{\text{in}} = f_{\text{IF}} = [4.65, 5.65]$ GHz and output frequency range $f_{\text{out}} = f_{\text{RF}} = [9.7, 13.8]$ GHz at the output RF port versus frequency. The desired output RF signal power levels ($\langle -1, 1 \rangle$, black) and the parasitic LO power levels ($\langle 0, 1 \rangle$, blue) are shown together with the transmission coefficient of the RF bandpass filter (wideband filter bank) and eight more spurs.

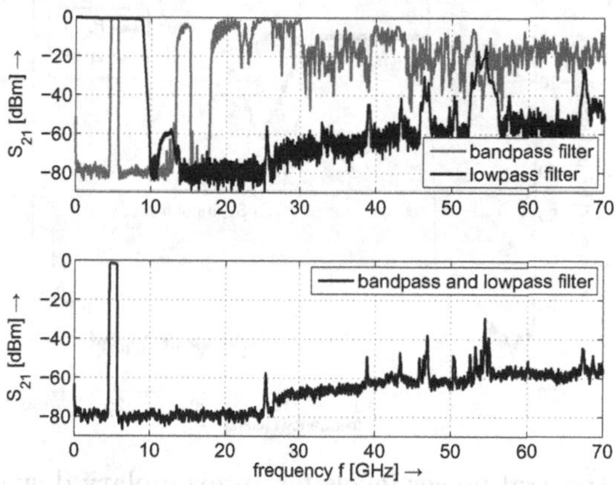

Figure 4.38: Transmission coefficients of narrowband IF bandpass filter (top, red) with passband center frequency $f_{\text{center}} = 5.15$ GHz ± 0.5 GHz and LPF with large bandwidth of stopband (top, blue). In cascade (bottom, black) a stopband rejection of 30 dB up to 70 GHz is achieved.

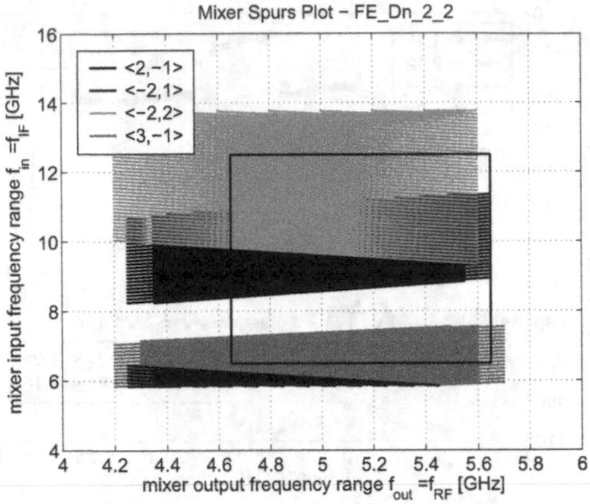

Figure 4.39: Mixer spurs plot for an exemplary downconversion case with input frequency range $f_{\mathrm{in}} = f_{\mathrm{RF}} = [6.5, 12.5]$ GHz, output frequency range $f_{\mathrm{out}} = f_{\mathrm{IF}} = [4.65, 5.65]$ GHz and four different spurs $\langle -2, 1\rangle$, $\langle -2, 1\rangle$, $\langle -2, 2\rangle$, $\langle 3, -1\rangle$. The LO frequency range is $f_{\mathrm{LO}} = [12.15, 18.15]$ GHz. Spurs inside the rectangle fall within the passband of the corresponding BPF (narrowband IF filter bank).

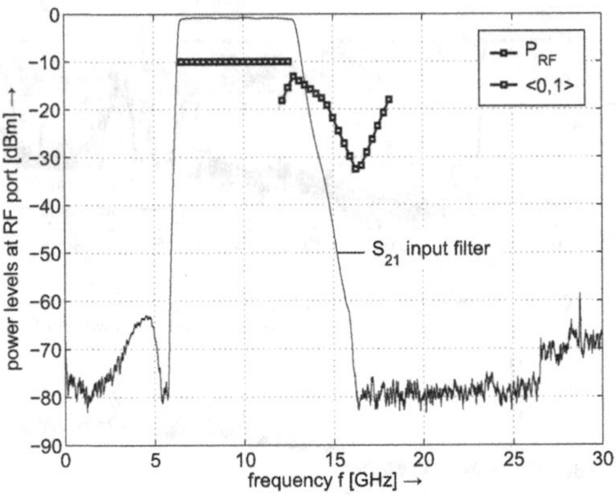

Figure 4.40: Spectral power levels for an exemplary downconversion case with input frequency range $f_{\mathrm{in}} = f_{\mathrm{RF}} = [6.5, 12.5]$ GHz and output frequency range $f_{\mathrm{out}} = f_{\mathrm{IF}} = [4.65, 5.65]$ GHz at the input RF port versus frequency. The desired input RF signal power levels (black) and the parasitic LO signal power levels (blue) are shown together with the transmission coefficient of the input BPF (wideband filter bank).

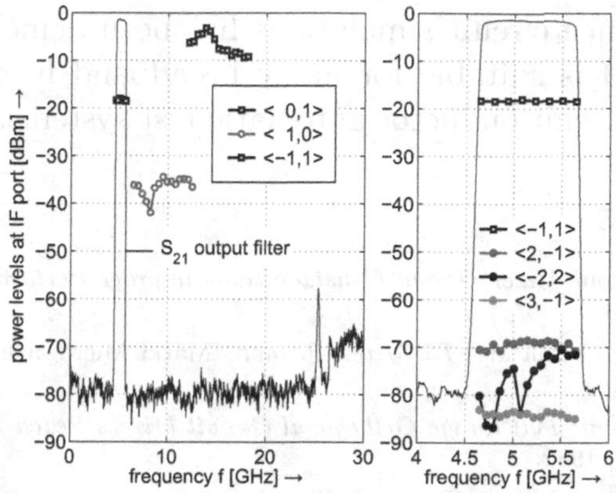

Figure 4.41: Spectral power levels for an exemplary downconversion case with input frequency range $f_{\text{in}} = f_{\text{RF}} = [6.5, 12.5]$ GHz and output frequency range $f_{\text{out}} = f_{\text{IF}} = [4.65, 5.65]$ GHz at the output IF port versus frequency. The desired output IF signal power levels (black) and the parasitic LO (blue) and RF (red) signal power levels are shown together with the transmission coefficient of the IF bandpass filter (narrowband IF filter bank).

Conclusion The design and construction of a hybrid broadband TBM based on commercial silicon crossed quad diodes has been presented. The mixer is realized on 10 mil SiO_2 substrate without bondwires or air bridges and an overall substrate size of (31.5×5.2) mm^2, which can easily be shrinked to (10.5×5.2) mm^2. A semi-rigid coaxial interface is used at the IF port but compared to other design approaches [30] the required assembly effort for usage in highly integrated planar front end modules is by far acceptable. Broadband planar baluns and the use of balanced eSTL and bSTL allow for covering an RF / LO frequency range of 1 to 45 / 50 GHz and at least 20 GHz IF bandwidth, if we claim for up- and downconversion loss values below 15 dB. This is achieved with local oscillator power levels $P_{\text{LO}} \in [14, 18]$ dBm and an input 1 dB compression point of 12 dBm. Reasonable agreement of the measurement and simulation results that are based on a co-simulation procedure between 3D EM simulations and nonlinear harmonic balance circuit simulations has been demonstrated. The proposed TBM is suitable for many broadband front end module applications of semiconductor automatic test systems.

References

[1] B. C. Henderson. *Mixer Design Considerations Improve Performance.* MSN, October Issue, 1981.

[2] F. Marki and C. Marki. *T3 Mixer Primer.* Marki Microwave, Inc., Application Note, 2012.

[3] B. C. Henderson. *Full Range Orthogonal Circuit Mixers Reach 2 to 26 GHz.* MSN, January Issue, 1982.

[4] P. Will. *Reactive Loads - The Big Mixer Menace.* Microwaves, April Issue, 1971.

[5] R. G. Hicks and P. J. Khan. *Numerical Analysis of Nonlinear Solid-State Device Excitation in Microwave Circuits.* IEEE Transactions on Microwave Theory & Techniques, Vol. 30, No. 3, pp. 251-259, 1982.

[6] A. R. Kerr. *A Technique for Determining the Local Oscillator Waveforms in a Microwave Mixer.* IEEE Transactions on Microwave Theory & Techniques, Vol. 23, No. 10, pp. 828-831, 1975.

[7] S. A. Maas. *Two-Tone Intermodulation in Diode Mixers.* IEEE Transactions on Microwave Theory & Techniques, Vol. 35, No. 3, pp. 307-314, 1987.

[8] S. A. Maas. *Nonlinear Microwave and RF Circuits.* Boston: Artech House, 2nd Edition, 2003. ISBN 1580534848.

[9] A. A. M. Saleh. *Theory of Resistive Mixers.* Cambridge, Massachusetts: MIT Press, 1971. ISBN 0262190931.

[10] D. E. Root, J. Verspecht, J. Horn, and M. Marcu. *X-Parameters: Characterization, Modeling, and Design of Nonlinear RF and Microwave Components.* Cambridge University Press, 2013. ISBN 0521193230.

[11] D. Vye. *Fundamentally Challenging Nonlinear Microwave Deisgn.* Microwave Journal, Vol. 53, No. 3, 2010.

[12] J. Verspecht and D. E. Root. *Polyharmonic Distortion Modeling.* IEEE Microwave Magazine, Vol. 7, No. 3, pp. 44-57, 2006.

[13] J. Wood and D. E. Root. *Fundamentals of Nonlinear Behavioral Modeling for RF and Microwave Design.* Boston: Artech House, 2005. ISBN 1580537758.

[14] H. C. Torrey and C. A. Whitmer. *Crystal Rectifiers.* New York: McGraw-Hill, MIT Radiation Laboratory Series, Vol. 15, 1948.

[15] R. V. Pound. *Microwave Mixers.* New York: McGraw-Hill, MIT Radiation Laboratory Series, Vol. 16, 1948.

[16] M. Sterns, R. Rehner, D. Schneiderbanger, S. Martius, and L.-P. Schmidt. *Broadband, Highly Integrated Receiver Frontend up to 67 GHz.* Proc. of the 41th European Microwave Conference, pp. 1063-1066, 2011.

[17] I. J. Bahl. *Fundamentals of RF and Microwave Transistor Amplifiers.* New York: John Wiley & Sons, 2009. ISBN 0470391669.

[18] H.-K. Chiou and H.-H. Lin. *A Miniature MMIC Double Doubly Balanced Mixer Using Lumped Dual Balun for High Dynamic Receiver Application.* IEEE Microwave & Guided Wave Letters, Vol. 7 , No. 8, pp. 227-228, 1997.

[19] D. Schneiderbanger, A. Cichy, R. Rehner, M. Sterns, S. Martius, and L.-P. Schmidt. *A Hybrid Broadband Millimeter-Wave Diode Ring Mixer with Advanced IF Extraction Technique.* Proc. of the 37th European Microwave Conference, pp. 656-659, 2007.

[20] D. Schneiderbanger, A. Cichy, R. Rehner, S. Martius, M. Sterns, and L.-P. Schmidt. *An Octave-Bandwidth Double-Balanced Millimeter-Wave Mixer.* Proc. of the 36th European Microwave Conference, pp. 1316-1319, 2006.

[21] G. D. Vendelin, A. M. Pavio, and U. L. Rohde. *Microwave Circuit Design Using Linear and Nonlinear Techniques.* New York: John Wiley & Sons, 2005. ISBN 0471414794.

[22] A. Milano. *2 to 18 GHz Triple Balanced Mixer for MMIC Implementation.* Proc. of the 27th European Microwave Conference, pp. 266-268, 1997.

[23] S. T. Choi, Y. J. Kim, Y. H. Kim, D. H. Lee, and C. T. Choi. *A Novel Uniplanar Ultra-Broadband Double Doubly Balanced Mixer for High Dynamic Range Applications.* Proc. of the 39th European Microwave Conference, pp. 65-68, 2009.

[24] M. N. Do, D. Langrez, J.-F. Villemazet, and J.-L. Cazaux. *Double and Triple Balanced Wideband Mixers Integrated in GaAs Technology.* Int. Workshop on Integrated Nonlinear Microwave and Millimeter-Wave Circuits, pp. 85-88, 2010.

[25] K. W. Kobayashi, R. Kasody, A. K. Oki, G. S. Dow, B. Allen, and D.C. Streit. *A Double-Double Balanced HBT Schottky Diode Broadband Mixer at X-band.* 16th Annual GaAs IC Symposium, pp. 315-318, 1994.

[26] W. Yun, V. Sundaram, and M. Swaminathan. *A Triple Balanced Mixer in Multi-Layer Liquid Crystalline Polymer (LCP) Substrate.* Proc. of the 57th Electronic Components and Technology Conference (ECTC), pp. 2000-2005, 2007.

[27] J. Eisenberg, J. Panelli, and W. Qu. *A New Planar Double-Double Balanced MMIC Mixer Structure.* IEEE MTT-S Int. Symp. Dig., Vol. 1, pp. 81-84, 1991.

[28] MITEQ Inc. *Triple Balanced Mixer TB0250LW1, conversion loss measurement data provided by Georg Nuetzl, EMCO Elektronik GmbH.* www.miteq.com.

[29] Marki Microwave Inc. *Triple Balanced Mixer M2-0240.* www.markimicrowave.com.

[30] Marki Microwave Inc. *Triple Balanced Mixer M2-0243.* www.markimicrowave.com.

[31] Marki Microwave Inc. *Triple Balanced Mixer M2-0440.* www.markimicrowave.com.

[32] Marki Microwave Inc. *Triple Balanced Mixer M2-0250.* www.markimicrowave.com.

[33] R. W. Klopfenstein. *A Transmission Line Taper of Improved Design.* Proc. of the IRE, Vol. 44 , No. 1, pp. 31-35, 1956.

[34] N. Marchand. *Transmission Line Conversion.* Electronics, Vol. 17, pp. 142-145, 1944.

[35] J. H. Cloete. *Exact Design of the Marchand Balun.* Proc. of the 9th European Microwave Conference, pp. 480-484, 1979.

[36] R. H. Pantell. *General Power Relationships for Positive and Negative Resistive Elements.* Proc. of the IRE, Vol. 46, No. 12, pp. 1910-1913, 1958.

[37] S. A. Maas. *Microwave Mixers.* Boston: Artech House, 1986. ISBN 0890061718.

[38] S. A. Maas. *The RF and Microwave Circuit Design Cookbook.* Boston: Artech House, 1998. ISBN 0890069735.

[39] B. C. Henderson. *Mixers in Microwave Systems (Part 1).* Watkins Johnson Communications, Inc., Tech-Note, Vol. 17, No. 1, 1990.

[40] B. C. Henderson. *Mixers in Microwave Systems (Part 2).* Watkins Johnson Communications, Inc., Tech-Note, Vol. 17, No. 2, 1990.

[41] B. C. Henderson. *Predicting Intermodulation Suppression in Double-Balanced Mixers.* Watkins Johnson Communications, Inc., Tech-Note, Vol. 10, No. 4, 1983.

[42] F. Marki and C. Marki. *Mixer Basics Primer.* Marki Microwave, Inc., Application Note, 2010.

[43] J. B. Cochrane and F. A. Marki. *Thin-Film Mixers Team Up To Block Out Image Noise.* Microwaves, March Issue, 1977.

[44] I. A. Glover, S. R. Pennock, and P. R. Shephard. *Microwave Devices, Circuits and Subsystems for Communications Engineering.* New York: John Wiley & Sons, 2005. ISBN 047189964X.

[45] J. Virtanen. *Harmonic Balance and Phase-Noise Analysis Methods in the APLAC Circuit Simulator.* Ph.D. Thesis, Helsinki University of Technology, Finland, 2005.

[46] B. Schueppert. *Ein Beitrag zur Analyse und Synthese realer Mikrowellenmischer unter besonderer Berücksichtigung integrierter Gegentaktstrukturen.* Ph.D. Thesis, Technical University Berlin, Germany, 1983.

[47] U. Tietze, C. Schenk, and E. Gamm. *Halbleiter-Schaltungstechnik.* Berlin: Springer, 2002. ISBN 3540428496.

[48] U. Tietze, C. Schenk, and E. Gamm. *Electronic Circuits.* Berlin: Springer, 2nd Edition, 2008. ISBN 978-3-540-00429-5.

[49] A. Milano. *Multioctave, Uniplanar, Triple Balanced Mixer for E. W. Systems.* Proc. of the 17th Conv. of Electrical and Electronics Engineers in Israel, pp. 380-383, 1991.

[50] M. Morgan and S. Weinreb. *A Millimeter-Wave Perpendicular Coax-To-Microstrip Transition.* IEEE MTT-S Int. Symposium Digest, Vol. 2, pp. 817-820, 2002.

5 Zero Bias Schottky Power Detectors

Scalar power measurements in the millimeter-wave frequency range and beyond are either performed with rather slow calorimetric devices or zero bias Schottky diode detectors. The latter are used for scattering parameter [1] and direction of arrival measurements [2, 3] utilizing the six-port receiver architecture. Focal plane arrays [4] and holographic imaging systems [5] are further fields of application. Within front end modules of semiconductor ATS and frequency extension modules of signal generators, diode detectors are used for power monitoring of forward and backward travelling waves to realize automatic level control loops and overload protection.

The first section includes theoretical foundations. Expressions for detector current β and voltage sensitivity γ, and the upper $\Delta P_{\text{sq}}, P_{\text{USL}}$ and lower boundary $P_{\text{NEP}}, P_{\text{TSS}}$ of the square law dynamic range are derived. This is followed by a section, which compares common detector architectures.

Two planar ultrawideband power detector designs operating from 1 to 40 GHz and 60 to 110 GHz [I8] are presented. The detectors are part of the signal generator frequency extension modules, described in chapter 6, to realize monitoring of the output power level and automatic level control (ALC).

5.1 Theoretical Foundations

We assume the nonlinear current-voltage dependency of the diode $i = g(v)$ to be infinitely differentiable and a sinusoidal input voltage time waveform with DC component $v(t) = V_0 + v_{RF}(t) = V_0 + \hat{v}\cos(\omega t), \omega =$

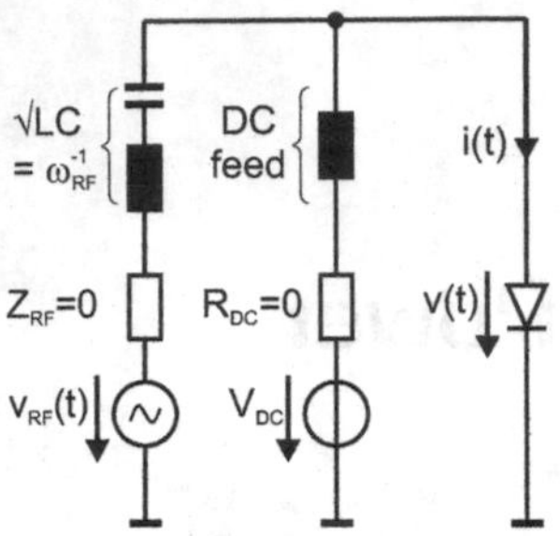

Figure 5.1: Simplified detector equivalent circuit.

$2\pi f$ according to Fig. 5.1. Then, we construct the Taylor[1] series around $v = V_0$. The DC current component $I_0 = g(V_0)$ corresponds to the applied DC bias voltage V_0, whereas the generated DC current component due to the applied RF signal is called I_{DC}.

$$
\begin{aligned}
i - I_0 &= \sum_{k=0}^{\infty} (v - V_0)^k \ \frac{g^{(k)}(V_0)}{k!} \\
i - I_0 &= (v - V_0)^1 \ \frac{g^{(1)}(V_0)}{1!} + \cdots + (v - V_0)^4 \ \frac{g^{(4)}(V_0)}{4!} + \ldots
\end{aligned}
\tag{5.1}
$$

After inserting $v(t)$ into Eq. (5.1), choosing $k \in \{0, 1, 2, 3, 4\}$ and factoring out the cosinus terms of the arising harmonic frequencies, the peak amplitudes of DC and harmonic frequency components are known as a function of $g^{(k)}$ and $v(t)$. The arguments of the derivatives $g^{(k)}(V_0)$ are omitted in Eq. (5.2).

[1]Brook Taylor (1685–1731), English mathematician.

$$i - I_0 = \underbrace{\left[\frac{\hat{v}^2}{4}g^{(2)} + \frac{3\hat{v}^4}{192}g^{(4)}\right]}_{I_{DC}} +$$

$$\underbrace{\left[\hat{v}g^{(1)} + \frac{\hat{v}^3}{8}g^{(3)}\right]}_{1^{\text{st}}\text{ harmonic}} \cos(\omega t) + \underbrace{\left[\frac{\hat{v}^2}{4}g^{(2)} + \frac{7\hat{v}^4}{192}g^{(4)}\right]}_{2^{\text{nd}}\text{ harmonic}} \cos(2\omega t)+$$

$$\underbrace{\left[\frac{\hat{v}^3}{24}g^{(3)}\right]}_{3^{\text{rd}}\text{ harmonic}} \cos(3\omega t) + \underbrace{\left[\frac{\hat{v}^4}{192}g^{(4)}\right]}_{4^{\text{th}}\text{ harmonic}} \cos(4\omega t) \tag{5.2}$$

With the first harmonic current from Eq. (5.2), the average real power of the first harmonic P_1 is known. In the sense of section 2.1, the scenario discussed here corresponds to a shunt mounted diode mixer (Z-mixer) with DC and small signal RF stimulus but without local oscillator signal. Hence, all idler impedances are zero (short circuit behaviour) and a discussion based on simple analytical 4th order Taylor series approximation is possible rather than single tone harmonic balance analysis (section 2.3).

$$\begin{aligned}
P_1 &= \frac{1}{T}\int_0^{T=1/f} \hat{v}\cos(\omega t)\left(g^{(1)}\hat{v} + g^{(3)}\frac{\hat{v}^3}{8}\right)\cos(\omega t)\,\mathrm{d}t \\
P_1 &= \frac{\hat{v}}{T}\left(g^{(1)}\hat{v} + g^{(3)}\frac{\hat{v}^3}{8}\right)\int_0^T \cos(\omega t)^2\,\mathrm{d}t \\
P_1 &= \frac{\hat{v}^2}{2}\left[g^{(1)} + g^{(3)}\frac{\hat{v}^2}{8}\right]
\end{aligned} \tag{5.3}$$

The current sensitivity (responsivity) β' is defined as the ratio of the arising DC current I_{DC} and first harmonic power level[2] P_1.

$$\beta' = \frac{I_{\mathrm{DC}}}{P_1} = \frac{1}{2}\frac{g^{(2)}\frac{\hat{v}^2}{2} + g^{(4)}\frac{\hat{v}^4}{32}}{g^{(1)}\frac{\hat{v}^2}{2} + g^{(3)}\frac{\hat{v}^4}{16}} \tag{5.4}$$

It is possible to factorize Eq. (5.4) into a square law term, the intrinsic low level current sensitivity β'_0, and a correction term. The latter depends on the applied input signal and covers deviation from square law behaviour due to generation of higher harmonics.

$$\begin{aligned} \beta' &= \frac{I_{\mathrm{DC}}}{P_1} = \beta'_0 \left[\frac{1+\Delta_1}{1+\Delta_2}\right] \\ \beta'_0 &= \frac{1}{2}\frac{g^{(2)}}{g^{(1)}} \quad \Delta_1 = \frac{\hat{v}^2}{16}\frac{g^{(4)}}{g^{(2)}} \quad \Delta_2 = \frac{\hat{v}^2}{8}\frac{g^{(3)}}{g^{(1)}} \end{aligned} \tag{5.5}$$

Using the simple Schottky junction model from Eq. (5.6) (compare section 2.2), derivation of β' as a function of the diode's equivalent circuit parameters is possible.

$$i = g(v) = I_{\mathrm{S}}\left[\exp\left(\frac{v}{\eta V_{\mathrm{T}}}\right) - 1\right] \quad V_{\mathrm{T}} = \frac{kT}{q_{\mathrm{e}}} \tag{5.6}$$

[2]When extracted from measurements, sensitivity is defined as the ratio of generated DC component to the available source power at the detector's input port.

The derivatives $g^{(m)}$ are summarized in Eq. (5.7).

$$
\begin{aligned}
g(V_0) &= I_S\left(\exp\left(\frac{V_0}{\eta V_T}\right)-1\right) = I_0 \\
g^{(1)}(V_0) &= \frac{I_S}{(\eta V_T)^1}\exp\left(\frac{V_0}{\eta V_T}\right) = \dots = \frac{I_0+I_S}{\eta V_T} = \frac{1}{R_j} \\
g^{(2)}(V_0) &= \frac{I_S}{(\eta V_T)^2}\exp\left(\frac{V_0}{\eta V_T}\right) = \dots = \frac{1}{R_j(\eta V_T)^1} \\
g^{(m)}(V_0) &= \frac{I_S}{(\eta V_T)^m}\exp\left(\frac{V_0}{\eta V_T}\right) = \dots = \frac{1}{R_j(\eta V_T)^{m-1}} \quad m \in \mathbb{N}^+
\end{aligned}
\tag{5.7}
$$

R_j is the detector diode's junction resistance at the applied DC voltage V_0 or $v = 0$ in case of zero bias operation and appears as part of the detector input resistance in equivalent circuits (Fig. 5.3).

$$
R_j = \left.\frac{dv}{di}\right|_{v=V_0} = \frac{\eta V_T}{I_S + I_0} \qquad R_j = \left.\frac{dv}{di}\right|_{v=0} = \frac{\eta V_T}{I_S} \tag{5.8}
$$

Inserting Eq. (5.7) into Eq. (5.5), detector sensitivity is known as a function of diode equivalent parameters.

$$
\begin{aligned}
\beta' &= \beta'_0\left[\frac{1+\Delta_1}{1+\Delta_2}\right] \\
\beta'_0 &= \frac{1}{2\eta V_T} \quad \Delta_1 = \frac{\hat{v}^2}{16(\eta V_T)^2} \quad \Delta_2 = \frac{\hat{v}^2}{8(\eta V_T)^2} = 2\Delta_1
\end{aligned}
\tag{5.9}
$$

The proposed detectors, both, drive high load resistances R_L and act as voltage sources. Hence, voltage sensitivity $\gamma' = V_{DC}/P_1 = \beta' R_j$ is the quantity of interest. In the sense of Fig. 5.2 (left side), β' and γ' correspond to the ratio of DC current resp. DC voltage and the first harmonic RF power $P_{R_j} = P_1$, absorbed in the diode's junction resistance R_j.

$$
\beta' = \frac{I_{DC}}{P_{R_j}} \quad \gamma' = \frac{V_{DC}}{P_{R_j}} \tag{5.10}
$$

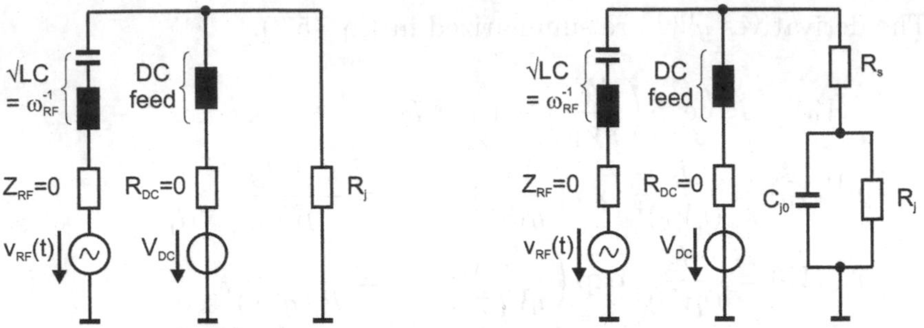

Figure 5.2: Simplified detector equivalent circuit at low input power levels with (right) and without (left) taking into account the lowpass effect from diode series resitance and junction capacitance.

To take into account the effects of series resistance R_s and junction capacitance C_j (Eq. (2.21) from subsection 2.2.1), the complex voltage divider on the right side of Fig. 5.2 is analyzed. The ratio of P_{R_j} to the total RF power P_{RF} is given by Eq. (5.11).

$$\frac{P_{R_j}}{P_{RF}} = \frac{1}{\left[1+\frac{R_s}{R_j}\right]} \frac{1}{\left[1+(f/f_c)^2\right]}$$
$$f_c = \frac{\sqrt{1+R_s/R_j}}{2\pi C_j \sqrt{R_s R_j}} \quad C_j = \frac{C_{j0}}{\sqrt{1+\frac{v}{\Phi_{bi}-V_T}}} \tag{5.11}$$

The value of junction capacitance C_j depends on the applied DC voltage. At zero bias, the maximum value C_{j0} has to be used. Hence, the current sensitivity of the detector in the presence of R_s and C_j becomes [6, 7]

$$\beta'' = \frac{I_{DC}}{P_{R_j}} \cdot \frac{P_{R_j}}{P_{RF}} = \beta_0' \left[\frac{1+\Delta_1}{1+\Delta_2}\right] \frac{1}{\left[1+\frac{R_s}{R_j}\right] \cdot \left[1+(f/f_c)^2\right]}. \tag{5.12}$$

As a final step, a non zero input reflection coefficient $S_{11}(f) > 0$ is considered. The general expression of current and voltage sensitivity is given by Eq. (5.13).

$$\beta = \underbrace{\beta_0' \frac{\left[1 - |S_{11}(f)|^2\right]}{\left[1 + \frac{R_S}{R_j}\right] \cdot \left[1 + (f/f_c)^2\right]}}_{\beta_0} \left[\frac{1 + \Delta_1}{1 + \Delta_2}\right]$$

$$\gamma = \underbrace{\underbrace{\overbrace{\frac{1}{2(I_S + I_0)}}^{R_j \beta_0'} \frac{1}{\left[1 + \frac{R_S}{R_j}\right]}}_{\gamma_0'} \frac{\left[1 - |S_{11}(f)|^2\right]}{\left[1 + (f/f_c)^2\right]}}_{\gamma_0} \left[\frac{1 + \Delta_1}{1 + \Delta_2}\right] \tag{5.13}$$

Fig. 5.3 shows Thévenin[3] and Norton[4] video equivalent circuits with the equivalent video resistance $R_\mathrm{V} = R_\mathrm{s} + R_\mathrm{j}$ and the load resistance R_L. R_L is either a real part of the detector or the input resistance of a cascaded video amplifier. In the above derivations, the load resistance R_L is assumed to be much greater than the video resistance $R_\mathrm{V} \ll R_\mathrm{L}$.

Upper Boundary of the Square Law Dynamic Range At low input power levels P_RF the detector shows square law behaviour. The output DC voltage has linear dependency of P_RF. It is $\gamma(f) \approx \gamma_0(f)$ in Eq. (5.13) and the correction term $[(1 + \Delta_1)/(1 + \Delta_2)]$ is negligible. The latter is called square law error ΔP_sq and quantifies the deviation from square law operation. The input power level, which leads to a specified square law error (e.g. 0.3 dB, 0.5 dB, 1 dB, 3 dB) is called the upper limit of square law operation P_USL. In case of power monitoring applications, ΔP_sq needs to be known at all frequencies and input power levels to calibrate the detector.

[3]Léon Charles Thévenin (1857–1926), French engineer.
[4]Edward Lawry Norton (1898–1983), American engineer.

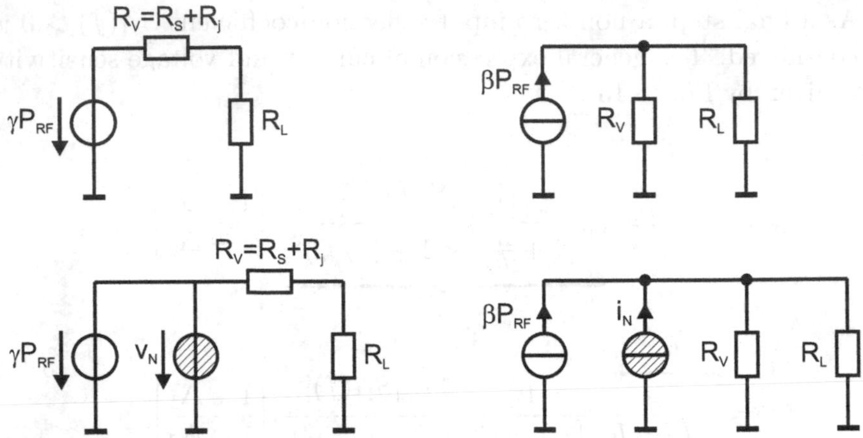

Figure 5.3: Thévenin (left) and Norton (right) video equivalent circuits with (bottom) and without (top) equivalent noise sources.

$$P_{\mathrm{USL}} \quad \rightarrow \quad \Delta P_{\mathrm{sq}} = \left[\frac{1+\Delta_1}{1+\Delta_2}\right] \tag{5.14}$$

Lower Boundary of Square Law Dynamic Range To evaluate the lower boundary of the square law dynamic range, we introduce equivalent noise sources, as shown in the bottom part of Fig. 5.3. In case of thermal noise[5] with one sided power spectral density $\mathrm{PSD}_i(f) = \frac{4kT}{R}$, $\mathrm{PSD}_v(f) = 4kTR$, the equivalent root mean square noise current I_{N} and voltage V_{N} of Fig. 5.3 are given by Eq. (5.15).

[5]Compare [8, 9] for consideration of shot noise $\mathrm{PSD}_i(f) = 2q_{\mathrm{e}}I$ and flicker noise $\mathrm{PSD}_i(f) = CI^b f^{-a}, C = \mathrm{const.}, a \approx 1, b \approx 2$.

$$\sqrt{\overline{i_{\mathrm{N}}^2(t)}} = I_{\mathrm{N}} = \sqrt{\left[\int\limits_{\Delta f} \mathrm{PSD}_i(f)\mathrm{d}f\right]} = \sqrt{\frac{4kT\Delta f}{R_{\mathrm{V}}}}$$
$$\sqrt{\overline{v_{\mathrm{N}}^2(t)}} = V_{\mathrm{N}} = \sqrt{\left[\int\limits_{\Delta f} \mathrm{PSD}_v(f)\mathrm{d}f\right]} = \sqrt{4kT\Delta f R_{\mathrm{V}}} \tag{5.15}$$

The detector's input power level P_{NEP}, which causes DC voltages V_{DC} equal to the noise voltage V_{N} is called noise equivalent power level P_{NEP}.

$$\begin{aligned} V_{\mathrm{DC}} &\overset{!}{=} V_{\mathrm{N}} \\ \gamma_0 P_{\mathrm{NEP}} &= \sqrt{4kT\Delta f R_{\mathrm{V}}} \\ &= \sqrt{4kT\Delta f R_{\mathrm{j}}\left(1+\frac{R_{\mathrm{s}}}{R_{\mathrm{j}}}\right)} \end{aligned} \tag{5.16}$$

Incorporating Eq. (5.13), the noise equivalent power level as a function of Schottky diode and detector parameters is given by Eq. (5.17).

$$P_{\mathrm{NEP}} = \underbrace{2\eta V_{\mathrm{T}}\sqrt{\frac{4kT\Delta f}{R_{\mathrm{j}}}}\sqrt{\left[1+\frac{R_{\mathrm{s}}}{R_{\mathrm{j}}}\right]}}_{P_{\mathrm{NEP}_0}} \frac{\left[1+(f/f_{\mathrm{c}})^2\right]}{\left[1-|S_{11}(f)|^2\right]}$$
$$P_{\mathrm{NEP}} = P_{\mathrm{NEP}_0}\frac{\left[1+(f/f_{\mathrm{c}})^2\right]}{\left[1-|S_{11}(f)|^2\right]} \tag{5.17}$$

In this sense, P_{NEP} belongs to a signal to noise ratio of SNR = 1 dB. It is more common to measure the so called tangential signal sensitivity (TSS) power level P_{TSS}, which belongs to SNR = 4 dB at video bandwidth VBW = 1 Hz. For TSS measurement the detector output is connected to a low noise video amplifier (Fig. 5.22) and an oscilloscope with defined video bandwidth in cascade. The input power level is periodically turned ON and OFF (square modulation) and lowered to the P_{TSS} level, where the lower noise peaks during the ON state approximately equal the upper noise peaks during the OFF state. Eq. (5.18) allows for straightforward calculation of P_{NEP}.

$$\begin{aligned} P_{\mathrm{TSS}} &= P_{\mathrm{NEP}} \cdot 2.5 \cdot \sqrt{\Delta f} \\ P_{\mathrm{TSS_{dB}}} &= P_{\mathrm{NEP_{dB}}} + 4\ \mathrm{dB} + 5 \log_{10}(\Delta f) \end{aligned} \tag{5.18}$$

5.2 Power Detector Architectures

Fig. 5.4 depicts two common power detector architectures. Although, both require the same number of components, realization at millimeter-wave frequency range differs. With variant #1 the incident RF power P_{RF} drives the diode toward a virtual short, realized by the parallel capacitor C_{p}. At DC the diode's anode is shorted by the RF choke L_{choke} and a DC voltage drop $V_{\mathrm{out}} = V_{\mathrm{DC}} \sim P_{\mathrm{RF}}$ across the load resistor R_{L} arises.

There are quasi-lumped element lowpass filters (compare filter designs from chapter 2) and resonator based bandstop filters [4, 10] with good transmission properties at DC and low frequencies together with short circuit reflection at high frequencies. Anyway, real short circuit behaviour $\Gamma = -1$ is possible at a single frequency only. Detector designs are sensitive to diode grounding, which limits these approaches to narrow bandwidth detector solutions. Metal insulator metal (MIM) capacitors or single layer capacitors (SLC), as utilized for RF grounding at gate and drain contacts of various amplifiers in chapter 6, constitute a promising realization possibility of C_{p}. But

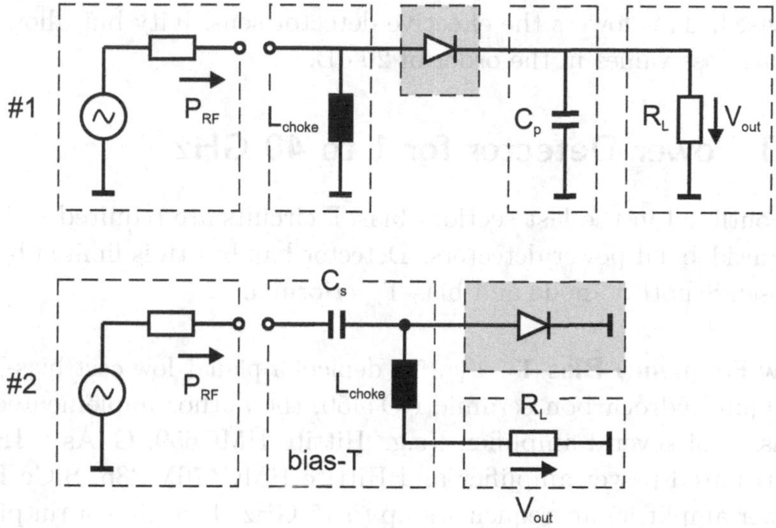

Figure 5.4: Two common power detector architectures.

MIM capacitors are not part of the standard PCB or thin-film process and the large minimum size of commercially available SLC's adds additional signal paths and increases the tendency to radiation of the EM wave rather than experiencing short circuit reflection.

Hence, when it comes to ultrawideband power detectors the variant #2 in Fig. 5.4 turns out advantageous. It utilizes a real DC coupled short circuit at the diode's cathode and a bias-T circuit to prevent RF power coupling to the load resistor R_L and DC voltage V_{out} coupling to the preceding system parts.

Two power detector designs, operating from 1 to 40 GHz and 60 to 110 GHz are presented in the next sections. These are dedicated to automatic level control, power monitoring and overload protection within front end modules and normally used in conjunction with loose directional couplers (10 to 20 dB coupling). If poor return loss behaviour of the detector is an issue of system design, combination with the author's DC to 110 GHz attenuator series from section 3.4 is advised. This lowers the effective detector sensitivity but allows for return loss values in the order of 20 dB.

5.3 Power Detector for 1 to 40 GHz

As outlined in the last section, bias-T circuits are required to build ultrawideband power detectors. Detector bandwidth is limited by the chosen Schottky diode and bias-T performance.

Low Frequency Bias-T Fig. 5.5 depicts a planar low cost bias-T on 200 µm hydrocarbon ceramic RO4450, the author implemented for biasing of several amplifiers (e.g. Hittite HMC659, GaAs pHEMT distributed power amplifier and Hittite HMC476MP86, SiGe HBT power amplifier) at frequencies up to 15 GHz. It utilizes a tin plated C = 10 nF chip ceramic capacitor GRM155R71C103KA01D [11] from Murata Manufacturing Co., Ltd. and a ferrite chip inductor $L = 560$ nH, $R = 0.92$ Ω, 0402AF-561XJLU [12] from Coilcraft. The maximum DC voltage is $V_{\max} = 16$ V due to the capacitor with a maximum DC current $I_{\max} = 200$ mA due to the coil.

A defected ground structure (DGS) according to Fig. 5.5 allows for compensation of the capacitor's width, which does not fit to the 50 Ω microstrip line (MSL). The shunt inductor is connected to the main path by high impedance MSL with 100 µm width, which is the minimum realizable strip width of the applied standard PCB process. Fig. 5.6 shows measured insertion loss, return loss and isolation performance versus frequency. These results include three SMA connectors. Insertion loss is −2.5 dB at 10 MHz and around 2 dB at 10 GHz.

High Frequency Bias-T To overcome these limitations, the multilayer ceramic capacitor C = 82 pF, $V_{\max}$ = 50 V, Opti-Cap®,

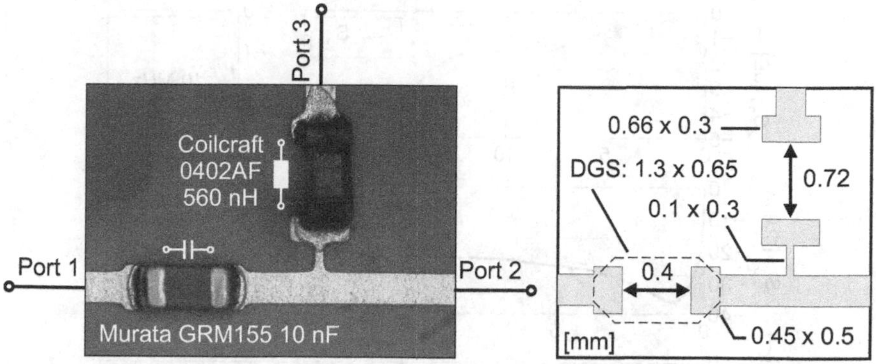

Figure 5.5: Photograph of the low frequency bias-T circuit, utilizing GRM155 series capacitor and 0402AF ferrite chip inductor.

P02BN820Z5S [13] from Dielectric Laboratories Inc. (DLI) is incorporated[6].

Fig. 5.7 shows the Opti-Cap® mounted on 5 mil alumina, including the dimensions of the DGS structure and two transitions from MSL to grounded coplanar waveguide (gCPW). Fig. 5.8 includes results from measurements with ground-signal-ground on-wafer probes after enhanced line-reflect-reflect-match calibration (eLRRM) [15].

Insertion loss values of 4 dB at 10 MHz and less than 2 dB up to 110 GHz are achieved, while return loss values are greater than 9 dB from 10 MHz to 110 GHz. These measurements have been carried out in the absence of metallic enclosure. Hardly any system operating up to 110 GHz comes without metallic housing to prevent radiation effects. Due to the large size of the Opti-Cap®, the maximum frequency within metallic channels is limited to frequencies less than 110 GHz by occurring higher order modes.

Fig. 5.9 shows the fully assembled Schottky power detector, which utilizes the Opti-Cap® series capacitor in conjunction with a shunt mounted ferrite filled conical coil, $L = 11$ µH, $R = 7\ \Omega$,

[6]The capacitor 550L from American Technical Ceramics (ATC) [14] has similar performance, but less maximum DC voltage capability $V_{\max} = 16$ V.

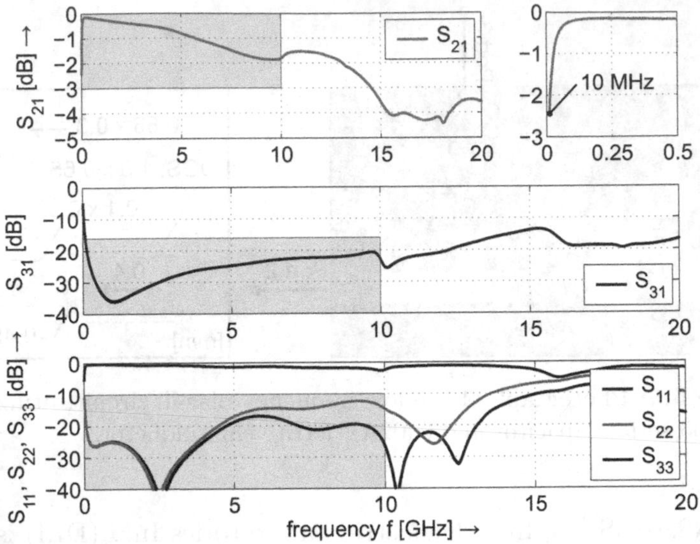

Figure 5.6: Measured scattering parameters of the low frequency bias-T circuit, utilizing GRM155 series capacitor and 0402AF ferrite chip inductor.

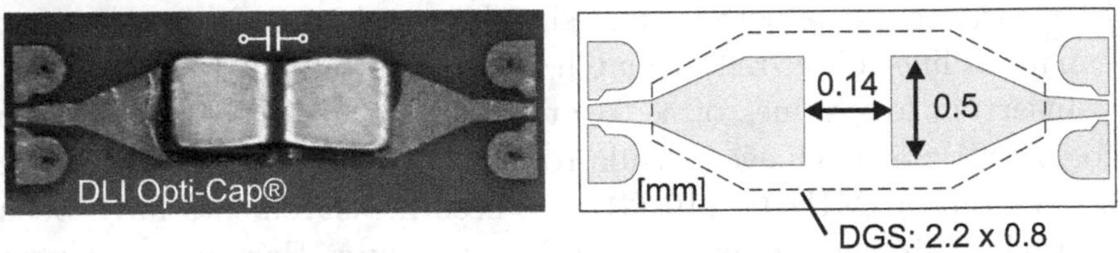

Figure 5.7: Photograph and dimensions of the Opti-Cap® on-wafer test structure.

506WLC110KG115B [16] from American Technical Ceramics (ATC). The coil tolerates a maximum DC current of $I_{\max} = 115$ mA and is conductively glued the Au plated Opti-Cap® rather than the alumina Au metallization to minimize interaction with the incident electromagnetic wave. A cylindrical piece of silicone gives mechanical stability.

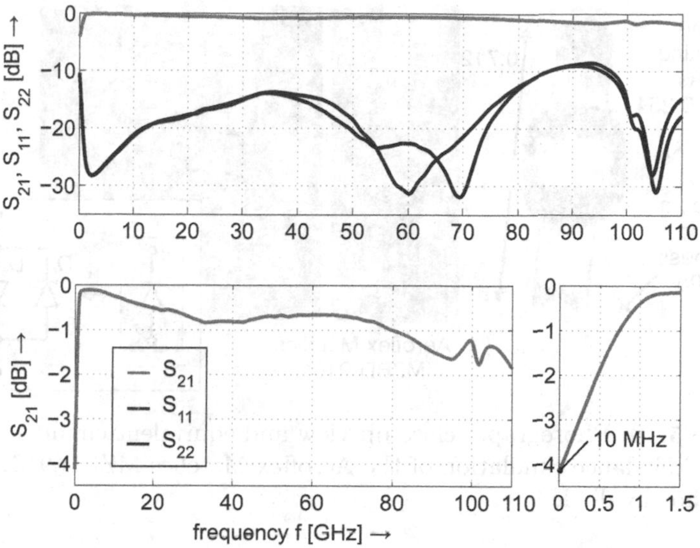

Figure 5.8: Measured scattering parameters of the Opti-Cap® series capacitor.

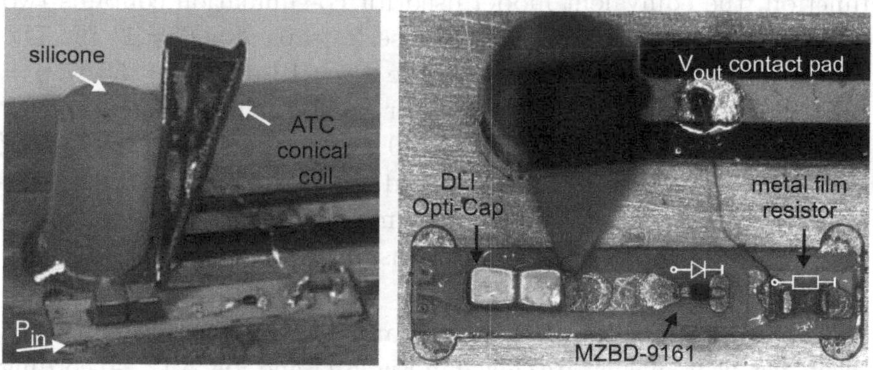

Figure 5.9: Photograph of the proposed planar power detector with assembled Opti-Cap®, conical coil, diode and load resistor R_L.

The detector follows architecture #2 of Fig. 5.4 with the Aeroflex Metelics, MZBD-9161 zero bias detector diode, which is a reproduction

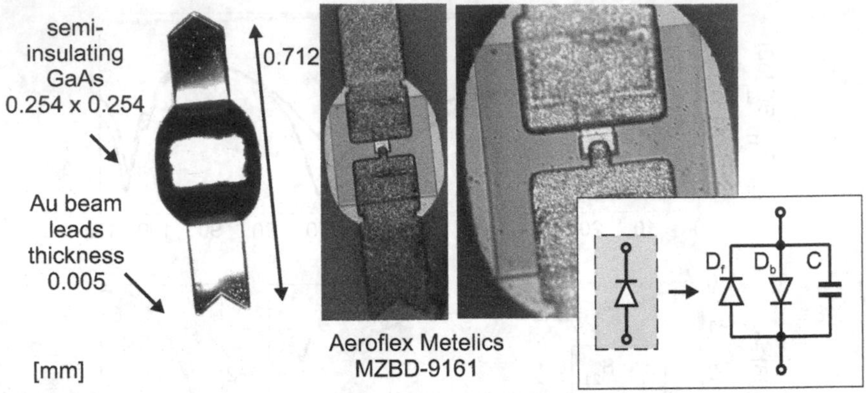

Figure 5.10: Photograph, close up view and equivalent circuit model for harmonic balance simulation of the Aeroflex Metelics MZBD-9161 zero bias Schottky diode.

of the famous Agilent HSCH-9161 diode. A detail view of the diode is given in Fig. 5.10. Although it features a single GaAs Schottky junction, the equivalent model used for co-simulation contains two antiparallel diodes D_f and D_b and a series capacitor $C = 11$ fF. The Schottky diode parameters of D_f are $I_S = 12$ µA, $R_s = 50\ \Omega, \eta = 1.2, C_{j0} = 0.03$ pF, $\Phi_{bi} = 0.26$ V. For D_b it is the same, except for $I_S = 84$ µA, $R_s = 10\ \Omega$ and $\eta = 40$. In analogy to the design of frequency multipliers in chapter 2 and mixers in chapter 4, the semi-insulating GaAs carrier and the first metallization layer of the diode are covered within 3D EM simulations.

Measurement Results Return loss and voltage sensitivity measurements versus frequency are shown in Fig. 5.11 and Fig. 5.12. According to Eq. (5.13), frequency dependency of sensitivity and return loss should be the same. This is not the case, due to electromagnetic radiation from the large coil and resistive losses. In the context of Eq. (5.13), S_{11} strictly covers the diode mismatch only.

The detector's square law voltage sensitivity averages out at 1 V/mW from 1 to 25 GHz and stays above 300 mV/mW up to 40 GHz.

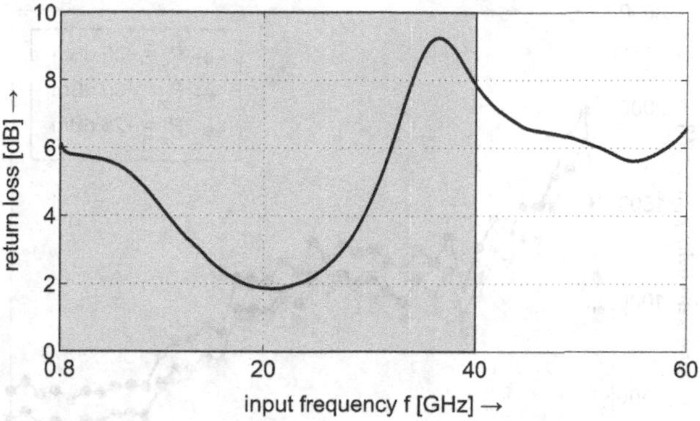

Figure 5.11: Measured return loss versus frequency.

Fig. 5.13 contains measured output voltage V_{out} versus input power level P_{in} at 15, 25, 37 GHz and 40 GHz, while all measurement data from 1 to 40 GHz are within the gray shaded area. The lowest input power levels are measured with 100 power line cycles (PLC), which corresponds to a video bandwidth of only 0.5 Hz. Within the source module of section 6.1, the detector works toward a low noise video amplifier LTC1037C (compare Fig. 5.22) with voltage gain ≈ 1000 and video bandwidth of 4 kHz. According to Eq. (5.18), the minimum measurable input power level shifts by the amount of $10 \log_{10} \left(\sqrt{\Delta f}\right) \approx +18$ dBm.

As it is shown in the next section (Table 5.1), the sensitivity strongly depends on the diode's junction capacitance and the achievable return loss (Eq. (5.13)). In case of the MZBD-9161 or HSCH-9161 diode with $C_{\text{j0}} = 35$ fF and the reasonable return loss values of 2 to 6 dB from 1 to 40 GHz, the analytical maximum sensitivity values are only 800 to 1700 mV/mW. This is in very good agreement with the measurement results of Fig. 5.12 and outperforms reported detectors like [17].

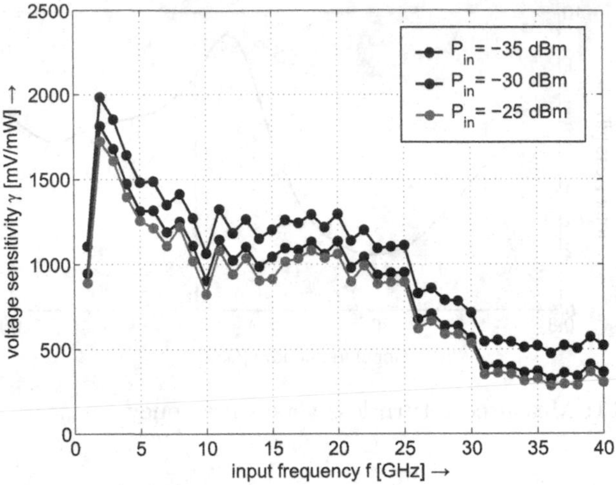

Figure 5.12: Measured square law voltage sensitivity γ versus frequency at three input power levels $P_{\text{in}} = -25, -30, -35$ dBm.

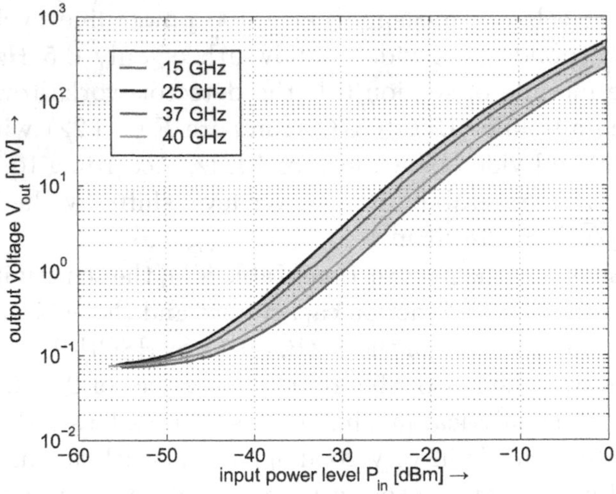

Figure 5.13: Measured DC output voltage V_{out} versus input power level P_{in} at $f = 15, 25, 37$ GHz and $f = 40$ GHz. Measurement data in the frequency range of 1 to 40 GHz lies within the gray shaded area.

5.4 Power Detector for 60 to 110 GHz

Recently detectors with antimony (Sb) heterostructure based backward diodes,[7] grown by molecular beam epitaxy (InAs/GaAlSb process) have been developed for passive W-band imaging cameras. These have superior noise properties [18] compared to GaAs Schottky diodes. At matched conditions $S_{11} = 0$ both, Sb and Schottky diodes would allow for power detectors with square law voltage sensitivities[8] γ_0' in the order of 20000 mV/mW to 40000 mV/mW (Eq. (5.13)). Nevertheless, Sb diodes exhibit much lower dynamic junction resistances R_j at zero bias and can therefore be matched more easily to circuits with 50 Ω input impedance. In [19] peak sensitivities of more than 10000 mV/mW at W-band frequencies have been achieved.

Anyway, the required monolithic process is proprietary and discrete Sb diodes are not commercially available. Therefore the focus is on a planar detector circuit with commercial GaAs zero bias Schottky diode (ZBD) from VDI to build up an ultrawideband $f \subset [60, 110]$ GHz detector for automatic level control and power monitoring purposes.

Fig. 5.14 shows a scanning electron micrograph, 3D EM model, circuit model parameters and the corresponding static current-voltage characteristic of the utilized diode. For comparison, a single Schottky junction of the UMS DBES105a diode reaches diode current of 1 mA at 750 mV. The model parameters are chosen to fit best to measured data and are used to describe the nonlinear Schottky junction, metal to depleted and undepleted epitaxial layer, within harmonic balance circuit simulations. Additional parasitics like pad to pad capacitance, anode finger inductance and the interaction of the semi-insulating GaAs carrier with the incident EM wave are considered within 3D EM fullwave simulations. The applied co-simulation procedure is the same as is in chapter 2 and chapter 4.

Several hybrid W-band Schottky diode detectors have been published. The detectors in [20] and [21] are based on the Agilent HSCH-9161 diode and exhibit sensitivities of 1000 mV/mW $< \gamma <$

[7] Backward diodes are a version of Esaki (tunnel) diodes.

[8] The term responsivity is used as a synonym of sensitivity by some authors.

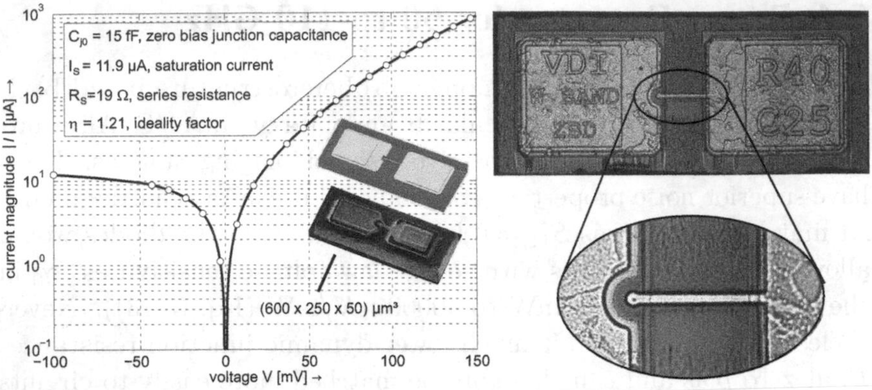

Figure 5.14: Photograph, 3D EM model, circuit model parameters and current versus voltage of the utilized VDI GaAs zero bias detector diode. Diode dimensions $(600 \times 250 \times 50)\ \mu m^3$.

3750 mV/mW in the frequency range of 75 GHz to 105 GHz with tangential signal sensitivity power levels of -68 dBm $\leq P_{\mathrm{TSS}} \leq -55$ dBm at video bandwidth VBW = 70 Hz. Datasheet information of the commercial detectors from Millitech Inc. DET-12 and DET-10 [22] guarantee sensitivities of $\gamma > 700$ mV/mW across the E-band and $\gamma > 1000$ mV/mW across the W-band with $P_{\mathrm{TSS}} \approx -45$ dBm at VBW = 1 kHz.

In the following the influence of return loss, frequency and junction capacitance on detector sensitivity is studied, based on analytical calculations. With Eq. (5.8) the video resistance in case of the VDI ZBD is $R_{\mathrm{V}} = R_{\mathrm{s}} + R_{\mathrm{j}} = 2871\ \Omega$. If the load resistance R_{L} is chosen much greater than the video resistance $R_{\mathrm{L}} \gg R_{\mathrm{V}}$, the square law voltage sensitivity of an ideal matched $S_{11} = 0$ detector is called γ_0'. Whereas γ_0 takes into account the lowpass effect of nonzero junction capacitance C_{j0} and the input return loss $S_{11} > 0$. Table 5.1 summarizes the theoretical square law sensitivity values γ_0' and γ_0 from Eq. (5.13) for several matching conditions $S_{11} \in \{-\infty, -8, -6, -4, -2\}$ dB at $f = 85$ GHz. It turns out, detectors with VDI ZBD ($C_{\mathrm{j0}} = 15$ fF)

Table 5.1: Analytical calculations of voltage sensitivities

[γ]=mV/mW	γ_0'	γ_0	γ_0	γ_0	γ_0	γ_0
$\lvert S_{11}\rvert^2$		**$\rightarrow \infty$ dB**	**-8 dB**	**-6 dB**	**-4 dB**	**-2 dB**
C_{j0} = 0 fF	41708	41708	35098	31232	25104	15392
C_{j0} = 15 fF °	41708	10154	8545	7604	**6112**	3747
C_{j0} = 35 fF †	41708	2328	1959	1743	**1401**	859

f = 85 GHz, I_S = 11.9 µA, η = 1.21, R_s = 19 Ω, $R_L/(R_L + R_V)$ = 0.9943

° VDI ZBD, † Agilent HSCH-9161 or Aeroflex Metelics MZBD-9161

can gain four times higher sensitivities than designs based on Agilent HSCH-9161 (C_{j0} = 35 fF).

The proposed planar power detector design is based on thin-film processed 5 mil Al_2O_3 substrate with 5 µm Au metallization. Fig. 5.15 shows a top view of the single layer structured substrate, which is soldered to Au plated brass housing using 62Sn-36Pb-2Ag alloy of solder paste with particle size of 15 to 30 µm. It further shows the assembled VDI ZBD and load resistor R_L. The substrate dimensions are (6.0 × 0.9) mm^2, whereas only the high frequency part with a length of 2.5 mm requires a metallic enclosure (1.0 × 0.635) mm^2 to avoid radiation losses. As depicted in Fig. 5.15 there is a grounded coplanar waveguide (gCPW) to microstrip line (MSL) transition at the detector input, which is intended to be used for bondwire intersubstrate connection within highly integrated front end modules (compare section 6.2). The input signal passes a DC blocking high pass filter (subsection 5.4) and drives the VDI ZBD toward a DC short circuit. The arising DC power from the rectification causes a voltage drop V_{out} across the load resistor R_L = 500 kΩ. An RF choke (subsection 5.4) consisting of NiCr thin-film sheet resistors prevents the incident power from coupling to R_L.

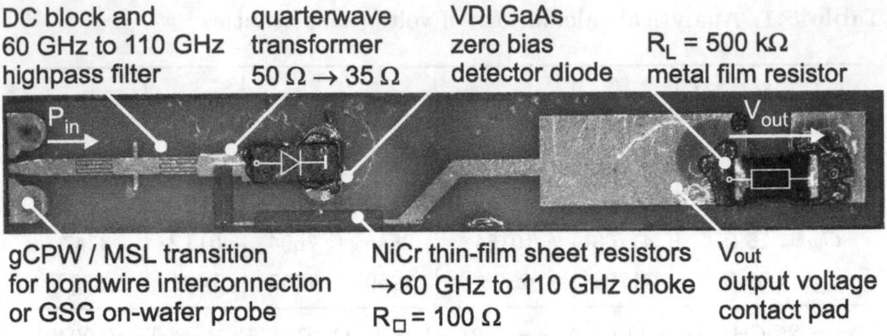

Figure 5.15: Photograph of the proposed planar power detector with assembled diode and load resistor R_L. Metallic enclosure (1.0×0.635) mm^2 is not depicted.

DC Block - Tapered Capacitor and Highpass Filter Metal insulator metal (MIM) capacitors are adequate candidates to realize DC blocks and millimeter wave short circuits, but are normally not part of the standard thin-film process. Interdigital capacitors are a straightforward alternative. The structure of Fig. 5.16 with 350 µm finger length and 10 µm spacing has an equivalent series capacitance $C_\mathrm{ESC} = 134$ fF at $f = 85$ GHz and a high series resonant frequency $f_\mathrm{SRF} \approx 152$ GHz.

$$C_\mathrm{ESC}(f_0) = \mathrm{Im}\{-Y_{21}(f_0)\}/\left(2\pi f_0\right) \quad C_\mathrm{ESC}(f_\mathrm{SRF}) = 0 \tag{5.19}$$

Fig. 5.16 compares simulated scattering parameters with results from on-wafer measurements after an eLRRM probe tip calibration [15] up to 110 GHz. The measurement data includes the parasitic influence of the probe pad structure. Following the suggestions in [23] a tapered geometry of the finger structure has been designed, that allows for return loss values greater than 10 dB at the input frequency range $f \in [60, 110]$ GHz. Utilizing three fingers with lengths $\ell \in [10, 330]$ µm, the equivalent series capacitance range of $C_\mathrm{ESC} \in [10, 385]$ fF can be covered. Within the proposed detector two three finger capacitors with $C_\mathrm{ESC}(85\ \mathrm{GHz}) = 126$ fF and $f_\mathrm{SRF} \approx 119$ GHz are connected by a small shunt capacitance with 50 Ω offset lengths. According to

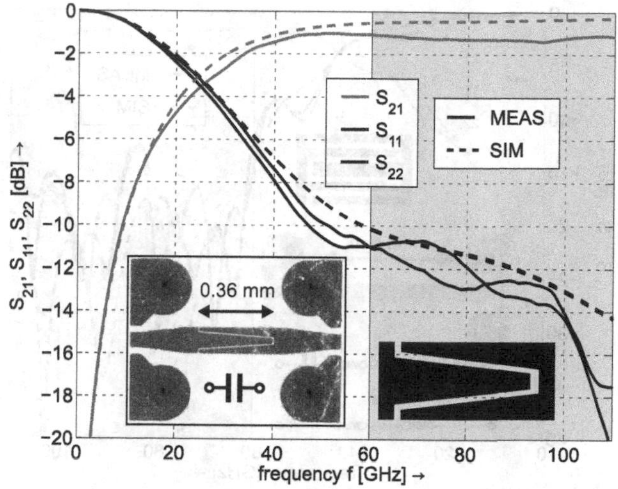

Figure 5.16: Simulated and measured scattering parameters of tapered series capacitor together with a photograph of an on-wafer test structure (length $\ell = 0.675$ mm).

Fig. 5.17 two high impedance ($Z \approx 90\ \Omega$) open stubs are used to realize the shunt capacitor with an equivalent parallel capacitance $C_{\mathrm{EPC}}(85\ \mathrm{GHz}) = 12$ fF. Simulation and measurement results validate the return loss improvement.

$$C_{\mathrm{EPC}} = \mathrm{Im}\{Z_{21}^{-1}(f_0)\}/\,(2\pi f_0) \tag{5.20}$$

The insertion loss in Fig. 5.17 is higher compared to Fig. 5.16 due to longer MSL feeding lines of the on-wafer test structure. The measured real part of propagation coefficient α is 0.27 dB/mm at $f = 110$ GHz.

RF Choke - Resistive Shunt Bypassing the DC power to the load resistor R_{L} without affecting the RF power flow to the Schottky junction requires a high impedance shunt path. The maximum realizable characteristic impedance with the utilized manufacturing technology is only 100 Ω. Size restrictions do not allow for the use of spiral

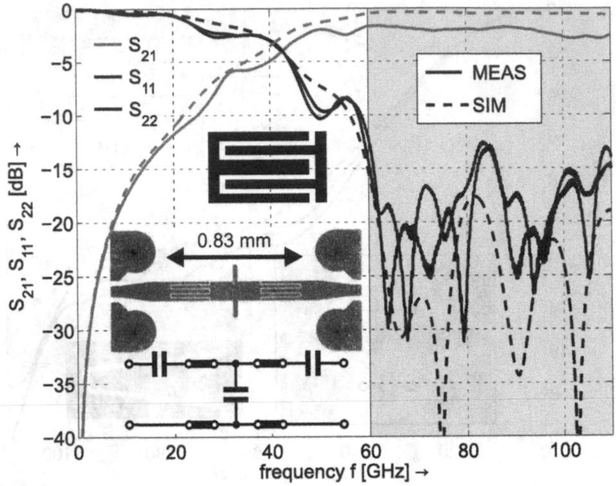

Figure 5.17: Simulated and measured scattering parameters of 60 GHz to 110 GHz highpass filter together with a photograph of an on-wafer test structure (length $\ell = 6.0$ mm).

inductors [B1]. Therefore NiCr thin-film sheet resistors constitute the most convincing RF choke design strategy. As shown in Fig. 5.15 the choke consists of two narrow sheet resistors connected in series.

$$R_{\text{choke}} = R_{\square} \cdot \left(\frac{\ell_1}{w_1} + \frac{\ell_2}{w_2} \right) = (2080 + 3900)\,\Omega = 5980\,\Omega \qquad (5.21)$$

Due to process variations of the sheet resistance $R_{\square}$ a beneficial higher total value of $R_{\text{choke}} = 7990\ \Omega$ has been measured. To establish reliable soldering of the diode a minimum pad width, that corresponds to $Z \approx 35\ \Omega$, is necessary. The transition is realized by a single stage quarterwave transformer ($\sqrt{50 \cdot 35}\ \Omega \approx 42\ \Omega$). Lowering the MSL impedance is unfavorable as we have to match the 50 Ω input impedance to the high dynamic junction resistance R_{j}, but the choke performance is improved in consequence of a higher impedance ratio between MSL and R_{choke}.

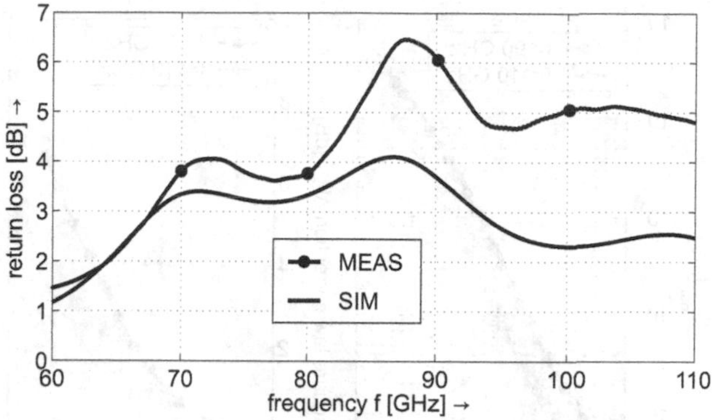

Figure 5.18: Simulated and measured return loss versus frequency.

Measurement Results Fig. 5.18 shows the detector return loss versus frequency from co-simulation and on-wafer measurement under small signal conditions. The discrepancy at $f > 85$ GHz between simulation and measurement can partly be explained by additional loss but the assembled detector seems to behave slightly better than predicted. E- and W-band frequency multipliers and variable waveguide attenuators have been used to guarantee precise power levels P_{in} at the detector input from 60 to 110 GHz. On-wafer probes instead of bondwires are used to couple the input power to the detector. Output voltage V_{out} measurements with a digital voltmeter at various input power levels P_{in} have been performed. For the sake of clarity Fig. 5.19 shows data of only two frequencies ($f = 90$ GHz and $f = 110$ GHz), whereas the measurement data from 60 to 110 GHz lies within the gray shaded area. The right side of Fig. 5.19 quantifies the detectors deviation from square law operation (Eq. (5.14)). The detector underestimates high input power levels by the plotted square law correction ΔP_{sq}, which is the necessary calibration information for power monitoring applications. The maximum tolerable ΔP_{sq} value is application specific and determines the upper limit of the square law dynamic range. In Fig. 5.21 four different upper square law limits P_{USL} are shown versus frequency. Fig. 5.20 compares the measured and simulated square law voltage sensitivity γ at three input power

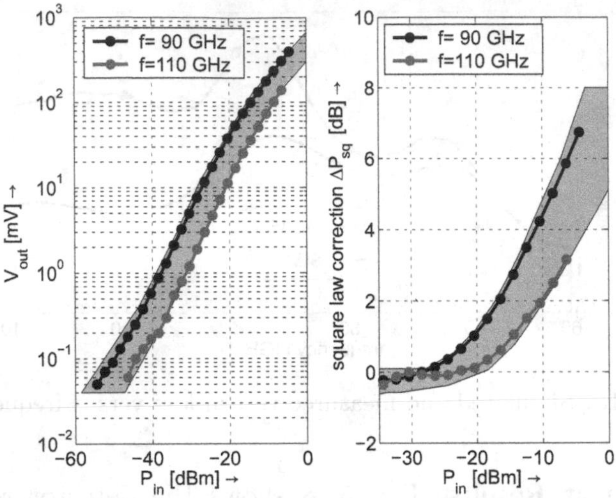

Figure 5.19: Measured DC output voltage V_{out} and square law correction ΔP_{sq} versus input power level P_{in} at $f = 90$ GHz and $f = 110$ GHz. Measurement data in the frequency range of 60 GHz to 110 GHz lies within the gray shaded area.

levels $P_{in} \in \{-25, -30, -35\}$ dBm versus frequency. γ values are between 2000 and 6000 mV/mW from 60 to 102 GHz and stay above 1000 mV/mW up to 110 GHz. Performance prediction from the applied co-simulation procedure fits well to the experimental data. The measured data outperforms the detectors mentioned at the beginning of this section. Fig. 5.20 further shows an analytically calculated γ using Eq. (5.13) with simulated return loss values from Fig. 5.18. Contrary to the power detector from section 5.3, the presented measurements have been carried out with metallic enclosure. Consequently, there is no radiation loss and frequency dependency of sensitivity and return loss are comparable.

To determine the lower boundary of the square law dynamic range the minimum detectable input power level P_{min} using a digital voltmeter has been measured versus frequency (Fig. 5.21). In addition the widely accepted tangential signal sensitivity power level P_{TSS} versus

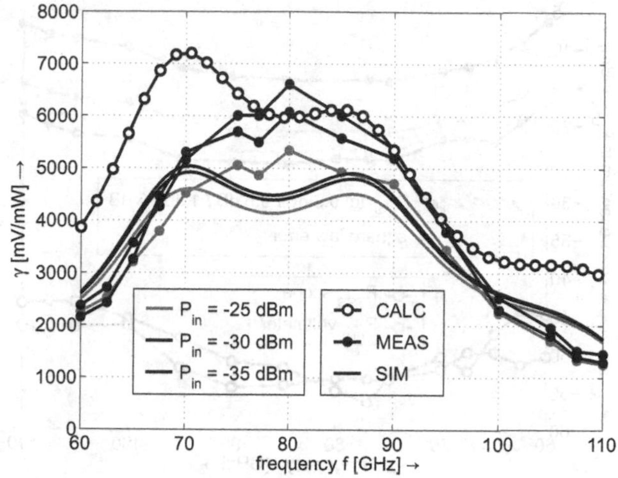

Figure 5.20: Square law voltage sensitivity γ versus frequency at three input power levels $P_{\text{in}} = -25, -30, -35$ dBm from measurement, simulation and analytical calculation.

frequency has been measured. For TSS measurement the detector output is connected to a low noise video amplifier (Fig. 5.22) with an oscilloscope in cascade (video bandwidth VBW = 4 kHz).

Fig. 5.23 illustrates an exemplary measured output voltage waveform $V_{\text{out}}(t)$ at two video bandwidths VBW = 20 MHz and VBW = 4 kHz, if a 75 GHz input signal with -50.3 dBm is periodically ($T \approx 5$ ms) turned ON and OFF. With VBW = 4 kHz the lower noise peaks during the ON state approximately equal the upper noise peaks during the OFF state, which means $P_{\text{TSS}} = -50.3$ dBm. The resulting P_{TSS} versus frequency is in the same order of magnitude as P_{min} (Fig. 5.21). Eq. (5.18) relates the measured P_{TSS} to the noise equivalent power P_{NEP} [6], that varies between 14.5 pW/$\sqrt{\text{Hz}}$ and 307.5 pW/$\sqrt{\text{Hz}}$ at the focused output frequency range $f \in [60, 110]$ GHz.

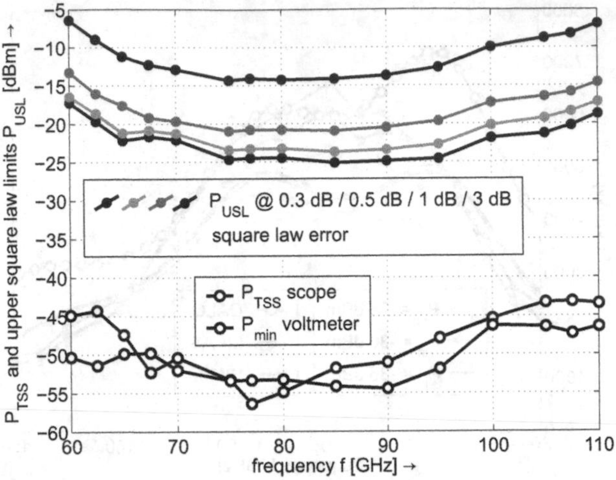

Figure 5.21: Tangential signal sensitivity power levels P_{TSS}, minimum detectable input power level using digital voltmeter P_{min} and upper square law power limits P_{USL} at several square law errors versus frequency.

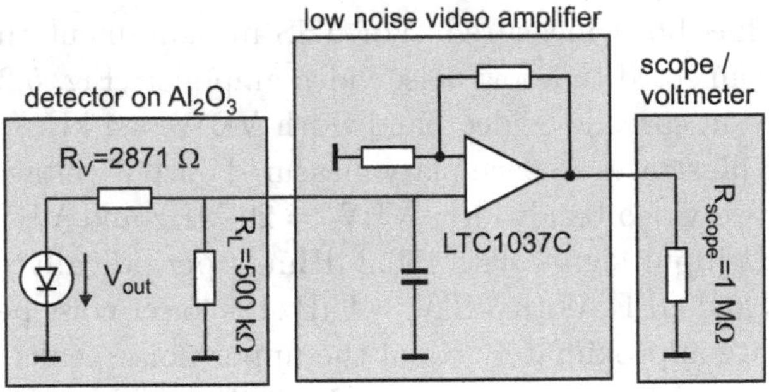

Figure 5.22: Video equivalent circuit of power detector, connected to low noise video amplifier and scope.

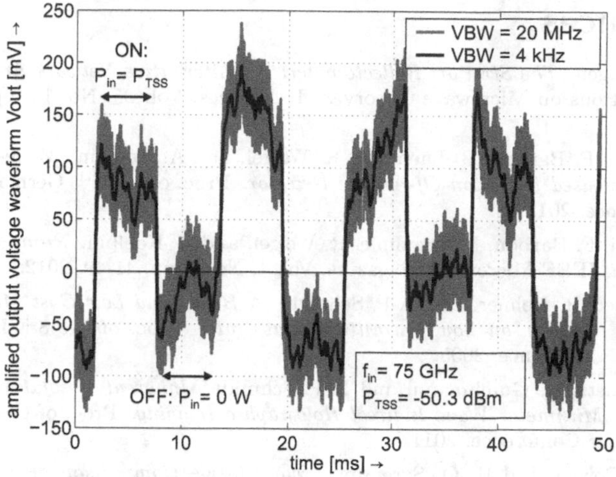

Figure 5.23: Measured output voltage waveform $V_{\mathrm{out}}(t)$ with ($P_{\mathrm{in}} = -50.25$ dBm) and without ($P_{\mathrm{in}} = 0$ W) detector input signal at $f =$ 75 GHz and two different video bandwidths VBW = 20 MHz and VBW = 4 kHz.

Conclusion The presented planar zero bias Schottky power detector covers an input frequency range of 60 to 110 GHz, which includes the entire E- and W-band and equals 59 % fractional bandwidth. Excellent square law sensitivity is achieved. The single layer planar design requires minimum strip widths and metal spacings of 10 µm which is non-standard but controllable. Laser trimming of the thin-film sheet resistors is not necessary.

The detector is ideally suited to be used in highly integrated planar front end modules of ATS, especially for power monitoring in conjunction with directional couplers at the output of signal generator frequency extension modules. In case of a maximum tolerable square law correction of 1 dB the linear dynamic range is at least 25 dB. The small size of (6.0×0.9) mm^2 can easily be shrinked to (2.5×0.9) mm^2, which makes the proposed detector attractive for array based E- and W-band imaging applications.

References

[1] G. F. Engen. *The Six-Port Reflectometer: An Alternative Network Analyzer.* IEEE Transactions on Microwave Theory & Techniques, Vol. 25, No. 12, pp. 1075-1080, 1977.

[2] G. Vinci, F. Barbon, B. Laemmle, R. Weigel, and A. Koelpin. *Wide-Range, Dual Six-Port based Direction-Of-Arrival Detector.* Proc. of the 7th German Microwave Conference, 2012.

[3] G. Vinci, F. Barbon, B. Laemmle, R. Weigel, and A. Koelpin. *Promise of a Better Position.* IEEE Microwave Magazine, Vol. 1, No. 7, pp. 41-49, 2012.

[4] J. Richter, R. Rehner, and L.-P. Schmidt. *A Broadband Low Cost Millimeterwave Slotline Detector on Low Permittivity Substrate.* Proc. of the 8th European Microwave Conference, 2005.

[5] J. Adametz, F. Gumbmann, and L.-P. Schmidt. *Inherent Resolution Limit Analysis for Millimeter-Wave Indirect Holographic Imaging.* Proc. of the 6th German Microwave Conference, 2011.

[6] A. M. Cowley and H. O. Sorensen. *Quantitative Comparison of Solid-State Microwave Detectors.* IEEE Transactions on Microwave Theory & Techniques, Vol. 14, No. 12, pp. 588-602, 1966.

[7] F. A. Benson. *Millimetre and Submillimetre Waves.* London: Iliffe Books Ltd., 1969. ISBN 0592027813.

[8] I. Bahl and P. Bhartia. *Microwave Solid State Circuit Design.* New York: John Wiley & Sons, 2003. ISBN 0471207551.

[9] G. D. Vendelin, A. M. Pavio, and U. L. Rohde. *Microwave Circuit Design Using Linear And Nonlinear Techniques.* New York: John Wiley & Sons, 1990. ISBN 0471414794.

[10] G. L. Matthaei, L. Young, and E. M. T. Jones. *Microwave Filters, Impedance Matching Networks, and Coupling Structures.* New York: McGraw-Hill, 1964. ISBN 0890060991.

[11] Murata Manufacturing Co. Ltd. *Murata GRM155R71C103KA01D Monolithic Chip Ceramic Capacitor.* www.murata.com.

[12] Coilcraft Inc. *Coilcraft 0402AF-561XJLU Ferrite Chip Inductor.* www.coilcraft.com.

[13] Dielectric Laboratories Inc. *DLI P02BN820Z5S Opti-Cap®.* www.dilabs.com.

[14] American Technical Ceramics. *ATC 550L Ultrabroadband Capacitor.* www.atceramics.com.

[15] L. Hayden. *An Enhanced Line-Reflect-Reflect-Match Calibration.* Proc. of the ARFTG Conference, pp. 143-149, 2006.

[16] American Technical Ceramics. *ATC 506WLC110KG115B Conical Coil.* www.atceramics.com.

[17] L. A. Tejedor-Alvarez, J. I. Alonso, and J. Gonzalez-Martin. *An Ultrabroadband Microstrip Detector up to 40 GHz.* Proc. of the 14th Conference on Microwave Techniques COMITE, 2008.

[18] H. P. Moyer, R. L. Bowen, J. N. Schulman, D. H. Chow, S. Thomas, J. J. Lynch, and K. S. Holabird. *Sb-Heterostructure Low Noise W-Band Detector Diode Sensitivity Measurements*. IEEE MTT-S Int. Microwave Symposium Digest, pp. 826-829, 2006.

[19] H. P. Moyer, J. N. Schulman, J. J. Lynch, J. H. Schaffner, M. Sokolich, Y. Royter, R. L. Bowen, C. F. McGuire, M. Hu, and A. Schmitz. *W-Band Sb-Diode Detector MMICs for Passive Millimeter Wave Imaging*. IEEE Microwave & Wireless Components Letters, Vol. 18, No. 10, pp. 686-688, 2008.

[20] L. Xie, Y. Zhang, Y. Fan, C. Xu, and Y. Jiao. *A W-Band Detector with High Tangential Signal Sensitivity and Voltage Sensitivity*. Proc. of the ICMMT Conference, pp. 582-531, 2010.

[21] K. Xu, Y. Zhang, L. Xie, and Y. Fan. *A Broad W-Band Detector Utilizing Zero-Bias Direct Detection Circuitry*. Proc. of the ICCP Conference, pp. 190-194, 2011.

[22] Millitech Inc. *Zero Bias Detectors DET-12 (E-Band) and DET-10 (W-Band)*. www.millitech.com.

[23] I. Huynen and G. Dambrine. *A Novel CPW DC-Blocking Topology with Improved Matching at W-Band*. IEEE Microwave & Guided Wave Letters, Vol. 8, No. 4, pp. 149-151, 1998.

[24] H. A. Watson. *Microwave Semiconductor Devices and their Circuit Applications*. New York: McGraw-Hill, 1969. ISBN 0070684758.

[25] P. L. D. Abrie. *The Design of Impedance-Matching Networks for Radio-Frequency and Microwave Amplifiers*. Boston: Artech House, 1985. ISBN 0890061726.

[26] J. N. Schulman and D. H. Chow. *Sb-Heterostructure Interband Backward Diodes*. IEEE Electron Device Letters, Vol. 21, No. 7, pp. 353-355, 2000.

[27] J. N. Schulman, D. H. Chow, C. W. Pobanz, H. L. Dunlap, and C. D. Haeussler. *Sb-Heterostructure Millimeter-Wave Zero-Bias Diodes*. Proc. of the 58th Device Research Conference, pp. 57 - 58, 2000.

[28] J. N. Schulman, D. H. Chow, and D. M. Jang. *InGaAs Zero Bias Backward Diodes for Millimeter Wave Direct Detection*. IEEE Electron Device Letters, Vol. 22, No. 5, pp. 200-202, 2001.

[29] J. N. Schulman, E. T. Croke, D. H. Chow, H. L. Dunlap, K. S. Holabird, M. Morgan, and S. Weinreb. *Quantum Tunneling Sb-Heterostructure Millimeter-Wave Diodes*. Internatinal Electron Devices Meeting IEDM, pp. 35.1.1-35.1.3 ,2001.

[30] J. N. Schulman, V. Kolinko, M. Morgan, C. Martin, J. Lovberg, S. Thomas, J. Zinck, and Y. K. Boegeman. *W-Band Direct Detection Circuit Performance with Sb-Heterostructure Diodes*. IEEE Microwave & Wireless Components Letters, Vol. 14, No. 7, 2004.

[31] N. Su, R. Rajavel, P. Deelman, J. N. Schulman, and P. Fay. *Sb-Heterostructure Millimeter-Wave Detectors with Reduced Capacitance and Noise Equivalent Power*. IEEE Electron Device Letters, Vol. 29, No. 6, pp. 536-539, 2008.

[32] J. J. Lynch, H. P. Moyer, J. H. Schaffner, Y. Royter, M. Sokolich, B. Hughes, Y. J. Yoon, and J. N. Schulman. *Passive Millimeter-Wave Imaging Module With Preamplified Zero-Bias Detection*. IEEE Transactions on Microwave Theory & Techniques, Vol. 56, No. 7, pp. 1592-1600, 2008.

[33] C.-Y. Huang, C.-M. Li, L.-Y. Chang, C.-C. Nien, Y.-C. Yu, and J.-H. Tarng. *A W-Band Monolithic HEMT Receiver for Passive Millimeter-Wave Imaging System.* Proc. of the Asia-Pacific Microwave Conference, 2011.

[34] J. L. Hesler and T. W. Crowe. *Responsivity and Noise Measurements of Zero-Bias Schottky Diode Detectors.* Proc. of the 18th Int. Symposium on Space Terahertz Technology, pp. 89-92, 2007.

6 Integrated Front End Assemblies

This chapter presents two signal generator frequency extension modules for the output frequency ranges 20 to 40 GHz and 60 to 110 GHz. Such source modules are required in automatic test systems to provide internal local oscillator signals and high frequency stimulus.

Although many necessary functionalities are available as microwave monolithic integrated circuits (MMIC), hybrid realization is preferred to combine the advantages of both, thin-film processed alumina Al_2O_3 or quartz SiO_2 and MMIC technology. There is no general guideline, ease of maintenance, module cost, and thermal engineering reasons determine which component is best integrated as an MMIC, hollow waveguide component, or hybrid planar device.

Such modules are either realized with frequency mixers, which easily allow for linear upconversion of single- and multi-tone signals at the expense of an additional required high power local oscillator (LO) signal, or with frequency multipliers. Utilizing frequency mixers, the module's output power level shows linear dependency on the input power level. This is different for modules incorporating nonlinear diode frequency multipliers, which require sufficiently high input power levels to saturate the diodes. Achieving high dynamic range in the latter case is possible with variable gain amplifiers or switched attenuators in cascade to the multiplier. To circumvent the difficulties of providing an additional LO signal, the proposed modules are based on frequency multipliers and commercial GaAs and InP amplifiers.

The presented systems utilize 1.85 mm / 1.00 mm coaxial to planar waveguide transitions and frequency tripler x3_HE_1 from chapter 2, two codirectional couplers with integrated 50 Ω termination from chapter 3 and tapered series capacitor and power detectors from chapter 5.

6.1 Planar Signal Source for 20 to 40 GHz

Fig. 6.1 and Fig. 6.2 show photographs of the fully assembled 20 to 40 GHz source module (dimensions: [38 × 58 × 9] mm^3) with its interconnection to the FR4 control board.

A block diagram is given in Fig. 6.3. Fig. 6.4 contains a detail view with all single components, including parts on alumina, quartz and MMICs.

Single Components The incident RF power at the fundamental frequency range of 10 to 20 GHz enters the module through a coaxial 1.85 mm (V) sparkplug connector. The transition with its parasitic capacitance is explained in chapter 2. Three double bondwires connect the first alumina substrate to the Hittite HMC441 [1] GaAs pHEMT medium power amplifier MMIC, which exhibits about 15 to 20 dB small signal gain from 10 to 20 GHz with drain voltage $V_D = 5$ V and drain current $I_D = 106$ mA (Fig. 6.5).

The amplified signal drives the Avago AMMC-6140 [2] GaAs pHEMT active frequency doubler MMIC. Fig. 6.6 illustrates measured output power levels versus frequency at various input power levels. At drain voltage $V_D = 4.5$ V and drain current $I_D = 39.5$ mA, the conversion loss is less than 5 dB for the envisaged output frequency range. It has less loss at higher frequencies and therefore compensates partly the frequency response of the HMC441 in Fig. 6.5.

An octave bandwidth fishgrate bandpass filter realized on 10 mil SiO_2 ensures fundamental signal rejection and to some extent rejection of the 4th harmonic after frequency doubling. The filter utilizes $N = 13$ parallel resonators in shape of short circuited stubs with quarterwave length and $N - 1$ impedance inverters (Fig. 6.7).

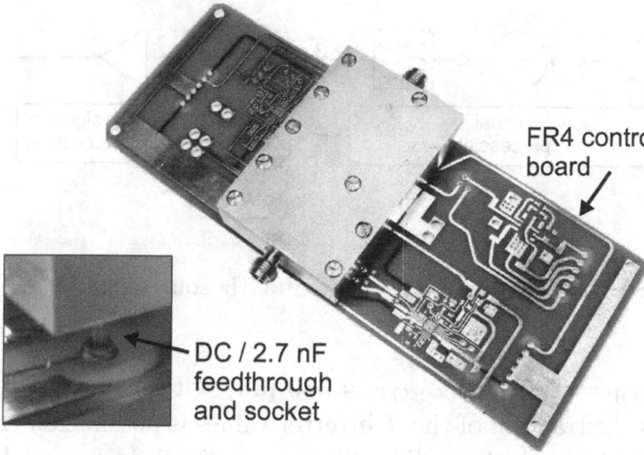

Figure 6.1: Photograph of the 20 to 40 GHz source module including the interconnection to the FR4 control board.

Figure 6.2: Photograph of the fully assembled 20 to 40 GHz source module.

It is essential for successful filter design to cover the bulky nature of the conical plated-through via holes (drill diameters at top and

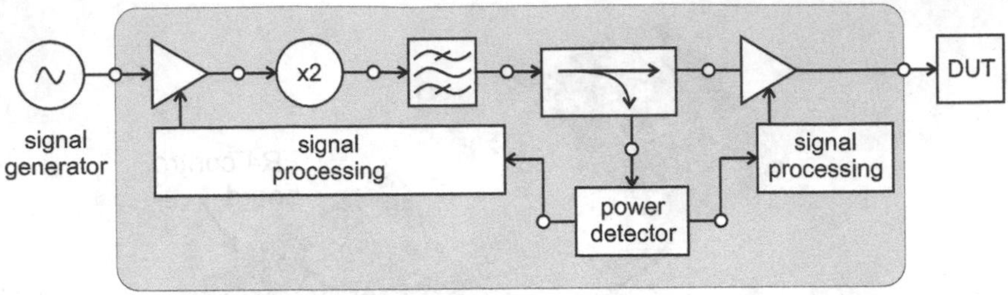

Figure 6.3: Block diagram of the 20 to 40 GHz source module.

bottom $\varnothing_{\mathrm{TOP}} = 200$ µm, $\varnothing_{\mathrm{BOT}} = 280$ µm) within 3D EM fullwave simulations. Extraction of the J-inverter values is possible on the basis of fast 2D EM simulations. The filter is synthesized from a lowpass prototype, following the theory of [3] (compare appendix A). Table 6.1 summarizes the filter parameters, which are characteristic impedances, strip widths and lengths of each resonator and impedance inverter. All resonator lengths ℓ_{res} correspond to a single short circuited stub and cover the distance from the center of the substrate to the edge of the via pad. Note, the first and last resonators are realized as single stubs rather than symmetrical double stubs.

Scattering parameter measurements with 1.85 mm coaxial testfixtures are compared to 3D EM simulations in Fig. 6.8. Return loss values average out at 15 dB. The filter's 3 dB passband covers 19 to 40 GHz. The reader should keep in mind, an octave bandwidth is realizable with quarterwave resonators only. The achieved bandwidth of passband of the halfwave resonator filter from chapter 2 is almost the limit for this filter class (70 to 110 GHz).

The fishgrate filter is connected to the codirectional coupler with integrated 50 Ω termination from chapter 3, which has simulated coupling values of 22.5 to 15 dB in the frequency range of 20 to 40 GHz. The termination is introduced in section 3.4.

The coupler drives the UMS CHA2097a [4] GaAs pHEMT three stage variable gain amplifier (VGA) MMIC in the main path and the power dector from chapter 3 in the coupled path. Fig. 6.9 includes measured VGA scattering parameters in high (solid lines, $V_{\mathrm{CTRL}} =$

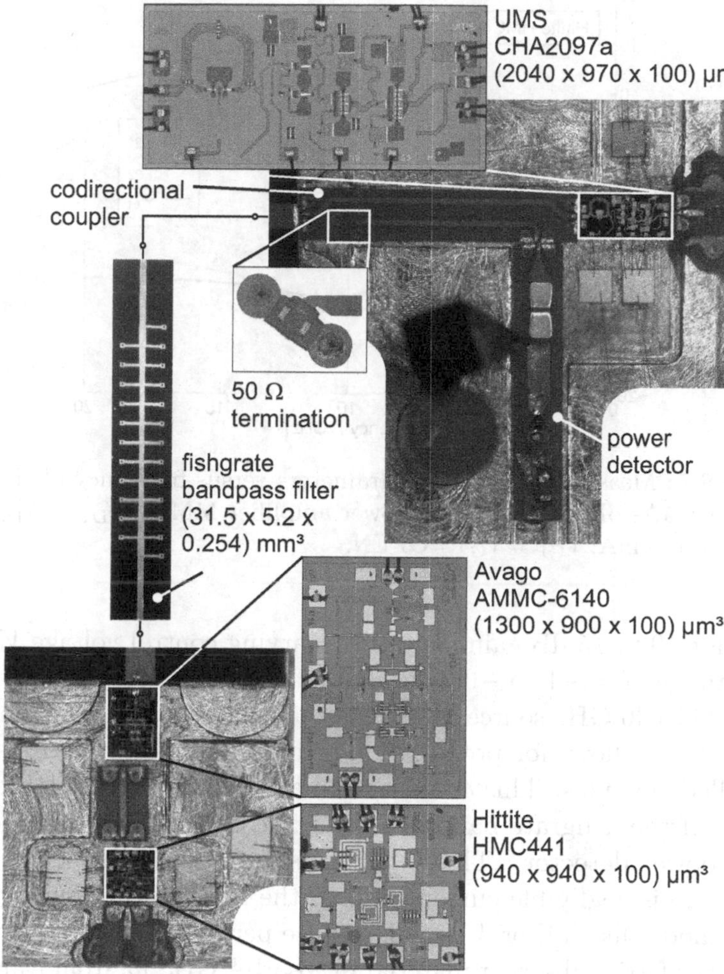

Figure 6.4: Detail view of 20 to 40 GHz source module, showing all single components including parts on alumina, quartz and MMICs.

−1 V) and low gain (dashed, $V_{\mathrm{CTRL}} = -0.3$ V) configuration. At drain voltage $V_{\mathrm{D}} = 3.5$ V and drain current $I_{\mathrm{D}} = 170$ mA, the VGA

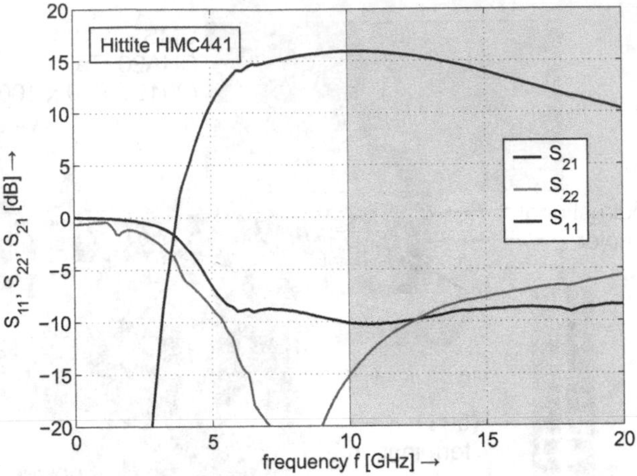

Figure 6.5: Measured scattering parameters versus frequency of Hittite HMC441 GaAs pHEMT medium power amplifier MMIC. $V_{D1} = V_{D2} =$ 5 V, $I_D = 106$ mA, $V_{G1} = V_{G2}$ =OPEN.

allows for 10 to 15 dB gain control by varying control voltage V_{CTRL} in the range of $= -1$ to -0.3 V.

The 20 to 40 GHz source module is the only design throughout the thesis, which allows for propagation of multiple hybrid modes along the milled channels. This could not be avoided due to the large size (width) of the fishgrate bandpass filter and the conical coil's height of the power detector. This is advantageous with respect to loss, since there is negligible current flow in the mechanical housing, but hybrid modes like TE or TM make device performance unpredictable. Hence, the ferrite absorber material Eccosorb® GDS [5] from Emerson & Cuming, which is a magnetically loaded, non-conductive silicone sheet[1] is glued to the inner surface of the top housing. Empirical analysis with several different sheet thicknesses is necessary to yield a successful module in single mode operation.

[1] $f = 3$ GHz, $\epsilon_r = 20, \tan\delta_e = 0.67, \mu_r = 3.5, \tan\delta_m = 0.4$.

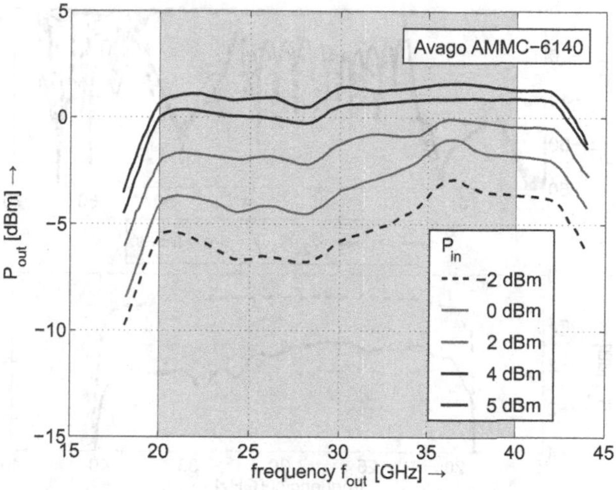

Figure 6.6: Measured output power level P_{out} versus output frequency f_{out} of Avago AMMC-6140 GaAs pHEMT active frequency doubler MMIC at various input power levels $P_{\mathrm{in}} = -2, 0, 2, 4, 5$ dBm. $V_{\mathrm{D}} = 4.5$ V, $I_{\mathrm{D}} = 39.5$ mA, $V_{\mathrm{G}} = -1.2$ V.

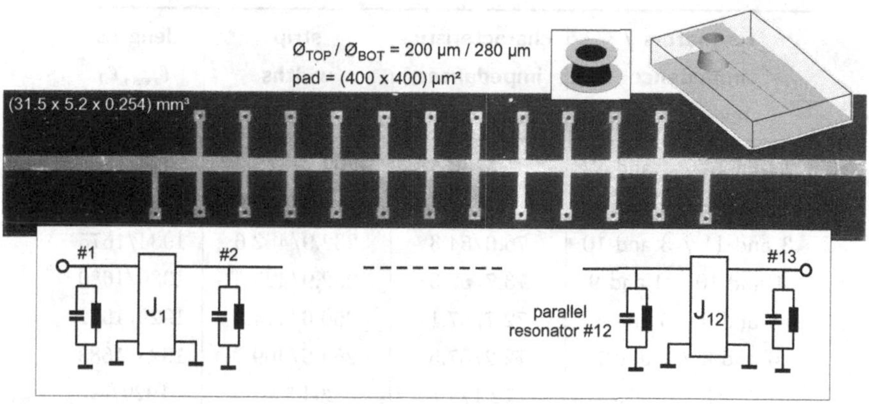

Figure 6.7: Photograph and equivalent circuit of 13th order fishgrate bandpass filter.

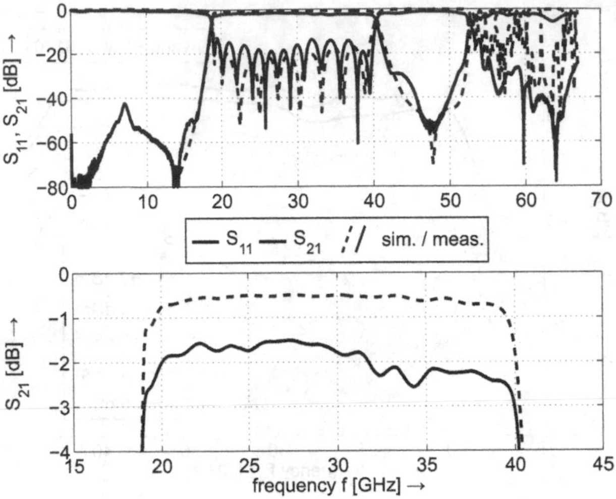

Figure 6.8: Simulated and measured scattering parameters versus frequency of 13th order fishgrate bandpass filter.

Table 6.1: 13th order quarterwave resonator bandpass filter (fishgrate)

resonators / impedance inverters J_i	characteristic impedances $Z_{res}, Z_J\,[\Omega]$	strip widths w_{res}, w_J [µm]	lengths ℓ_{res}, ℓ_J [µm]
1 and 13 / 1 and 12	81.5/46.8	202.7/579.9	2028/1658
2 and 12 / 2 and 11	80.1/49.4	210.6/531.0	1939/1664
3 and 11 / 3 and 10	76.6/54.3	232.9/452.6	1934/1675
4 and 10 / 4 and 9	73.7/56.3	252.9/425.6	1930/1680
5 and 9 / 5 and 8	72.7/57.1	260.6/414.1	1929/1682
6 and 8 / 6 and 7	72.2/57.5	263.6/409.7	1928/1683
7 / –	72.1/–	264.5/–	1929/–

filter order $N = 13$, fractional bandwidth FBW $= 67\,\%$,
center frequency $f_c = \sqrt{20 \cdot 40}$ GHz $= 28.3$ GHz.

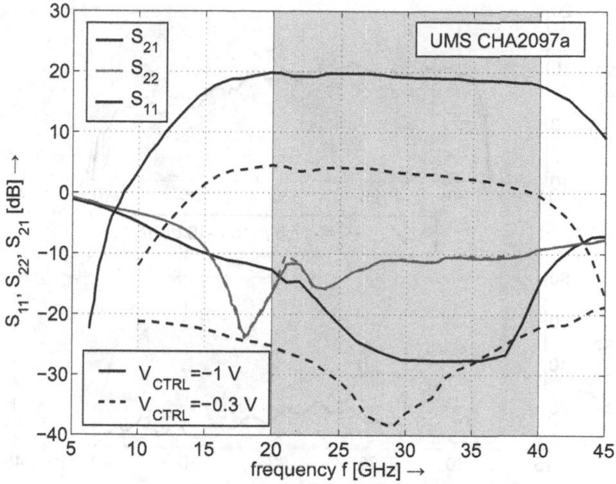

Figure 6.9: Measured scattering parameters versus frequency of UMS CHA2097a GaAs pHEMT three stage variable gain amplifier in high (solid, $V_{\mathrm{CTRL}} = -1$ V) and low gain (dashed, $V_{\mathrm{CTRL}} = -0.3$ V) configuration. $V_{\mathrm{D}} = 3.5$ V, $I_{\mathrm{D}} = 170$ mA, $V_{\mathrm{G}} = -0.1$ V.

Source Module Performance

Simultaneously measured output power level [6] of the module and output voltage of the detector are shown in Fig. 6.10 for three different input power levels. Information of Fig. 6.10 already provides power monitoring feature and can be used to build an automatic level control (ALC). The latter does not involve the VGA, which is placed after the loop to gain maximum output power level. Hence, an ALC controls the HMC441 and regulates the VGA.
According to spectral measurements at the module's output (Fig. 6.11), the 3rd harmonic rejection is 14 dB at worst.

Measured output power level versus input power level is shown in Fig. 6.12. All measurement data in the output frequency range of 20 to 40 GHz are within the gray shaded area. Fig. 6.13 and Fig. 6.14 illustrate the relationship of dependency between the output power level and VGA control voltage V_{CTRL}. Such modules normally operate at a fixed input power level which can be chosen between -20

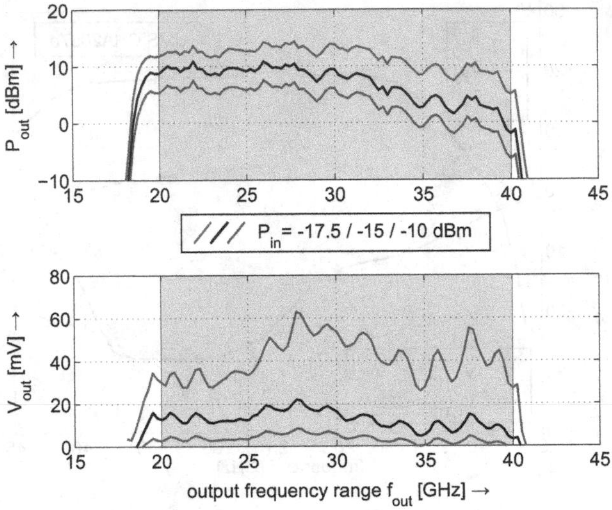

Figure 6.10: Measured output power level P_{out} (top) and output DC voltage V_{out} (bottom) versus output frequency f_{out} at several input power levels $P_{\text{in}} = -17.5, -15, -10$ dBm.

to -10 dBm. The VGA then allows for gain control of 10 to 15 dB. If the input power level is varied in addition, the dynamic range is about 25 dB. The maximum output power level varies between 5 and 14 dBm from 20 to 40 GHz.

The measured return loss at the coaxial interfaces after SOLT calibration is higher than 10 dB at the input port and varies between 5 and 20 dB at the output port (Fig. 6.15).

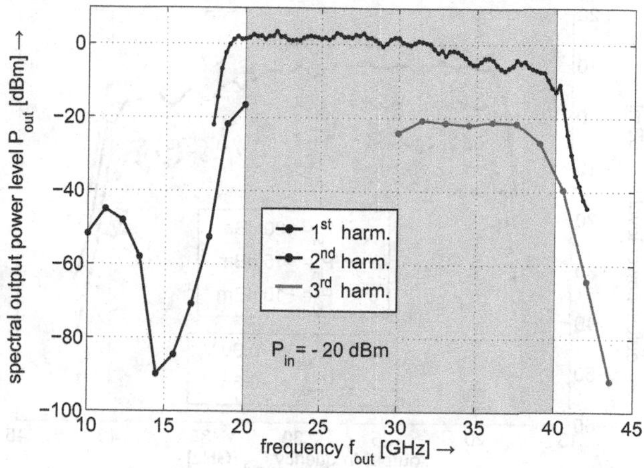

Figure 6.11: Measured spectral output power levels up to the 3rd harmonic versus output frequency f_{out} at input power level $P_{\text{in}} = -20$ dBm.

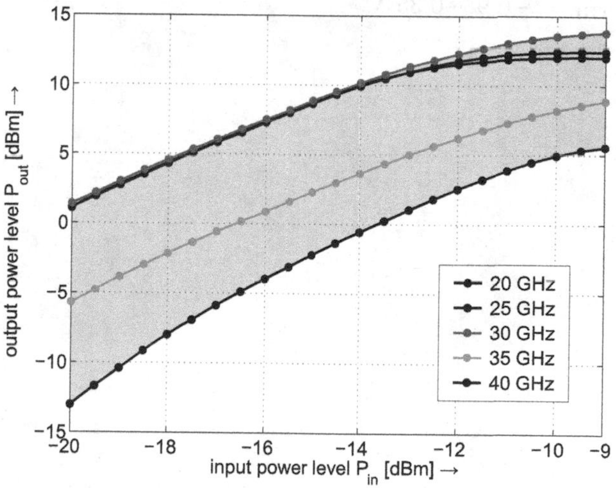

Figure 6.12: Measured output power level versus input power level at various output frequencies $f_{\text{out}} = 20, 25, 30, 35, 40$ GHz.

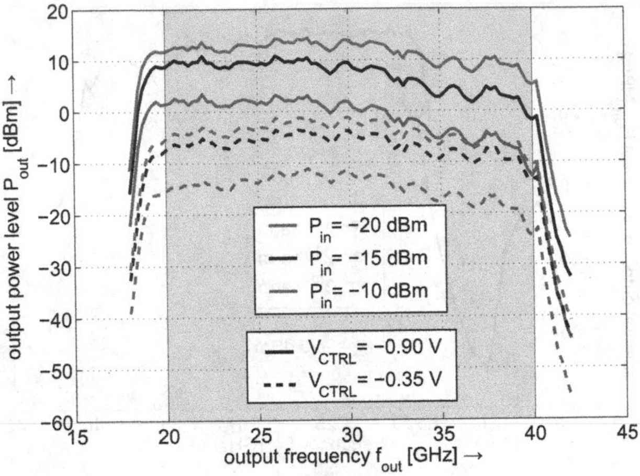

Figure 6.13: Measured output power level versus output frequency at various input power levels $P_{\text{in}} = -20, -15, -10$ dBm and two control voltages $V_{\text{CTRL}} = -0.9, -0.35$ V.

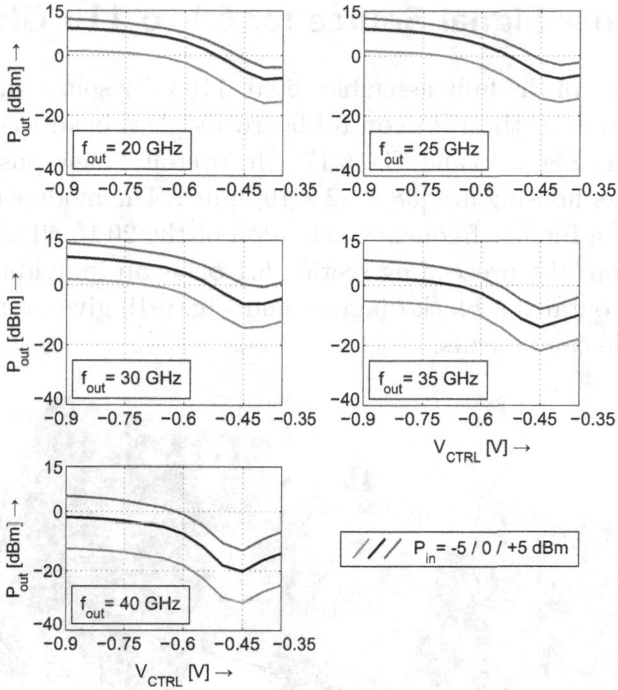

Figure 6.14: Measured output power level versus control voltage at various output frequencies $f_{\text{out}} = 20, 25, 30, 35, 40$ GHz.

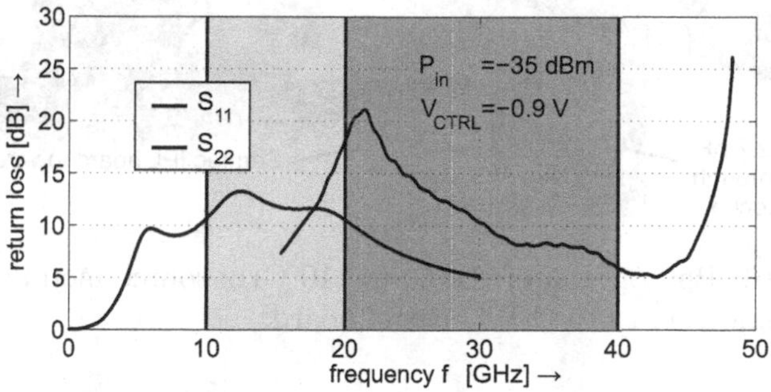

Figure 6.15: Measured return loss versus frequency of 20 to 40 GHz source module.

6.2 Planar Signal Source for 60 to 110 GHz

Photographs of the fully assembled 60 to 110 GHz source module and its connection to the FR4 control board by glass bead feedthroughs are shown in Fig. 6.16 and Fig. 6.17. The overall dimensions of the Au plated brass housing are [38 × 32 × 10] mm^3. The module can either be used as a further frequency extension of the 20 to 40 GHz source module from the preceeding section 6.1 or as an individual system. Fig. 6.18 contains a block diagram and Fig. 6.19 gives a detail view of all single components.

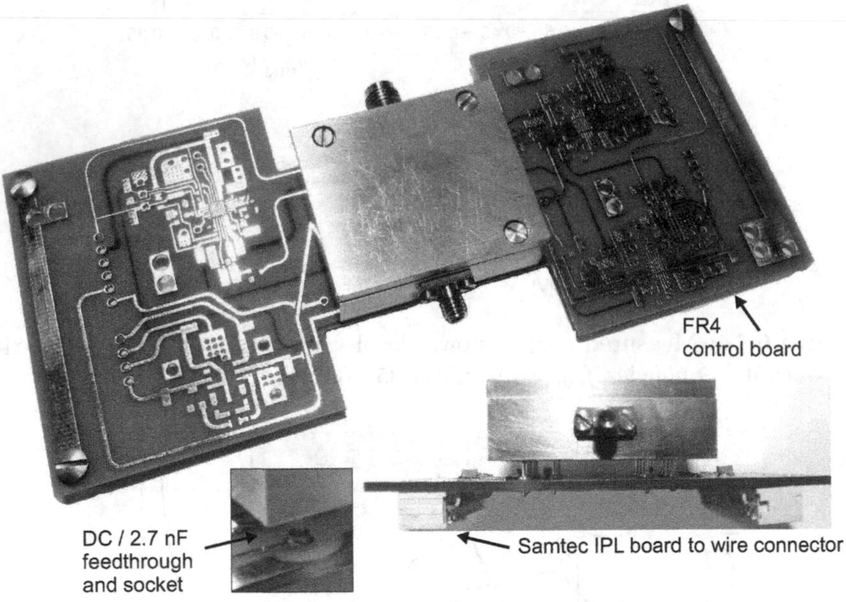

Figure 6.16: Photograph of the 60 to 110 GHz source module including the interconnection to the FR4 control board.

Figure 6.17: Photograph of the fully assembled 60 to 110 GHz source module.

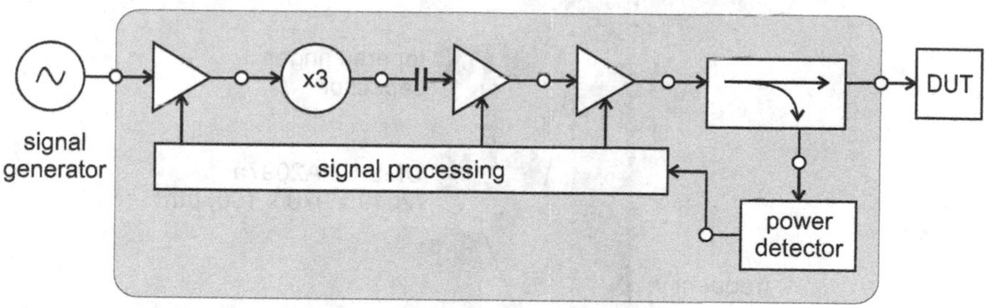

Figure 6.18: Block diagram of the 60 to 110 GHz source module.

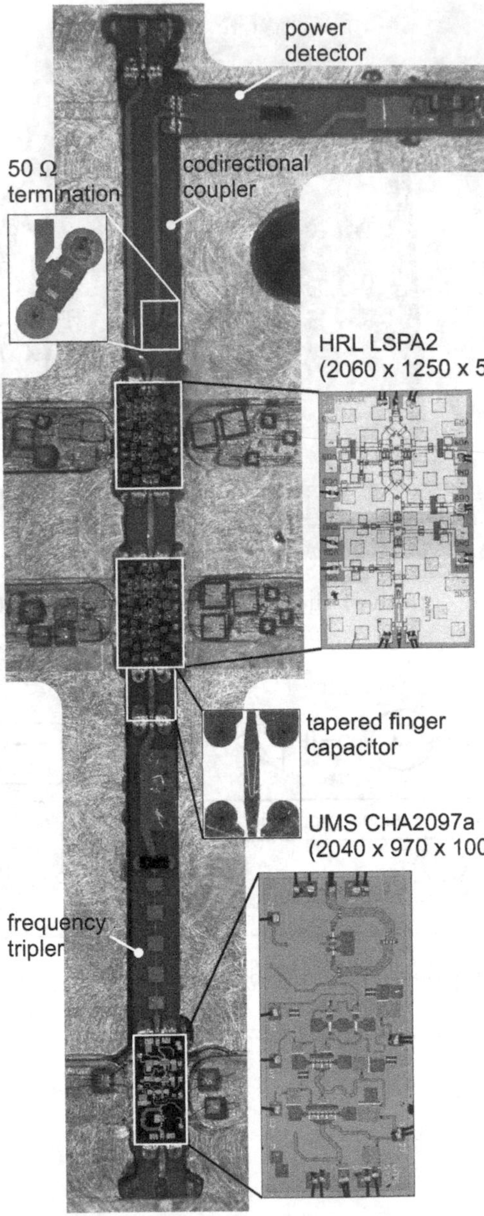

Figure 6.19: Detail view of 60 to 110 GHz source module, showing all single components including parts on alumina and MMICs.

Single Components

The dimensions of the milled channel are [1000 × 635] μm^2 and guarantee single mode operation up to the maximum frequency of 110 GHz. The channel width is widened only to allow for assembly of the MMICs and single layer capacitors at amplifiers gate and drain contacts. This is realized by an internal screwed on top plate, as indicated in Fig. 6.17. Appropriate milling compensates for the different heights of alumina (5 mil) and MMICs (100 µm, 50 µm). Manual assembly allows for distances between two single components of only 25 to 50 µm.

Bondwire modelling is a big topic for semiconductor packaging [B1] industry. It has been shown by many different research groups (e.g. IBM), that accurate modelling of very complex bondwire configurations based on 3D EM fullwave simulation is possible. The point is, 3D models have to be based on the actually manufactured bondwires, which is out of proportion to the benefit. In the author's opinion, it is good practice to just figure out the minimum number of bondwires or minimum bondwire diameter and geometrical limits to keep return loss values of the transition greater than 15 to 20 dB. Grid and ribbon bonds have better high frequency performance in general, but are not part of the available technology. Conventional wedge bonds in 3 × 2 configuration are used for the entire module. The 3D EM model from Fig. 6.20 is used to study the performance of two different bondwire heights for intersubstrate gaps between 25 and 150 µm. The results are shown in Fig. 6.21. It turns out, with a bondwire height of only 25 µm, bridging distances up to 50 µm do not affect system performance up to 110 GHz. If bondwire height equals 100 µm, intersubstrate distance should be kept below 50 µm. Hence, there is no need for considering bondwire interconnects in the case at hand. This holds true for the developed bondpad design of Fig. 6.20 and the utilized 5 mil alumina substrate.

The incident RF power at fundamental frequency range of 20 to 36.7 GHz enters the module in coaxial mode by a 1.85 mm sparkplug connector (compare chapter 2). After amplification by the UMS CHA2097a [4] GaAs pHEMT variable gain amplifier (VGA) MMIC, the signal enters frequency tripler $x3\_HE_1$ from chapter 2. The

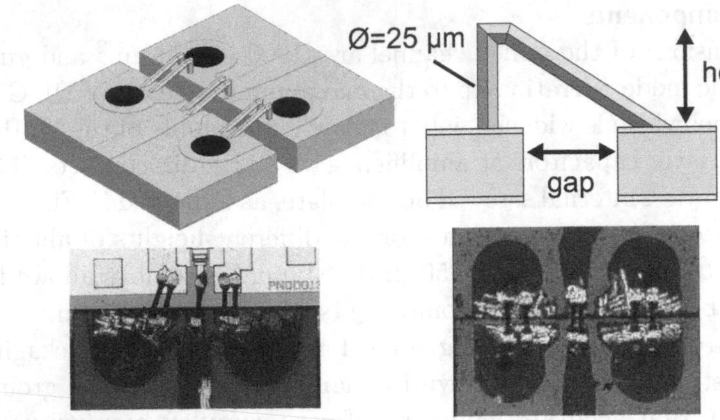

Figure 6.20: 3D EM model and photographs of bondwire interconnections.

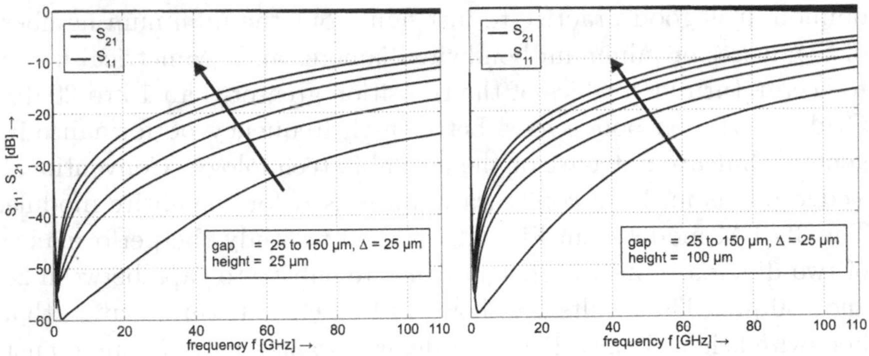

Figure 6.21: Simulated scattering parameters of bondwire interconnections as a function of gap width. Bondwire heights are 25 µm (left) and 100 µm (right).

hybrid HE_1 waveguide at the output of the tripler has short circuit behaviour at DC, which is changed to an open circuit by the tapered finger capacitor from section 5.4. The latter does also improve the fundamental rejection of x3_HE_1. Two cascaded HRL LSPA2 [7] InP HEMT three stage power amplifier MMICs build the heart of the

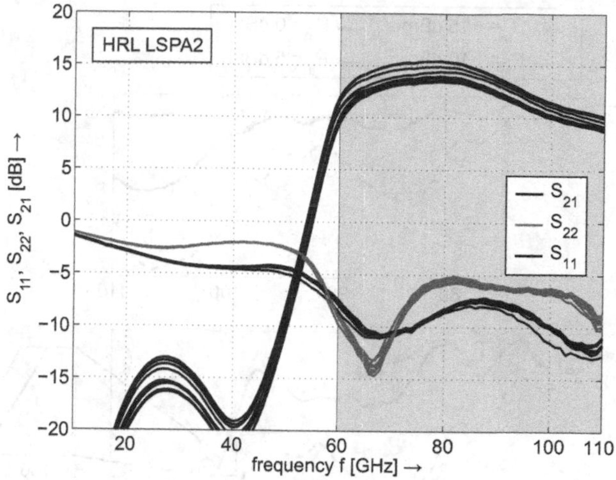

Figure 6.22: Measured scattering parameters of ten HRL LSPA2 InP HEMT three stage power amplifier MMICs from the same wafer run. $V_G = -0.18$ V, $V_D = 2$ V, $I_D = 200$ mA.

module. Each of them provides 8 to 15 dB gain with a maximum input power level of 5 dBm. To the author's knowledge, this is the only commercially available amplifier covering the focused frequency range of 60 to 110 GHz. Fig. 6.22 shows on-wafer scattering parameter measurements of ten LSPA2 amplifiers from the same wafer run. The amplifier operates at drain voltage $V_D = 2$ V with drain currents from $I_D = 112$ to 200 mA.

The codirectional coupler from section 3.3 with integrated 50 Ω termination passes the amplified signal to a transition from microstrip line (MSL) to 1.00 mm coaxial connector. The transition is explained in chapter 2. The coupled arm (simulated coupling of 15 to 10 dB) is connected to the power detector from section 5.4.

Source Module Performance Results from simultaneously measured output power level P_{out} [6] and detector output voltage V_{out} versus output frequency f_{out} at four input power levels P_{in} are shown in Fig. 6.23. At about 105 GHz it looks like different output power

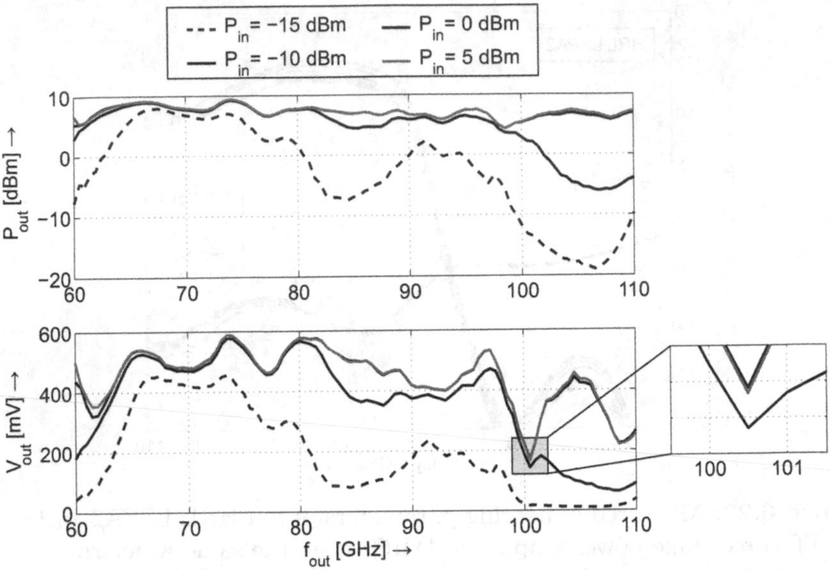

Figure 6.23: Measured output power level P_{out} (top) and output DC voltage V_{out} (bottom) versus output frequency f_{out} at several input power levels $P_{\text{in}} = -15, -10, 0, 5$ dBm.

levels are mapped to the same detector output voltage, but the enlarged view shows the detector allows for differentiation and therefore power monitoring / automatic level control. From 60 to about 95 GHz, saturation of the frequency tripler at a module input power level of $P_{\text{in}} = -10$ dBm is already in high gear, whereas at least $P_{\text{in}} = -5$ dBm is necessary for saturation up to 110 GHz. This fits to the results of chapter 2 and is even more obviously in Fig. 6.24, which contains measured output power levels versus input power levels at various output frequencies.

Spectral measurements (Fig. 6.25) at an input power level $P_{\text{in}} =$ 0 dBm have been performed. The module achieves worst case rejections of 33 dBc, 19 dBc, 30 dBc and 21 dBc for fundamental signals, 2nd, 3rd, 4th and 5th harmonic, respectively.

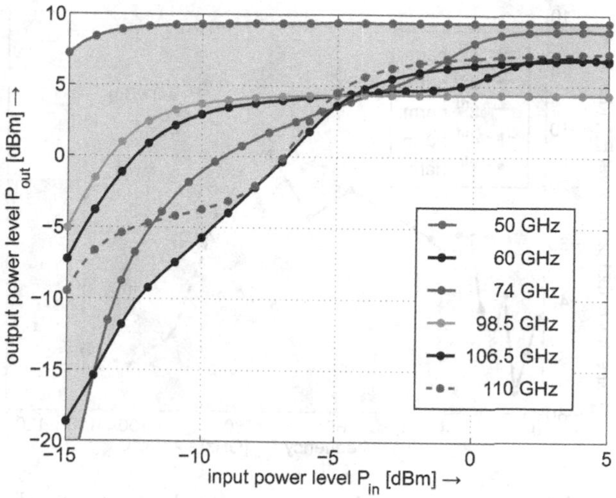

Figure 6.24: Measured output power level versus input power level at various output frequencies $f_{out} = 50, 60, 74, 98.5, 106.5, 110$ GHz.

In general, upconverters and local oscillator modules based on frequency multipliers are driven with fixed input power level in saturation region. Variable output power levels are realized with step attenuators in cascade. In this sense, the variable gain amplifier CHA2097a within the 20 to 40 GHz source module from the preceding section is a kind of step attenuator. In the present case, the only way to establish different output power levels is varying the tripler's input power level. This is not the best solution for several reasons. Below saturation, the output power level is very sensitive to changes of the input power level. This is illustrated in Fig. 6.26 and Fig. 6.27. Fig. 6.26 shows measured output power levels versus output frequency at three input power levels and high (solid) and low (dashed) gain configuration of the VGA. Fig. 6.27 contains output power levels versus VGA control voltage at three different output frequencies. Furthermore, with too low P_{in} the diodes of the tripler exhibit high impedance and therefore change the tripler's spurious tone rejection and return loss performance. Consequently the source module will change its performance

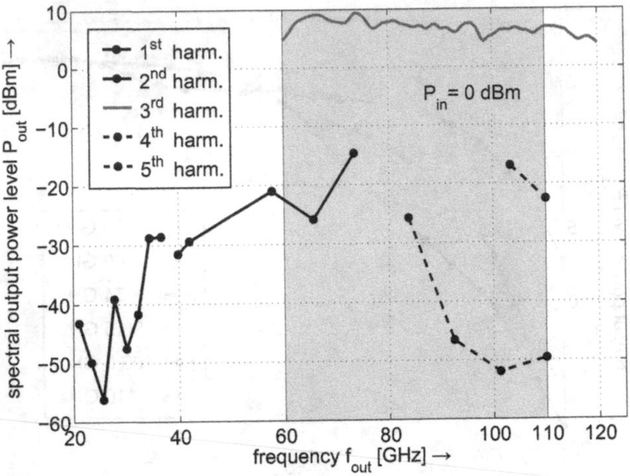

Figure 6.25: Measured spectral output power levels up to the 5th harmonic versus output frequency f_{out} at input power level $P_{\mathrm{in}} = 0$ dBm.

with varying input power level and datasheet information becomes hard to specify. Anyway, in conjunction with an automatic level control, this practice allows for varying the module's output power level from -10 dBm to the maximum level $P_{\max}(f)$. Maximum output power levels in the range of 4.3 to 9.5 dBm are achieved.

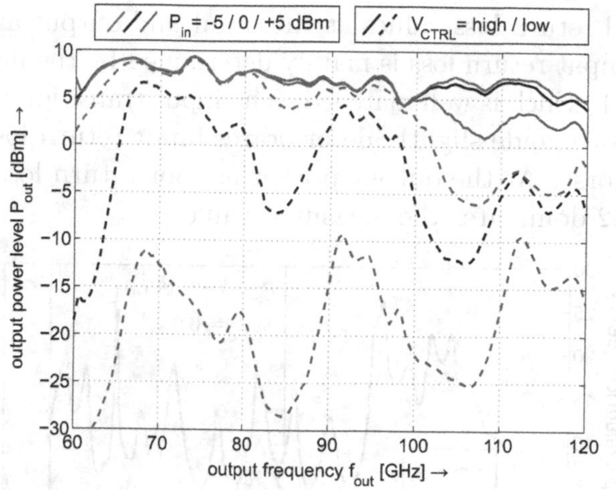

Figure 6.26: Measured output power level versus output frequency at various intput power levels $P_{\text{in}} = -5, 0, 5$ dBm and two control voltages V_{CTRL}.

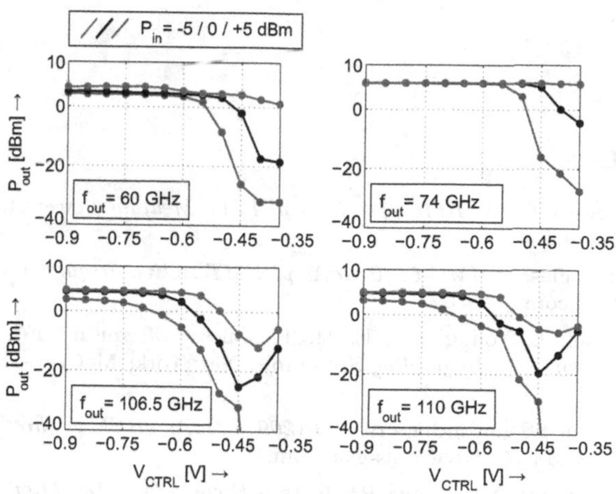

Figure 6.27: Measured output power level versus control voltage at various output frequencies $f_{\text{out}} = 60, 74, 106.5, 110$ GHz.

Measured return loss values at the input and output are shown in Fig. 6.28. Input return loss is mainly determined by the first amplifier (CHA2097a), which is well matched. The input transition from coaxial to planar waveguide slightly deteriorates input return loss values to 13 dB at worst. At the output port, the poor return loss behaviour from LSPA2 dominates the measured value.

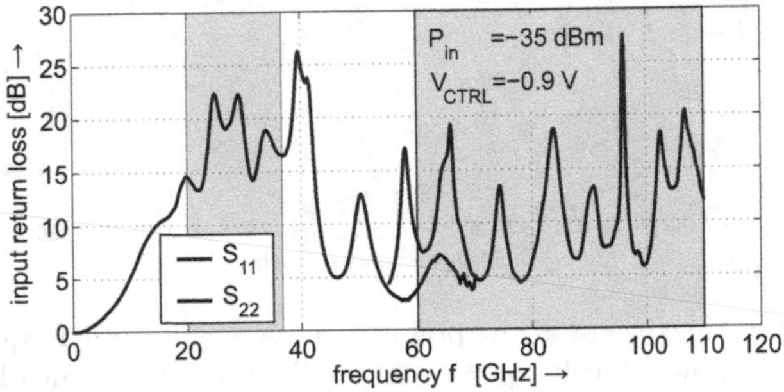

Figure 6.28: Measured return loss versus frequency of 60 to 110 GHz source module.

References

[1] Hittite Microwave Corp. *HMC441 GaAs pHEMT Medium Power Amplifier MMIC.* www.hittite.com.

[2] Avago Technologies. *AMMC-6140 GaAs pHEMT Active Frequency Doubler MMIC.* www.avagotech.com.

[3] G. L. Matthaei, L. Young, and E. M. T. Jones. *Microwave Filters, Impedance Matching Networks, and Coupling Structures.* New York: McGraw-Hill, 1964. ISBN 0890060991.

[4] United Monolithic Semiconductors. *CHA2097a GaAs pHEMT Three Stage Variable Gain Amplifier MMIC.* www.ums-gaas.com.

[5] Emerson & Cuming Microwave Products. *Eccosorb® GDS Microwave Absorber.* www.eccosorb.eu.

[6] Rohde & Schwarz GmbH & Co. KG. *Operating Manual of NRP-Z51/-Z52/-Z55/-Z56/-Z57/-Z58 Power Sensors.* www.rohde-schwarz.com.

[7] HRL Laboratories LLC. *InP HEMT Three Stage Power Amplifier MMIC.* www.mmics.hrl.com.

7 Summary

1 Synthetic Instruments

Synthetic Automatic Test Systems allow for coverage of a dedicated number of required measurements with a minimum of redundant hardware components. Chapter 1 introduces history and **fundamental idea** of synthetic instruments. Block diagrams of synthetic and traditional instruments are shown and **critical millimeter-wave components** for synthetic automatic test systems are identified. Within the scope of this work, innovative ultrawideband components and modules for synthetic instruments are developed. All designs are based on commercially available materials and technologies and are also interesting for usage within traditional instruments.

2 Resistive Diode Frequency Multipliers

In chapter 2, a complete **modelling procedure** for discrete **Schottky diodes** is illustrated, which includes the semiconductor and linear parts of the diode. On the basis of effective material parameters a clear connection to results from 3D EM simulation is established. One section deals with single tone large signal analysis of nonlinear circuits, as it is used by harmonic balance circuit simulators. Different diode configurations, optimum embedding impedances and the author's frequency **multiplier design flow** are explained. Experimental results are presented. In section 2.5 an octave bandwidth frequency **tripler for 20 to 40 GHz** based on bilateral finlines is presented. Conventional PCB processing on woven fiberglass reinforced, ceramic filled, PTFE based composite material AD600 is used. The tripler achieves 16 to 20 dB conversion loss from 21 to 40 GHz and constitutes a hybrid alternative to the integrated circuit TGC1430G. Using

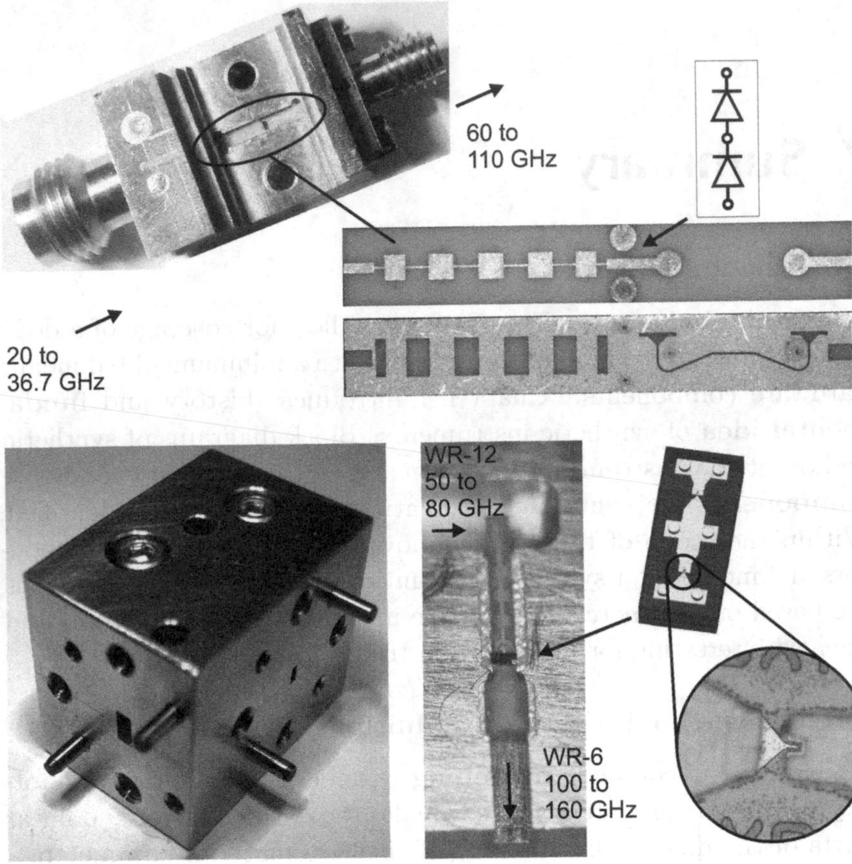

thin-film processed 5 mil alumina, two frequency doublers for **50 to 110 GHz** and three triplers for **60 / 67 to 110 GHz** are developed. One frequency doubler operates with conversion loss values below 15 dB from 50 to 89 GHz and below 22 dB from 50 to 110 GHz. One frequency tripler achieves conversion loss values around 18 dB from 60 to 95 GHz and below 22 dB from 60 to 110 GHz. Hence, with an input power level of 18 dBm, the multipliers deliver output power levels in the range of -5 to 5 dBm. Section 2.7 illustrates a frequency doubler and frequency tripler for **D-band and Y-band**, respectively. These multipliers are manufactured with conventional PCB technology on pure PTFE. The doubler operates from 100 to 160 GHz with conversion loss values from 15 to 20 dB. At its WR-6 output interface, output power levels in the range of -5 to 5 dBm are measured. The output power levels of the Y-band tripler are in the range of -20 to -11 dBm from 180 to 230 GHz. In addition, the design equations of Dolph-Chebyshev waveguide tapers and an uncertainty analysis of spectral power measurements above 50 GHz are outlined.

3 Planar Directional Couplers and Filters

In the first section of chapter 3 the modal and nodal scattering and impedance network parameters of coupled transmission lines according to the **general waveguide circuit theory** (GWCT) are introduced. This analysis is based on results, which are known from 2D EM Eigenmode analysis. The second section deals with **backward wave directional couplers**, which are required for separation of forward and backward travelling waves within testsets of (synthetic) vector network analyzers. The coupler's directivity values need to be high from very low frequencies in the kHz or MHz range up to the maximum frequencies 50 / 67 / 70 GHz. Directivity is directly related to the phase velocity deviations of the propagating common / even and differential / odd modes. Theoretical and experimental investigations on backward wave couplers with **dielectric overlay technique (stripline)** and **wiggly-line technique** to equalize the even and odd mode phase velocities are presented. Stripline couplers are realized by multilayer PCB technology on cost effective hydrocarbon laminates. Subsection 3.2.2 includes results of different coupler designs. Up to

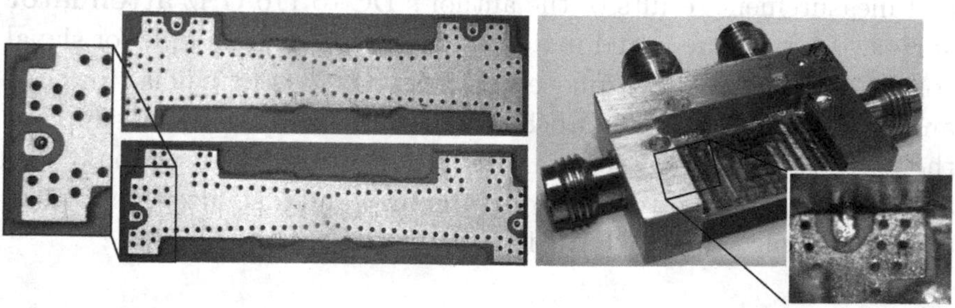

50 GHz directivity values greater than 10 dB are achieved. One coupler operates with 5.5 dB directivity from 10 MHz to 69.5 GHz. Subsection 3.2.3 applies wiggly-line technique to equalize phase velocities of coupled microstrip lines on thin-film processed 10 mil alumina. The synthesis procedure of nonuniform transmission line couplers and especially application of wiggly-line technique are outlined. A manufactured wiggly-line coupler achieves 8 dB directivity from 10 MHz to 48.4 GHz at an average coupling value of 16.6 dB ± 2.4 dB. Section 3.3 is dedicated to **codirectional couplers**. Beside the synthesis procedure, simulated and measured results (5 to 67 GHz) of various codirectional couplers on thin-film processed alumina are given. Two of the presented couplers are used within the modules of chapter 6. Couplers

for planar integration require internal nonreflective impedance terminations. Several 50 Ω **terminations** based on nickel chrome (NiCr) sheet resistors are presented in section 3.4. From DC to 100 GHz return loss values greater than 20 dB are achieved. Up to 110 GHz return loss values are below 15 dB. It further includes simulation and measurement results of the author's DC to 110 GHz **attenuator series**. These are required in almost every front end module for signal conditioning. Equalized phase velocities are also beneficial for edge and broadside **coupled line bandpass filters (BPF)** to suppress the parasitic second passband. This is demonstrated in section 3.5 by applying wiggly-line technique to the first and last filter element of an edge coupled line BPF on 10 mil alumina.

4 Triple Balanced Mixers

Chapter 4 presents the author's investigations on **triple balanced mixers (TBM)**, which are the only mixers providing overlapping RF, LO and IF frequency ranges together with large IF bandwidth and enhanced RF large signal handling capability. The first section highlights major differences and similarities between frequency multipliers and mixers. This is followed by a section about analytical and numerical methods for mixer analysis. Different methods are listed together with their strengths and weaknesses. With the results of single tone large signal analysis (harmonic balance) from chapter 2, section 2.3, as a starting point, **large signal / small signal (LSSS) mixer analysis** based on conversion matrices is derived. The minimum achievable conversion loss at different embedding impedances and also optimum embedding impedances are calculated for a simplified case, which includes nonzero diode series resistances. Different mixer configurations (SBM, DBM and TBM) are compared from a small signal analysis point of view and regarding the mixers' spurious tone behaviour, following a method from Henderson. Section 4.4 presents a **planar TBM realization** on SiO_2 using commercial silicon crossed quad diodes on thin-film processed 10 mil SiO_2. The TBM operates

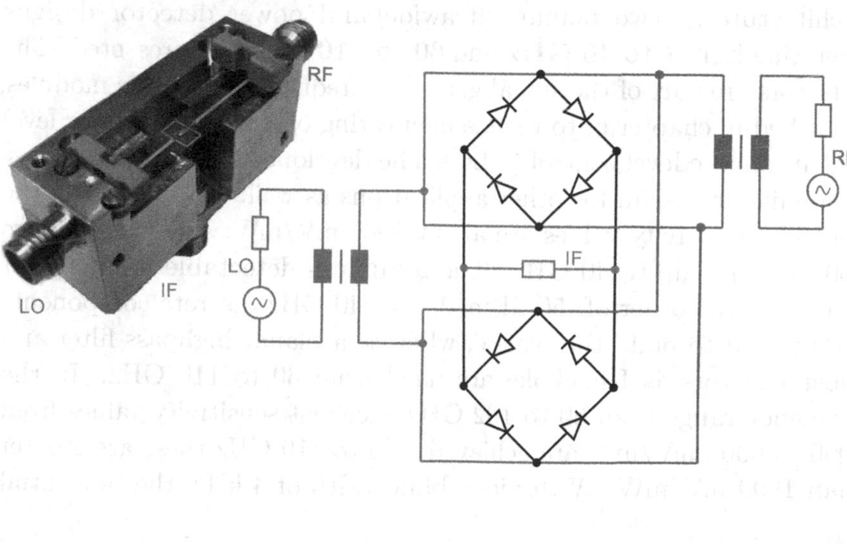

from **1 to 45 / 50 GHz** at RF / LO and achieves at least 20 GHz IF bandwidth with conversion loss values below 15 dB. Simulation and measurement results of the component are presented. Measurement results of the TBM within an automatic test system front end module are included.

5 Zero Bias Schottky Power Detectors

The first section of chapter 5 introduces **theoretical foundations** of zero bias **Schottky power detectors**. Expressions for detector current β and voltage sensitivity γ, and the upper $\Delta P_{\mathrm{sq}}, P_{\mathrm{USL}}$ and lower boundary $P_{\mathrm{NEP}}, P_{\mathrm{TSS}}$ of the square law dynamic range are derived. This is followed by a section which compares common detector

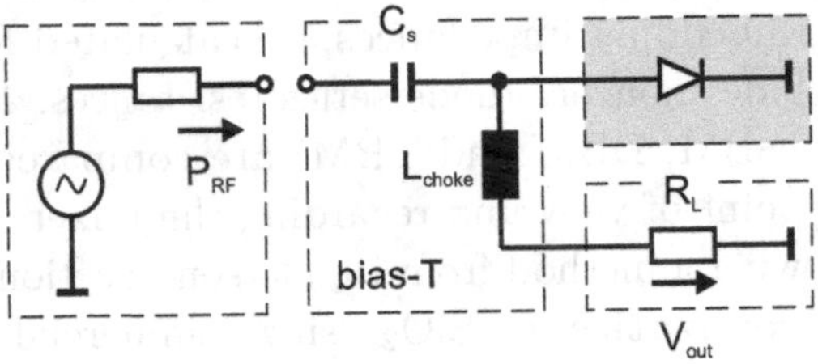

architectures. Two planar ultrawideband power detector designs operating from **1 to 40 GHz** and **60 to 110 GHz** are presented. The detectors are part of the signal generator frequency extension modules, described in chapter 6, to realize monitoring of the output power level and automatic level control (ALC). The developed ultrawideband bias-T circuits are useful for other applications as well. Up to 25 GHz the detector sensitivity values are about 1000 mV/mW. These stay above 300 mV/mW up to 40 GHz. The minimum detectable input power level is in the order of -50 dBm. Up to 40 GHz discrete components are utilized to build the bias-T, whereas a planar highpass filter and sheet resistors as RF choke are used from 60 to 110 GHz. In the frequency range from 60 to 102 GHz excellent sensitivity values from 2000 to 6000 mV/mW are achieved. Up to 110 GHz these are greater than 1000 mV/mW. With video bandwidth of 4 kHz, the tangential signal sensitivity power levels are in the range of -42 to -55 dBm. An analytical study is presented, that illustrates the dependency of Schottky junction capacitance on the maximum achievable sensitivity.

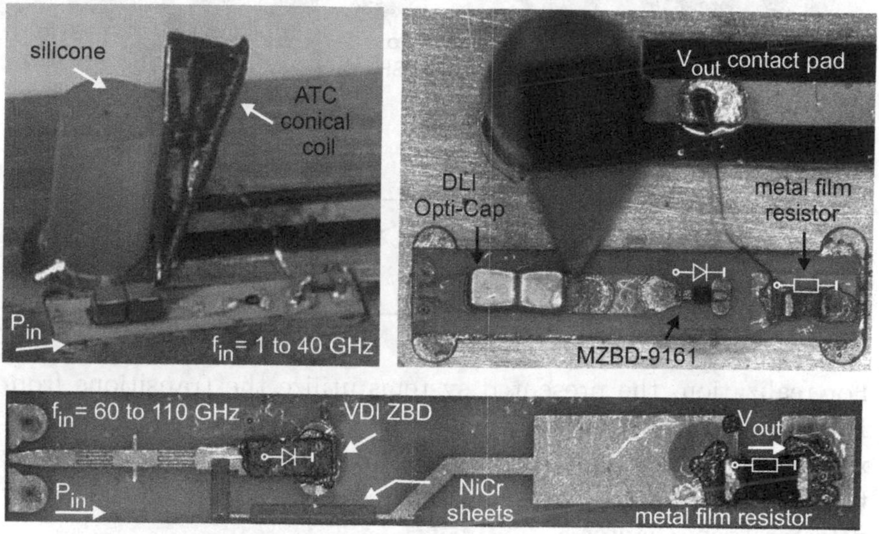

6 Integrated Front End Assemblies

Chapter 6 presents two **signal generator frequency extension modules** for the output frequency ranges **20 to 40 GHz** and **60 to 110 GHz**. Such source modules are required in automatic test systems to provide internal local oscillator signals and high frequency stimulus. Although many necessary functionalities are available as microwave monolithic integrated circuits (MMIC), hybrid realization is preferred to combine the advantages of both, thin-film processed alumina Al_2O_3 or quartz SiO_2 and MMIC technology. The modules can operate individually or in cascade, building a frequency sixtupler for 60 to 110 GHz output frequency range. The first module achieves maximum output power levels in the order of 10 dBm from 20 to 40 GHz. Using different control voltages allows for an output dynamic range of about 15 dB at a fixed input power level. The maximum output power level from 60 to 110 GHz is in the range of 5 to 10 dBm.

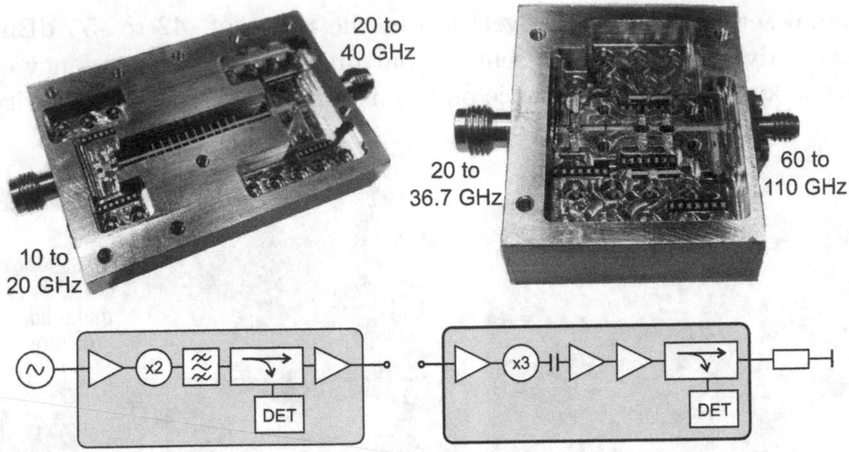

For realization, the presented systems utilize the transitions from 1.85 / 1.00 mm coaxial to planar waveguide and frequency tripler x3_HE$_1$ of chapter 2, two codirectional couplers with integrated 50 Ω termination from chapter 3 and tapered series capacitor and power detectors from chapter 5.

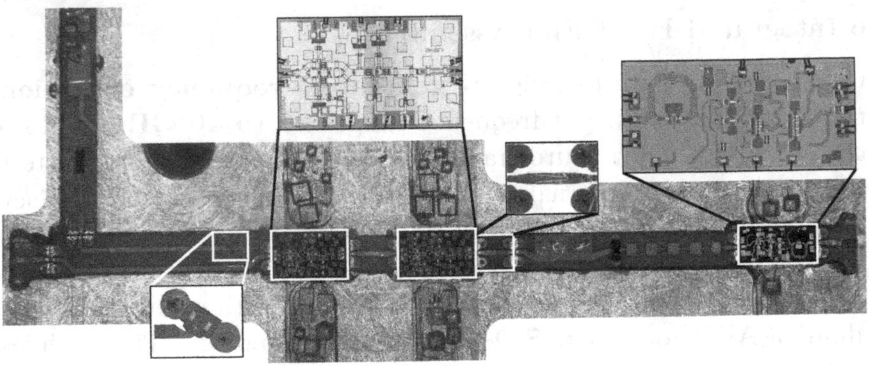

A Filter Synthesis

The appendix includes equations for **synthesis** of **Chebyshev** lowpass (LPF) and bandpass **filters** (BPF) in a very compact form. The filters are either based on **g** parameters of a lowpass prototype filter or **G** parameters of a prototype quarterwave transformer.

8 Zusammenfassung

1 Synthetische Instrumente

Synthetische Testsystem-Architekturen erlauben es eine dedizierte Anzahl von erforderlichen Messungen mit einem Minimum an redundanter Hardware abzudecken. In Kapitel 1 wird der **Grundgedanke dieser Instrumentenklasse** erklärt. Der Aufbau von synthetischen und traditionellen Instrumenten wird in Kapitel 1 anhand von Blockdiagrammen gezeigt. Im Rahmen dieser Arbeit wurden **innovative Breitbandkomponenten und -module** für synthetische Automatische Testsysteme im Millimeterwellenbereich entwickelt. Alle Entwicklungsarbeiten basieren auf kommerziell verfügbaren Materialen und Technologien und sind gleichwohl für den Einsatz in konventionellen Systemen der Messtechnik geeignet.

2 Resistive Diodenvervielfacher

In Kapitel 2 werden **resistive Diodenvervielfacher** auf der Basis von diskreten Schottky Dioden behandelt. Verfahren zur **Modellierung diskreter Schottky Dioden** werden ausführlich dargestellt. Durch die Verwendung effektiver Materialparameter wird eine Verknüpfung mit 3D elektromagnetischer Feldsimulation hergestellt. Die Methode der harmonischen Balance mit einem Großsignal zur Analyse von Vervielfacherschaltungen wird erläutert. Im Abschnitt 2.5 wird ein Frequenzverdreifacher, basierend auf bilateralen Flossenleitungen, vorgestellt. Dabei kommt gewöhnliche Leiterplattentechnologie in Verbindung mit dem Teflon-Keramik Verbundmaterial AD600 zum Einsatz. Der Verdreifacher erzielt Konversionsverluste im Bereich 16 bis 20 dB von **21 bis 40 GHz** und stellt eine hybride Alternative zum integrierten Schaltkreis TGC1430G dar. Unter Verwendung einer Dünnfilmtechnik und 5 mil Keramiksubstraten wurden zwei Verdoppler für den Frequenzbereich **50 bis 110 GHz** und drei Verdreifacher für den Frequenzbereich **60 / 67 bis 110 GHz** entwickelt. Ein Verdoppler erreicht Konversionsverluste unterhalb von 15 dB von 50 bis 89 GHz und unterhalb von 22 dB von 50 bis 110 GHz. Ein Verdreifacher erzielt Konversionsverluste um 18 dB von 60 bis 95 GHz und unterhalb von 22 dB von 60 bis 110 GHz. Damit sind bei einer Eingangsleistung von 18 dBm, Ausgangsleistungen im Bereich -5 bis 5 dBm möglich. Der Abschnitt 2.7 stellt einen Verdoppler

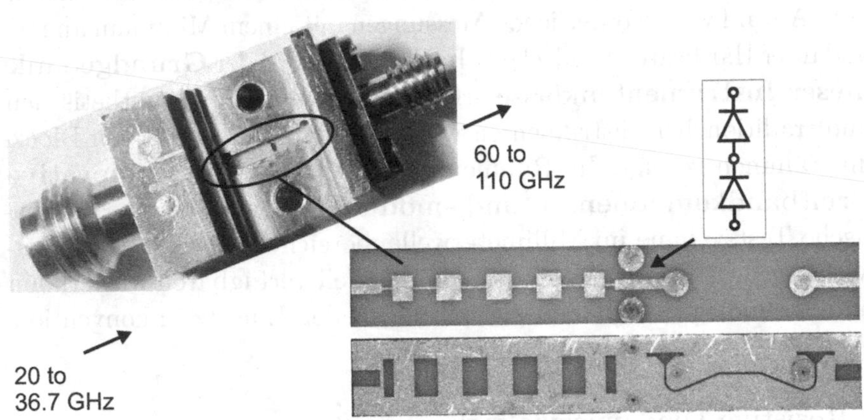

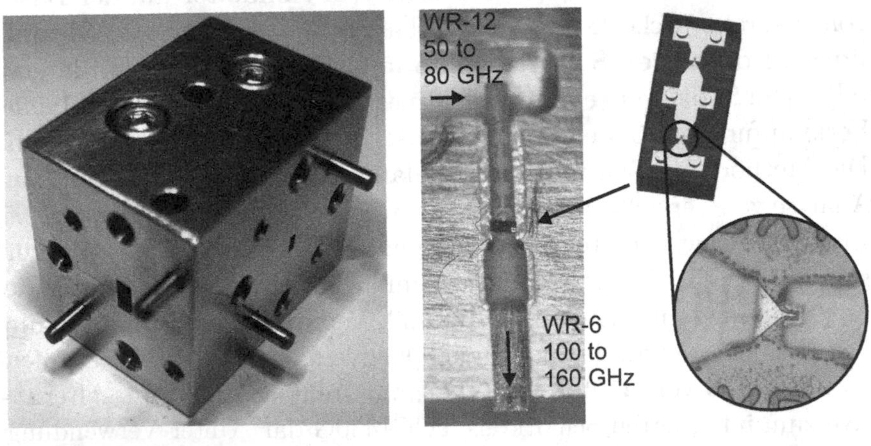

für das **D-Band** und einen Verdreifacher für das **Y-Band** vor. Die Schaltungen wurden mit gewöhnlicher Leiterplattentechnologie auf purem Teflon gefertigt. Der Verdoppler arbeitet mit Konversionsverlusten von 15 bis 20 dB im Frequenzbereich 100 bis 160 GHz. Im WR-6 Ausgangshohlleiter des D-Band Verdopplers werden Leistungen im Bereich von -5 bis 5 dBm erreicht. Die Ausgangsleistungen des Y-Band Verdreifachers liegen im Bereich -20 bis -11 dBm von 180 bis 230 GHz.

3 Planare Richtkoppler und Filter

Das Kapitel 3 beinhaltet eine ausführliche Herleitung der modalen und nodalen Streu- und Impedanzparameter. Die Ableitungen beruhen auf der **generalisierten Wellenleiter-Schaltungstheorie (GWCT)**. Alle Betrachtungen beziehen sich auf Ergebnisse, wie sie aus einer 2D elektromagnetischen Feldsimulation der jeweiligen modalen Leitungskonfiguration bekannt sind. Im Abschnitt 3.2 wird die Synthese von **Rückwärtswellenkopplern** mit ungleichförmigem Querschnitt über der Kopplerlänge behandelt. Diese werden im Testset von (synthetischen) Vektor-Netzwerkanalysatoren zur Wellentrennung benötigt. Entscheidend für die Streuparametermessung sind Koppler mit direktionaler Wirkung (Direktivität) von sehr niedrigen Frequenzen (kHz- oder MHz-Bereich) bis zu den Maximalfrequenzen 50 / 67 / 70 GHz. Bei Architekturen, die auf gekoppelten Leitungen beruhen, ist die Direktivität mit dem Phasengeschwindigkeitsunterschied der beiden ausbreitungsfähigen Wellentypen (gerade und ungerade) verknüpft. Experimentelle Ergebnisse mit zwei unterschiedlichen Methoden zum Phasengeschwindigkeitsangleich werden vorgestellt.
Gekoppelte Streifenleitungen mit quasi-homogenem Querschnitt lassen sich durch Anwendung von Mehrlagen-Leiterplattentechnik mit kostengünstigen Hydrokarbon Laminaten herstellen. Der Abschnitt 3.2.2 enthält Ergebnisse von verschiedenen Kopplervarianten. Bis 50 GHz können Direktivitätswerte oberhalb von 10 dB erzielt werden. Ein Koppler weist Direktivitätswerte oberhalb von 5.5 dB von 10 MHz bis 69.5 GHz auf. Im Abschnitt 3.2.3 wird durch Ein-

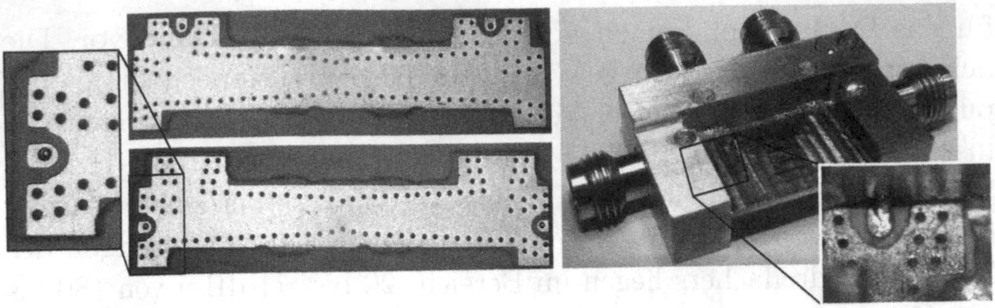

führung einer **gezackten Leitungsführung** ein Angleich der Phasengeschwindigkeiten erreicht. Dies erlaubt die Verwendung einer Dünnfilmtechnik auf zweilagigen 10 mil Keramiksubstraten. Das implementierte Syntheseverfahren wird vorgestellt. Ein gefertigter Koppler erreicht Direktivitätswerte oberhalb von 8 dB von 10 MHz bis 48.4 GHz bei einem mittleren Kopplungswert von 16.6 dB ± 2.4 dB. Der Ab-

schnitt 3.3 beschäftigt sich mit dem **Vorwärtskoppler**. Die Synthese wird beschrieben und experimentelle Ergebnisse auf Keramiksubstraten im Frequenzbereich 5 bis 67 GHz sind enthalten. Ferner werden zwei Vorwärtskoppler vorgestellt, welche zur Überwachung der Ausgangsleistung in den Modulen von Kapitel 6 eingesetzt werden. Die vorgenannten Koppler profitieren von einem integrierten reflexionsarmen Leitungsabschluß an einem der vier Tore. Im Abschnitt 3.4 werden verschiedene **reflexionsarme 50 Ω Abschlüsse** präsentiert, welche mit Nickel-Chrom Schichtwiderständen realisiert wurden. Von 0 bis 100 GHz sind die Rückflußdämpfungswerte größer als 20 dB. Bis 110 GHz liegen Sie oberhalb von 15 dB. Darüberhinaus wird

eine **Serie von Dämpfungsgliedern** (0 bis 110 GHz) vorgestellt, wie man sie in nahezu allen synthetischen Modulen zur Signalkonditionierung benötigt. Möglichst identische Phasengeschwindigkeiten des geraden und ungeraden Wellentyps sind auch für **Bandpass-Filter** wichtig, welche auf gekoppelten Leitungen und **Leitungsresonatoren mit Halbwellenlänge** basieren. Im Abschnitt 3.5 wird die erfolgreiche Anwendung der gezackten Leitungsführung zur Unterdrückung des parasitären Durchlaßbereiches von derartigen Bandpass-Filtern gezeigt.

4 Dreifach balancierte Mischer

Das Kapitel 4 behandelt den **dreifach balancierten Diodenmischer (TBM)**, welcher überlappende HF, LO und ZF Frequenzbereiche zusammen mit einer großen ZF Bandbreite erlaubt und über ein ausgezeichnetes HF Großsignalverhalten verfügt. Der Unterabschnitt 4.2.1

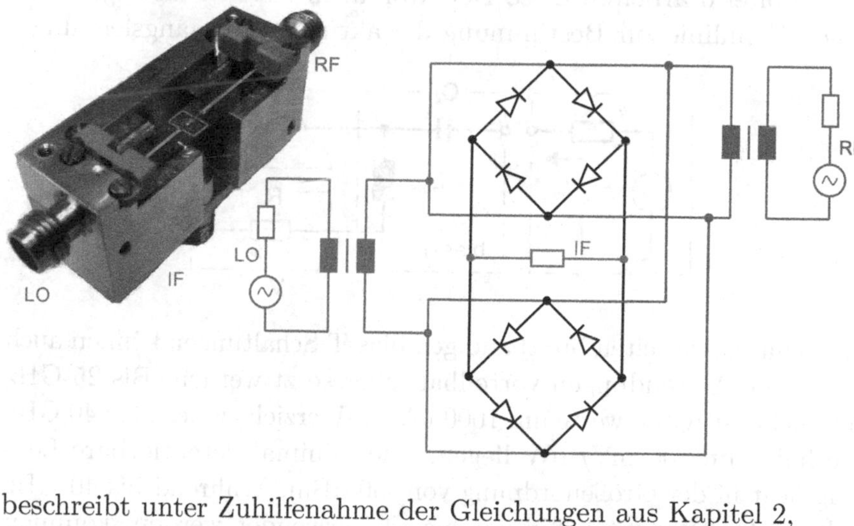

beschreibt unter Zuhilfenahme der Gleichungen aus Kapitel 2, Abschnitt 2.3, ausführlich die **Großsignal-Kleinsignal-Methode (LSSS)** zur Mischeranalyse mit Konversionsmatrizen. Das Klein- und Großsignalverhalten von einfach, zweifach und dreifach balancierten Mischern wird im Abschnitt 4.3 gegenübergestellt. Der Abschnitt 4.4

illustriert den Aufbau und die Synthese eines **planaren dreifach balancierten Mischers.** Es kommen gekreuzte Silizium Schottky Diodenringe auf einem 10 mil Quartzsubstrat (Dünnfilmtechnik) zum Einsatz. Bei einem Konversionsverlust unterhalb von 15 dB erreicht der Mischer eine LO / HF Bandbreite von 1 bis 45 / 50 GHz und mindestens 20 GHz ZF Bandbreite. Neben Simulations- und Messdaten, wird auch der Einsatz des Mischers in einem Modul eines Automatischen Testsystems beschrieben.

5 Schottky Leistungsdetektoren ohne Vorspannung

Leistungsdetektoren auf der Basis von **Schottky Dioden** werden im Kapitel 5 behandelt. Im Abschnitt 5.2 werden zwei gängige Realisierungsformen diskutiert. Zwei breitbandige Detektoren auf 5 mil Keramiksubstrat (Dünnfilmtechnik) für die Frequenzbereiche **1 bis 40 GHz** und **60 bis 110 GHz** werden vorgestellt. In den Modulen von Kapitel 6 arbeiten diese Detektoren im Bereich ihrer quadratischen Kennlinie zur Bestimmung der aktuellen Ausgangsleistung.

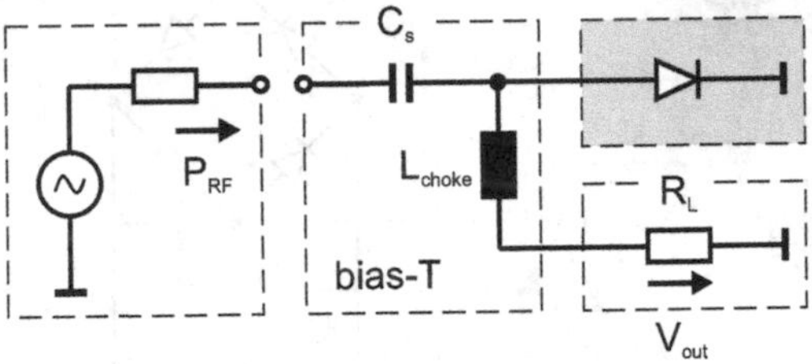

Die dafür entwickelten breitbandigen bias-T Schaltungen können auch für andere Anwendungen vorteilhaft eingesetzt werden. Bis 25 GHz werden Sensitivitätswerte um 1000 mV/mW erzielt, welche bis 40 GHz oberhalb von 300 mV/mW liegen. Die minimal detektierbare Leistung liegt in der Größenordnung von -50 dBm. Während bis 40 GHz diskrete Komponenten für das bias-T verwendet werden, kommen von 60 bis 110 GHz ein planares Hochpass-Filter und ein HF-Block basierend auf Nickel-Chrom Schichtwiderständen zum Einsatz. Im Frequenzbereich von 60 bis 102 GHz werden exzellente Sensitivitätswerte im Bereich 2000 bis 6000 mV/mW erreicht. Diese bleiben bis

110 GHz oberhalb von 1000 mV/mW. Bei einer Videobandbreite von 4 kHz liegen die zur tangentialen Sensitivität gehörenden Leistungspegel im Bereich -42 bis -55 dBm. Eine analytische Berechnung verdeutlicht den Zusammenhang zwischen dem Kapazitätswert des Schottky Übergangs und der erreichbaren maximalen Sensitivität.

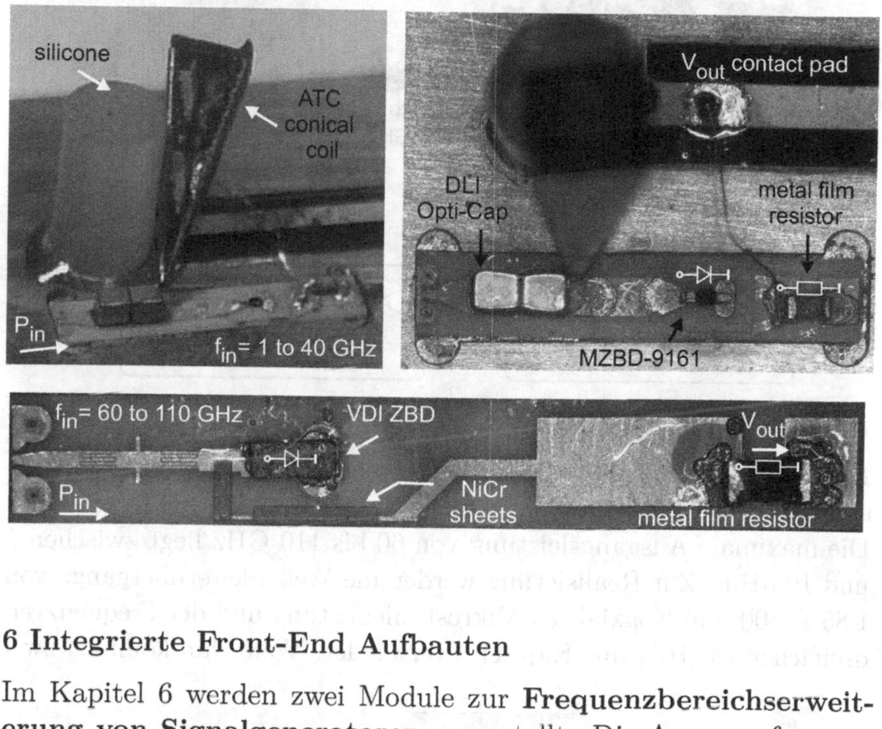

6 Integrierte Front-End Aufbauten

Im Kapitel 6 werden zwei Module zur **Frequenzbereichserweiterung von Signalgeneratoren** vorgestellt. Die Ausgangsfrequenzbereiche sind **20 bis 40 GHz** und **60 bis 110 GHz**. Derartige Module benötigt man in Automatischen Testsystemen zur Bereitstellung von Testsignalen im Sendepfad und als internes Lokaloszillatorsignal. Viele erforderliche Subkomponenten solcher Quellmodule sind bereits als monolithisch integrierte (MIC) Schaltkreise verfügbar. Eine hybride Realisierung wurde gewählt um die Vorteile der MIC und Dünnfilmtechnologie zu kombinieren. Es gibt keine generelle Regel dafür, welche Subkomponenten man besser als IC, Al_2O_3 oder SiO_2 integriert. Im Hinblick auf die Modulwartung, die Modulkosten und die thermischen Verhältnisse existiert eine Optimalkonfiguration. Die Module können einzeln betrieben werden oder in Kaskade als Frequenzversechsfacher von 10 bis 20 GHz auf 60 bis 110 GHz. Das erste

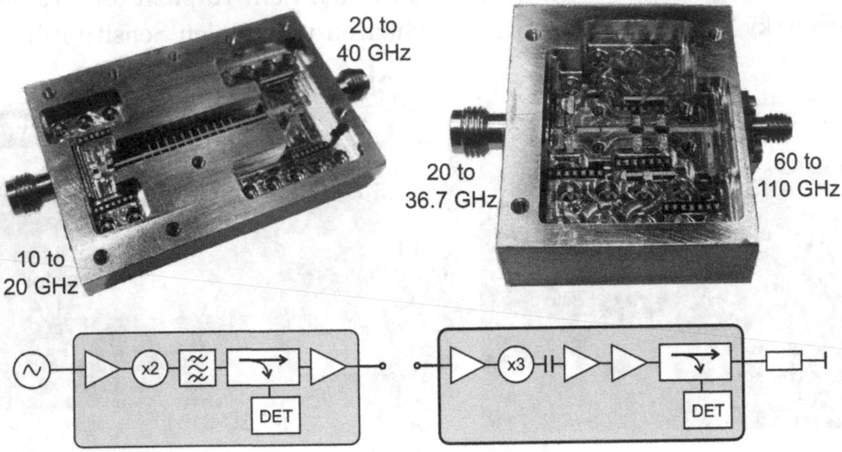

Module erreicht maximale Ausgangsleistungen in der Größenordnung 10 dBm von 20 bis 40 GHz, wobei durch Veränderung der Kontrollspannungen ein Signaldynamikbereich von etwa 15 dB möglich ist. Die maximale Ausgangsleistung von 60 bis 110 GHz liegt zwischen 5 und 10 dBm. Zur Realisierung werden die Wellenleiterübergänge von 1.85 / 1.00 mm Koaxial- zu Mikrostreifenleitung und der Frequenzverdreifacher x3_HE$_1$ aus Kapitel 2 verwendet. Weiterhin kommen zwei

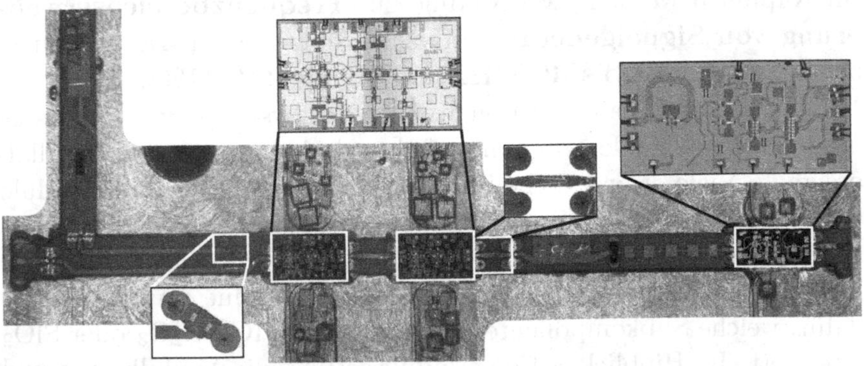

Vorwärtskoppler mit integriertem 50 Ω Abschluß aus Kapitel 3 und eine planare Serienkapazität sowie die beiden Leistungsdetektoren aus Kapitel 5 zum Einsatz.

A Filter Synthese

Gleichungen für den **Entwurf von Chebyshev Tiefpass- und Bandpass-Filterstrukturen** werden im Anhang der Arbeit in kompakter Form zusammengefasst. Die vorgestellten Filterstrukturen basieren entweder auf **g** Parametern einer Tiefpass-Prototypen-Filterstruktur oder auf **G** Parametern eines Prototypen-Lambda-Viertel-Transformators.

A Filter Synthesis

This appendix includes equations for synthesis of Chebyshev[1] lowpass (LPF) and bandpass filters (BPF) in a very compact form. The filters are either based on **g** parameters of a lowpass prototype filter [1] or **G** parameters of a prototype quarterwave transformer [2][2].

Chebyshev LPF and BPF

In order to design a Chebyshev LPF or BPF it is necessary to determine a prototype LPF with normalized 3 dB cut-off frequency $\omega = 1$ and its minimum passband ripple r_{dB} in decibel, which corresponds to the minimum passband return loss $\mathrm{RL}_{\mathrm{min}} > 0$.

$$\mathrm{RL}_{\mathrm{min\,dB}} = -10\log_{10}\left(\frac{1}{|S_{11\mathrm{min}}|^2}\right) = -10\cdot\log_{10}\left(1 - 10^{-0.1\ r_{\mathrm{dB}}}\right) \tag{A.1}$$

Alternatively, the minimum passband return loss $\mathrm{RL}_{\mathrm{min}} > 0$ is specified, that corresponds to the minimum passband ripple $r_{\mathrm{dB}} > 0$.

$$r_{\mathrm{dB}} = -10\cdot\log_{10}\left(1 - 10^{-0.1\ \mathrm{RL}_{\mathrm{min\,dB}}}\right) \tag{A.2}$$

In addition, the stopband attenuation $|S_{21\,\mathrm{dB}}(f_{\mathrm{stop}})|^2$ in decibel at the stopband frequency f_{stop} has to be specified (Fig. A.1).

$$|S_{21}(f_{\mathrm{stop}})| = -10\log_{10}|S_{21\,\mathrm{dB}}(f_{\mathrm{stop}})|^2 \tag{A.3}$$

---

[1] Pafnuty Lvovich Chebyshev (1821–1894), Russian mathematician.
[2] More sophisticated, modern filter synthesis can be found in [3–6].

As a consequence of Eq. (A.1 to A.3) the minimum filter order N is calculated from Eq. (A.4).

$$N \geq \left(\frac{\operatorname{acosh}\left(\frac{\sqrt{|S_{21}(f_{\text{stop}})|-1}}{10^{-0.1\ r_{\text{dB}}-1}} \right)}{\operatorname{acosh}\left(\frac{f_{\text{stop}}}{f_0} \right)} \right) \tag{A.4}$$

The N poles p_k of the Chebyshev prototype LPF transfer function are given by Eq. (A.5) with the abbreviations ε and η.

$$\begin{aligned} p_k &= -\sin\left(\frac{(2k-1)\pi}{2N}\right)\eta + j\cos\left(\frac{(2k-1)\pi}{2N}\right)\cosh\left(N^{-1}\operatorname{asinh}\left(\varepsilon^{-1}\right)\right) \\ k &= 1,\ldots,N \\ \varepsilon &= \sqrt{10^{0.1\ r_{\text{dB}}}-1} \qquad \eta = \sinh\left(N^{-1}\operatorname{asinh}\left(\varepsilon^{-1}\right)\right) \end{aligned} \tag{A.5}$$

Eq. (A.5) allows for calculation of the prototype LPF's scattering parameters $S_{21} = S_{12}$ and $S_{11} = S_{22} = \sqrt{1 - S_{21}}$.

$$S_{21}(p) = \frac{\prod_{k=1}^{N} \sqrt{\eta^2 + \sin^2\left(\frac{k\pi}{N}\right)}}{\prod_{k=1}^{N} (p - p_k)} \tag{A.6}$$

Lowpass Filters (LPF)

The scattering parameters of the prototype LPF from Eq. (A.6) are transformed to an arbitrary 3 dB cut-off frequency $f_0 = f_c$ by applying the lowpass transform of Eq. (A.7).

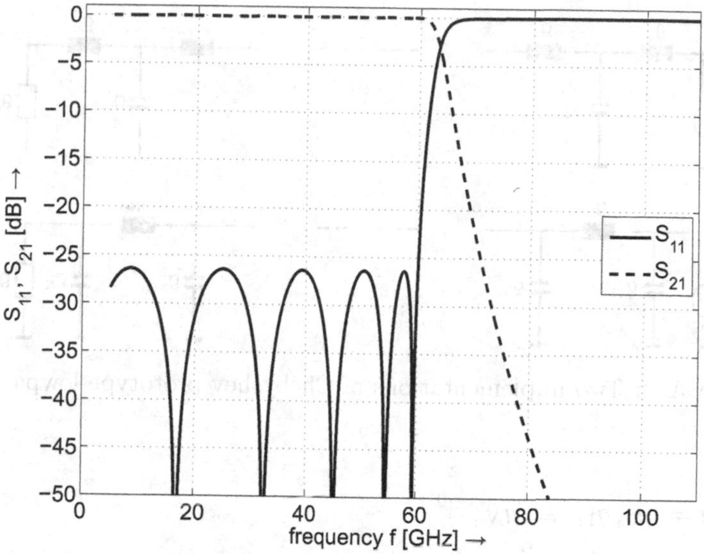

Figure A.1: Calculated scattering parameters of a Chebyshev lowpass filter with $N = 11, \mathrm{RL}_{\mathrm{min\,dB}} = 26$ dB, $r_{\mathrm{dB}} = 0.01$ dB, $f_0 = 60$ GHz, $f_{\mathrm{stop}} = 72$ GHz, and $|S_{21\,\mathrm{dB}}(f_{\mathrm{stop}})|^2 = 25$ dB.

$$
\begin{aligned}
&p \to j\frac{f}{f_0} \\
&10\log_{10}|S_{21}(f)|^2 = 10\log_{10}\left|S_{21}\left(p \to j\frac{f}{f_0}\right)\right|^2 \qquad \text{(A.7)} \\
&S_{11}(f) = \sqrt{1 - S_{21}(f)}
\end{aligned}
$$

Fig. A.1 illustrates exemplary lowpass filter scattering parameters. The resulting **g** parameters are given by Eq. (A.8). There are two different ways of implementation, which are shown in Fig. A.2.

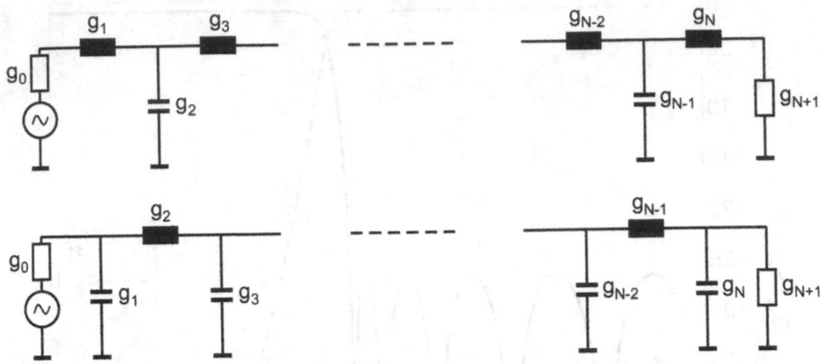

Figure A.2: Two implementations of Chebyshev prototype lowpass filter.

$$
\begin{aligned}
\mathbf{g} &= [g_0, g_1, \ldots, g_{N+1}]^{\mathrm{T}} \\
g_0 &= 1 \quad g_1 = \frac{2}{\gamma} \sin\left(\frac{\pi}{2N}\right) \\
g_k &= \frac{4 \sin\left(g_{2k-1} \frac{\pi}{2N}\right) \sin\left((2k-3)\frac{\pi}{2N}\right)}{g_{k-1}\left(\gamma^2 + \left(\sin\left((k-1)\frac{\pi}{N}\right)\right)^2\right)} \quad k = 2, \ldots, N \\
g_{N+1} &= \begin{cases} 1 & N \text{ odd}, Z_\mathrm{s} = Z_\ell \\ \coth^2\left(\frac{\beta}{4}\right) & N \text{ even}, Z_\mathrm{s} \neq Z_\ell \end{cases} \\
\beta &= \ln\left(\coth\left(\frac{r_{\mathrm{dB}}}{17.37}\right)\right) \quad \gamma = \sinh\left(\frac{\beta}{2N}\right)
\end{aligned}
\tag{A.8}
$$

Bandpass Filters (BPF)

In analogy to the lowpass transform of Eq (A.7), Eq (A.9) constitutes the corresponding bandpass transform from a LPF with normalized cut-off frequency $\omega = 2\pi f = 1$ to the center frequency f_0 with fractional bandwidth FBW. Fractional bandwidth is the ratio of bandwidth $\Delta f = f_2 - f_1$ to either the center frequency f_0 or the geometric average frequency $\sqrt{f_1 f_2}$.

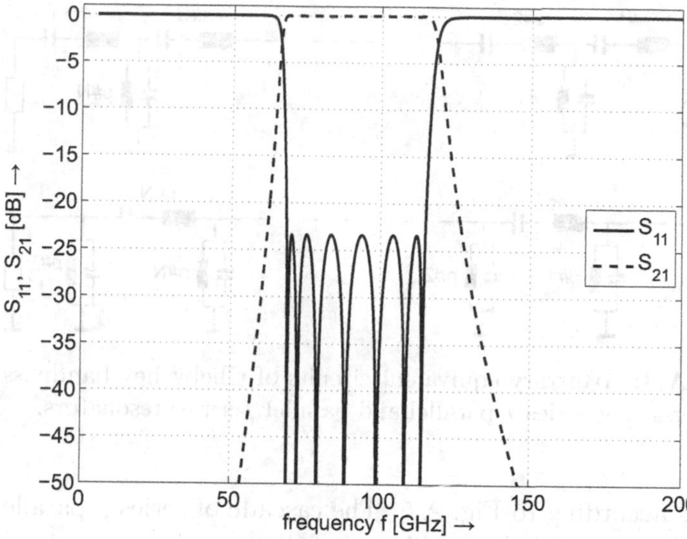

Figure A.3: Calculated scattering parameters of a Chebyshev bandpass filter with $N = 7$, $\mathrm{RL}_{\mathrm{min\,dB}} = 23$ dB, $r_{\mathrm{dB}} = 0.02$ dB, $f_{\mathrm{stop}} = 125$ GHz, $f_1 = 68$ GHz, $f_2 = 115$ GHz, $\sqrt{f_1 f_2} = 88$ GHz, FBW $= 53$ %, and $|S_{21\,\mathrm{dB}}(f_{\mathrm{stop}})|^2 = 20$ dB.

$$\begin{aligned}
p &\to \frac{j}{\mathrm{FBW}}\left(\frac{f}{f_0} - \frac{f_0}{f}\right) \\
S_{21\,\mathrm{dB}} &= 10\log_{10}|S_{21}(f)|^2 \\
&= 10\log_{10}\left|S_{21}\left(p \to \frac{j}{\mathrm{FBW}}\left(\frac{f}{f_0} - \frac{f_0}{f}\right)\right)\right|^2 \\
S_{11}(f) &= \sqrt{1 - S_{21}(f)}
\end{aligned} \tag{A.9}$$

Fig. A.3 illustrates exemplary bandpass filter scattering parameters.

In general, Chebyshev BPF consist of alternating series / parallel resonators pared with parallel / series resonators, according to Fig. A.4. To simplify BPF design, especially at high frequencies, all series / parallel resonators are exchanged with admittance J or impedance K

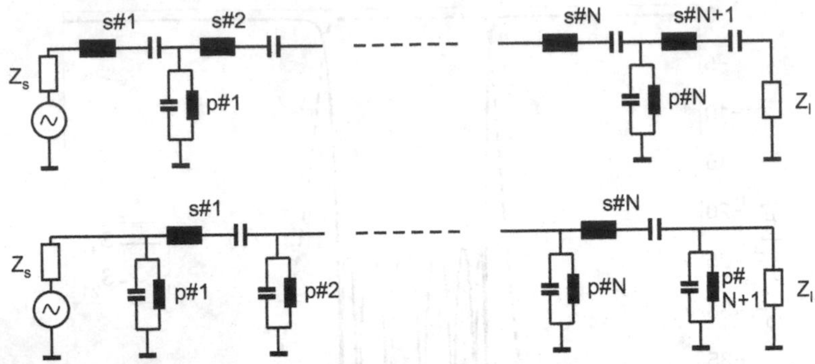

Figure A.4: Arbitrary equivalent circuits of Chebyshev bandpass filters with alternating series / parallel and parallel / series resonators.

inverters according to Fig. A.5. The cascade of series / parallel resonator with inverter behaves like a parallel / series resonator. Hence, only a single resonator type is required. For a given Chebyshev prototype LPF **g**, the required inverter values are either given by Eq. (A.10) or Eq. (A.11), [1]. Several realizations of J and K inverters have been reported. Transmission lines with quarterwave length are simple inverter realizations. Lumped element equivalent circuits of inverters are given in the next subsection.

$$\begin{gathered}\frac{J_1}{Y_0} = \sqrt{\frac{\pi \mathrm{FBW}}{2 g_1 g_2}} \quad \frac{J_{N+1}}{Y_0} = \sqrt{\frac{\pi \mathrm{FBW}}{2 g_{N+1} g_{N+2}}} \\ \frac{J_k}{Y_0} = \frac{\pi \mathrm{FBW}}{2\sqrt{g_k g_{k+1}}} \quad k = 2, \ldots, N\end{gathered} \tag{A.10}$$

$$\begin{gathered}\frac{K_1}{Z_0} = \sqrt{\frac{\pi \mathrm{FBW}}{2 g_1 g_2}} \quad \frac{K_{N+1}}{Z_0} = \sqrt{\frac{\pi \mathrm{FBW}}{2 g_{N+1} g_{N+2}}} \\ \frac{K_k}{Z_0} = \frac{\pi \mathrm{FBW}}{2\sqrt{g_k g_{k+1}}} \quad k = 2, \ldots, N\end{gathered} \tag{A.11}$$

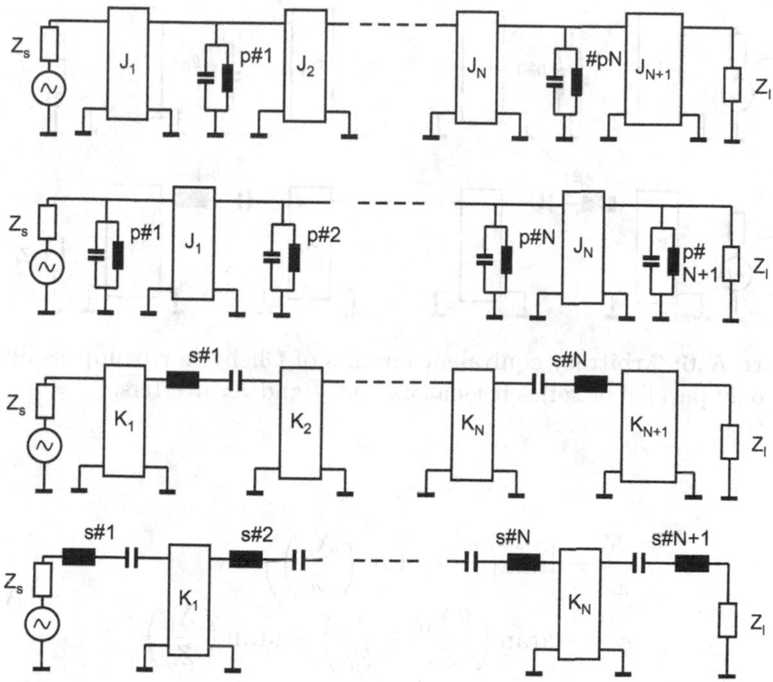

Figure A.5: Arbitrary equivalent circuits of Chebyshev bandpass filters with only parallel or series resonators and J or K inverters.

Fig. A.6 shows another class of BPF using both, K and J inverters with a single type of resonator. Fig. A.6 includes two of four possible configurations only.

Equivalent Circuits of K and J Inverters

Fig. A.7 shows lumped element equivalent circuits of impedance (left) and admittance (right) inverters [5]. The impedance inverter value normalized to the reference impedance K/Z_0 of the T circuit (X_s, X_p) in the upper left side of Fig. A.7 is given by Eq. (A.12). Whereas $\phi/2$ is the length of the feeding transmission line at the inverter input and output.

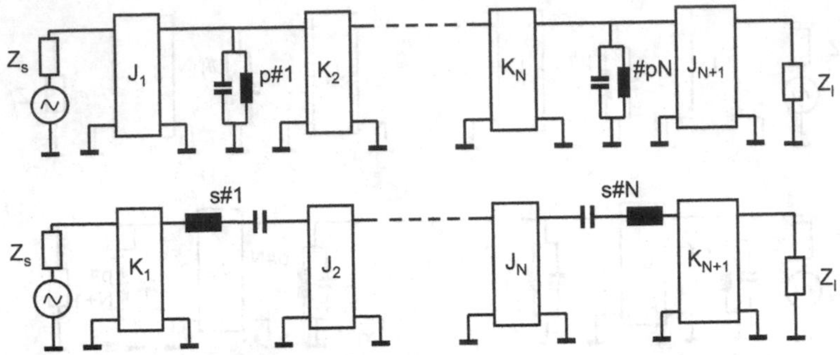

Figure A.6: Arbitrary equivalent circuits of Chebyshev bandpass filters with only parallel or series resonators and J and K inverters.

$$\begin{aligned}\frac{K}{Z_0} &= \left|\tan\left(\frac{\phi}{2} + \operatorname{atan}\left(\frac{X_\mathrm{s}}{Z_0}\right)\right)\right| \\ \phi &= -\operatorname{atan}\left(\frac{2X_\mathrm{p}}{Z_0} + \frac{X_\mathrm{s}}{Z_0}\right) - \operatorname{atan}\left(\frac{X_\mathrm{s}}{Z_0}\right)\end{aligned} \tag{A.12}$$

The parallel inductor ($X_\mathrm{p} > 0, \phi < 0$) or parallel capacitor ($X_\mathrm{p} < 0, \phi > 0$) are simplified versions of the T circuit inverter. Eq. (A.13) includes the corresponding normalized inverter value.

$$\begin{aligned}\frac{K}{Z_0} &= \tan\left|\frac{\phi}{2}\right| \\ \phi &= -\operatorname{atan}\left(\frac{2X_\mathrm{p}}{Z_0}\right) \\ \left|\frac{X_\mathrm{p}}{Z_0}\right| &= \frac{\frac{K}{Z_0}}{1-\left(\frac{K}{Z_0}\right)^2}\end{aligned} \tag{A.13}$$

The π circuit ($B_\mathrm{s}, B_\mathrm{p}$) in the upper right side of Fig. A.7 behaves analogously to the T circuit. The normalized admittance inverter value is given by Eq. (A.14).

$$
\begin{aligned}
\frac{J}{Y_0} &= \left|\tan\left(\frac{\phi}{2} + \operatorname{atan}\left(\frac{B_\mathrm{p}}{Y_0}\right)\right)\right| \\
\phi &= -\operatorname{atan}\left(\frac{2B_\mathrm{s}}{Y_0} + \frac{B_\mathrm{p}}{Y_0}\right) - \operatorname{atan}\left(\frac{B_\mathrm{p}}{Y_0}\right)
\end{aligned}
\tag{A.14}
$$

The series capacitor ($B_\mathrm{s} > 0, \phi < 0$) or series inductor ($B_\mathrm{s} < 0, \phi > 0$) are simplified versions of the π circuit inverter. Eq. (A.15) includes the corresponding normalized inverter value.

$$
\begin{aligned}
\frac{J}{Y_0} &= \tan\left|\frac{\phi}{2}\right| \\
\phi &= -\operatorname{atan}\left(\frac{2B_\mathrm{s}}{Y_0}\right) \\
\left|\frac{B_\mathrm{s}}{Y_0}\right| &= \frac{\frac{J}{Y_0}}{1-\left(\frac{J}{Y_0}\right)^2}
\end{aligned}
\tag{A.15}
$$

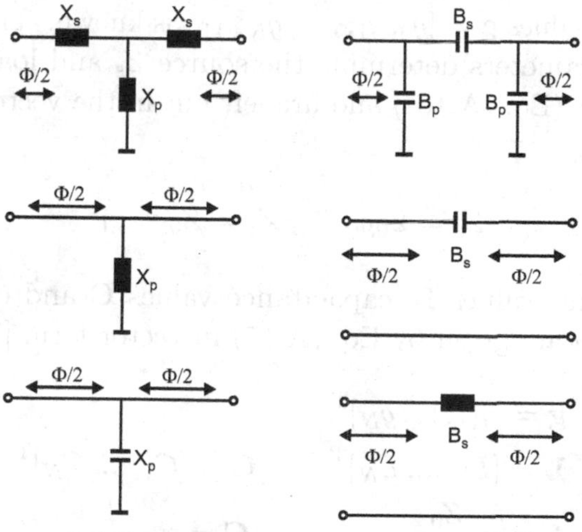

Figure A.7: Equivalent circuits of impedance (left) and admittance (right) inverters.

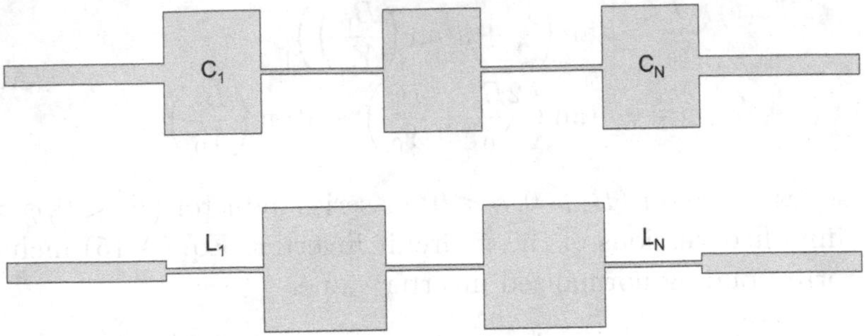

Figure A.8: Top view of exemplary microstrip line stepped impedance LPF realization $N = 5$.

Stepped Impedance LPF

Fig. A.8 shows the top view of exemplary microstrip line stepped impedance LPF realizations. From Eq. (A.8) the $\mathbf{g}$ parameter set with $(N+2)$ values $\mathbf{g} = [g_0, g_1, \dots, g_{N+1}]^{\mathrm{T}}$ is known. The first g_0 and last g_{N+1} parameters determine the source Z_{s} and load impedance Z_ℓ of the LPF (Eq. (A.16)) and are left out in the vector notation of Eq. (A.17).

$$Z_{\mathrm{s}} = Z_0 g_0 \qquad Z_\ell = Z_0 g_{N+1} \tag{A.16}$$

The inductance values $\mathbf{L}$, capacitance values $\mathbf{C}$ and corresponding lengths $\boldsymbol{\ell}_{\mathbf{L}}, \boldsymbol{\ell}_{\mathbf{C}}$ are given by Eq. (A.17) in vector form [7].

$$\begin{aligned}
&\mathbf{g} = [g_1, \dots, g_N]^{\mathrm{T}} \\
&\mathbf{L} = [L_1, ..., L_N]^{\mathrm{T}} && \mathbf{C} = [C_1, ..., C_N]^{\mathrm{T}} \\
&\mathbf{L} = \mathbf{g}\frac{Z_0}{2\pi f_{\mathrm{c}}} && \mathbf{C} = \mathbf{g}\frac{1}{Z_0 2\pi f_{\mathrm{c}}} \\
&\boldsymbol{\ell}_{\mathbf{L}} = \mathbf{L}\frac{c_0}{\sqrt{\epsilon_{\mathrm{reff\,L}}} Z_{0L}} && \boldsymbol{\ell}_{\mathbf{C}} = \mathbf{C}\frac{c_0 Z_{0C}}{\sqrt{\epsilon_{\mathrm{reff\,C}}}}
\end{aligned} \tag{A.17}$$

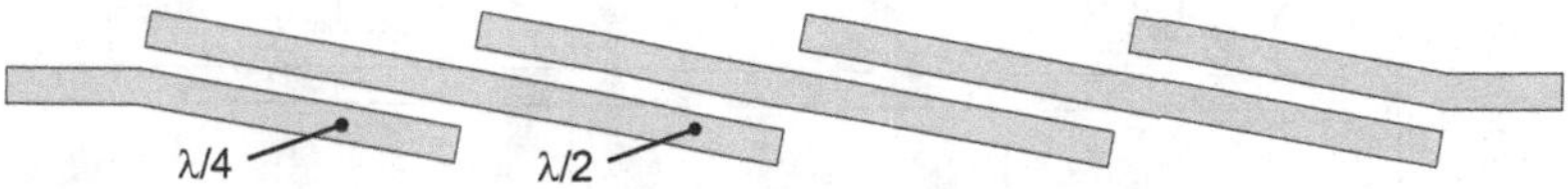

Figure A.9: Top view of exemplary edge coupled microstrip line BPF with $N = 3$ halfwavelength series resonators.

Edge Coupled Microstrip Line BPF

An edge coupled microstrip line BPF with N halfwavelength $\lambda/2$ series resonators consists of $(N+1)$ coupled quarterwave sections acting as inverters. With the **g** parameter set ($N+2$ values) from Eq. (A.8), the $(N+1)$ normalized inverter values J_k/Y_0 are known from Eq. (A.10).

The required even and odd mode impedances Z_{0ek}, Z_{0ok} of each inverter are given by Eq. (A.18), [7].

$$\begin{aligned} Z_{0ek} &= \frac{1}{Y_0}\left(1 + \frac{J_k}{Y_0} + \left(\frac{J_k}{Y_0}\right)^2\right) \\ Z_{0ok} &= \frac{1}{Y_0}\left(1 - \frac{J_k}{Y_0} + \left(\frac{J_k}{Y_0}\right)^2\right) \end{aligned} \tag{A.18}$$

And the inverter lengths ℓ_k, $k = 1 \ldots N+1$ are chosen to equal a quarter of the geometric average of the even and odd mode effective wavelengths $1/4\sqrt{\lambda_e \lambda_o}$.

$$\ell_k = \lambda_{f0}/4 = \frac{1}{4}\sqrt{\lambda_e \lambda_o} = \frac{\lambda_0}{4\sqrt{\sqrt{\epsilon_{re}\epsilon_{ro}}}} \tag{A.19}$$

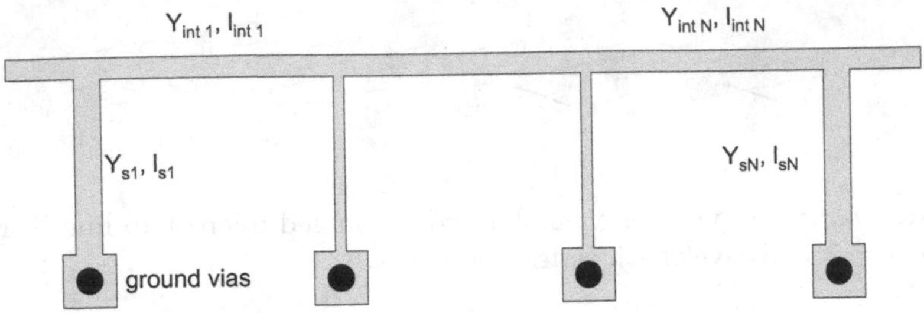

Figure A.10: Top view of exemplary microstrip line BPF with $N = 4$ short circuited quarterwave stublines (resonators) and $(N-1)$ quarterwave interconnection lines as inverters.

Quarterwave Short Circuited BPF

Fig. A.10 depicts the top view of an exemplary microstrip line BPF with $N = 4$ short circuited quarterwave stubs (parallel resonators) and $(N-1)$ quarterwave interconnection lines as inverters.

This type of filter is also known as fishgrate BPF. The filter parameters are derived from $(N+2)$ $\mathbf{g}$ parameters (Eq. (A.8)) using the J inverter model [1] of Eq. (A.20).

$$\begin{aligned} \frac{J_1}{Y_0} &= g_1\sqrt{\frac{hg_2}{g_3}} \qquad \frac{J_{N-1}}{Y_0} = g_1\sqrt{\frac{hg_2g_{N+2}}{g_1g_N}} \\ \frac{J_k}{Y_0} &= \frac{hg_1g_2}{\sqrt{g_{k+1}g_{k+2}}} \qquad k = 2,\ldots,N-2 \end{aligned} \tag{A.20}$$

The scaling parameter h (initial value $h = 2$) is used to adjust the filter parameters until practical realizable values are found. Eq. (A.21) includes N stub admittances $\mathbf{Y}_\mathrm{s}$ and lengths $\boldsymbol{\ell}_\mathbf{s}$ together with $(N+1)$ interconnection line admittances $\mathbf{Y}_\mathrm{int}$ and lengths $\boldsymbol{\ell}_\mathrm{int}$. The latter include the source $Y_{\mathrm{int}0} = Y_\mathrm{s} = g_0/Y_0$ and load impedances $Y_{\mathrm{int}N} = Y_\ell = Y_0/g_{N+1}$.

$$\begin{aligned} \mathbf{Y}_\mathrm{s} &= [Y_{\mathrm{s}1}, \ldots, Y_{\mathrm{s}N}]^\mathrm{T} & \boldsymbol{\ell}_\mathbf{s} &= [\ell_{\mathrm{s}1}, \ldots, \ell_{\mathrm{s}N}]^\mathrm{T} \\ \mathbf{Y}_\mathrm{int} &= [Y_{\mathrm{int}0}, \ldots, Y_{\mathrm{int}N}]^\mathrm{T} & \boldsymbol{\ell}_\mathrm{int} &= [\ell_{\mathrm{int}1}, \ldots, \ell_{\mathrm{int}N-1}]^\mathrm{T} \end{aligned} \tag{A.21}$$

The remaining $(N-1)$ components of $\mathbf{Y}_\mathrm{int}$ are determined by the J_k inverter values.

$$Y_k = J_k \qquad k = 1 \ldots n-1 \tag{A.22}$$

Whereas the N components of $\mathbf{Y}_s$ are known from Eq. (A.23).

$$\begin{aligned} \frac{Y_1}{Y_0} &= g_1 g_2 (1 - h/2) \tan\Theta + \left(N_1 - \frac{J_1}{Y_0}\right) \\ \frac{Y_N}{Y_0} &= (g_{N+1} g_{N+2} - g_1 g_2 h/2) \tan\Theta + \left(N_{n-1} - \frac{J_{N-1}}{Y_0}\right) \\ \frac{Y_k}{Y_0} &= \left(M_{k-1} + M_k - \frac{J_{k-1}}{Y_0} - \frac{J_k}{Y_0}\right) \quad k = 2, \ldots, N-1 \\ M_k &= \sqrt{\left(\frac{J_k}{Y_0}\right)^2 + \left(\frac{h g_1 g_2 \tan\Theta}{2}\right)^2} \quad k = 1 \ldots N-1 \\ \Theta &= \frac{\pi}{2}(1 - \mathrm{FBW}/2) \quad \mathrm{FBW} = \frac{f_2 - f_1}{\sqrt{f_1 f_2}} \end{aligned} \tag{A.23}$$

The individual lengths of all resonators are $\lambda/4 = \frac{c_0}{\sqrt{\epsilon_\mathrm{reff}} f_0}$. Consideration of the parasitic influence of short circuited vias is mandatory.

Halfwave BPF and Quarterwave BPF with Lumped Element Inverters

Fig. A.11 illustrates the top view of exemplary microstrip line BPF with $N = 2$ halfwavelength series resonators and $(N+1)$ lumped inverters (top) and BPF with $N = 4$ quarterwavelength series resonators and $(N+1)$ lumped inverters (bottom). In case of the quarterwavelength resonators, impedance and admittance inverters are used alternately. In both cases, the impedance of the resonators $Z_{\mathrm{res}k}$ equals the reference impedance Z_0. For a given $\mathbf{g}$ parameter set from Eq. (A.8),

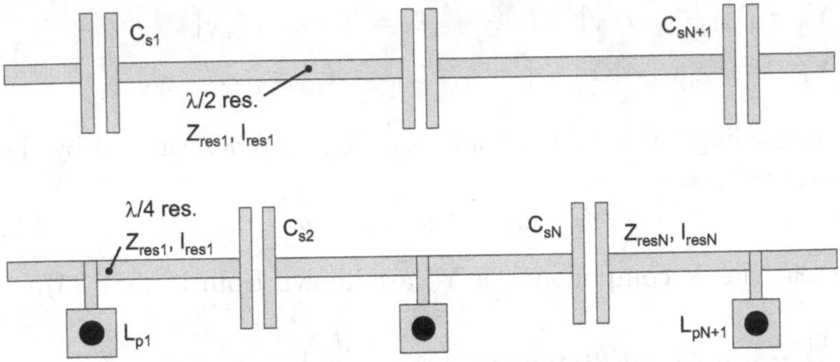

Figure A.11: Top view of exemplary microstrip line BPF with $N = 2$ halfwavelength series resonators and $(N + 1)$ lumped inverters (top) and BPF with $N = 4$ quarterwavelength series resonators and $(N + 1)$ lumped inverters (bottom).

the inverter values are calculated using Eq. (A.10, A.11). Inserting the inverter values into Eq. (A.12 to A.15) leads to the corresponding lumped element values. It is necessary to calculate the individual phase shift ϕ_k, incorporated with each lumped inverter. These values have to be used to correct the halfwave or quarterwave resonator lengths [1].

Distributed LPF based on Quarterwave Prototype

Contrary to the filters of the preceding sections, the distributed LPF of this section is based on a quarterwave transformer prototype $\mathbf{G} = [G_1, , \ldots, G_N]^{\mathrm{T}}$ rather than on a lumped lowpass prototype filter and its **g** parameters. Choosing prototype filter order N, the G_k, $k = 1 \ldots N$ parameters are given by Eq. (A.24), which is a closed form expression from Rhodes ([8], [2] page 146 et seq.). The exact theory from Young [9–11] and Levy [12] could be used equally.

$$G_k = A_k \left(\frac{2\sin((2k-1)\frac{\pi}{2N})}{\sin((2k+1)\frac{\pi}{2N})} - \frac{\alpha}{4}\left(\frac{\eta^2 + \sin^2(k\frac{\pi}{n})}{\sin((2k+1)\frac{\pi}{2n})} + \frac{\eta^2 + \sin^2(k-1)\frac{\pi}{n}}{\sin((2k-3)\frac{\pi}{2n})} \right) \right) \tag{A.24}$$

Whereas A_k is given by Eq. (A.25) with the last term $\eta^2 + \sin^2(0)$, replaced by η, e.g. $A_2 = \frac{\eta}{\eta^2+\sin^2(\pi/N)}$.

$$\begin{aligned} A_r &= \frac{\{\eta^2 + \sin^2((r-2)\frac{\pi}{N})\}\{\eta^2 + \sin^2((r-4)\frac{\pi}{N})\}\dots}{\{\eta^2 + \sin^2((r-1)\frac{\pi}{N})\}\{\eta^2 + \sin^2((r-3)\frac{\pi}{N})\}\dots} \qquad r = 1 \dots N \\ \varepsilon &= \sqrt{10^{0.1\ r_{\mathrm{dB}}} - 1} \qquad \eta = \sinh\left[N^{-1}\mathrm{asinh}\left(\varepsilon^{-1}\right)\right] \\ \alpha &= \sin(\mathrm{FBW}) \end{aligned} \tag{A.25}$$

For a specified passband ripple r_{dB} or minimum passband return loss $\mathrm{RL}_{\mathrm{min\,dB}}$ and stopband rejection $|S_{21}(f_{\mathrm{stop}})|$, the required minimum filter order N is known from Eq. (A.4). The fractional bandwidth of the quarterwave transformer prototype and therefore the LPF's 3 dB cut-off frequency are determined by the bandwidth scaling factor $\alpha < 1$. The G_k parameters are the normalized admittances of the quarterwave prototype. The prototype consists of $(2N-1)$ cascaded transmission lines with impedances equal to

$$G_k^{-1} Z_0, \qquad k = 1, \dots, N-1, N, N-1, \dots, 1 \tag{A.26}$$

and quarterwavelength at the desired center frequency as shown in Fig. A.12. In fact, it is two transformers, from Z_{s} to the impedance[3] at the center section and back to Z_ℓ, connected together. The source Z_{s} and load impedances Z_ℓ are given by Eq. (A.27).

[3]The ratio of the center section's impedance to Z_{s} is greater than 1.

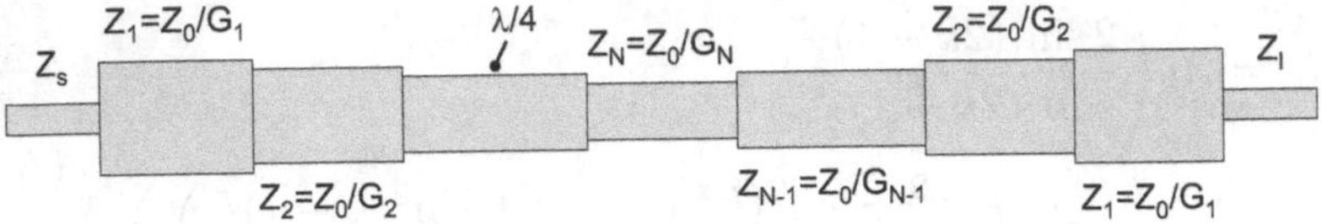

Figure A.12: Quarterwave transformer prototype ($N = 4$), consisting of $(2N-1)$ cascaded commensurate transmission lines with quarterwavelength.

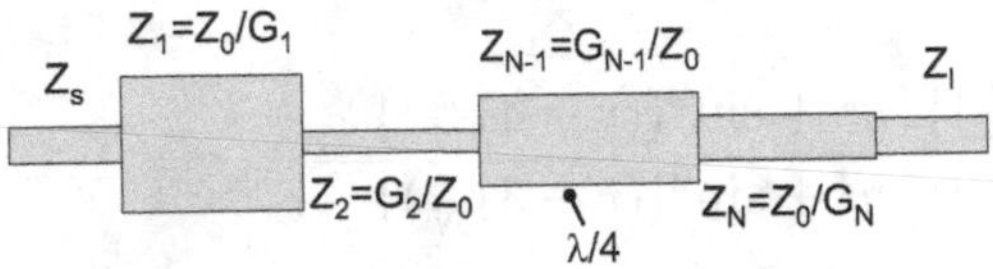

Figure A.13: Distributed lowpass prototype derived from quarterwave transformer prototype.

$$Z_s = Z_0 \qquad Z_\ell = \begin{cases} Z_0 & N \text{ odd} \\ Z_0 \frac{\sqrt{1+\varepsilon^2}+\varepsilon}{\sqrt{1+\varepsilon^2}-\varepsilon} & N \text{ even} \end{cases} \tag{A.27}$$

The quarterwave transformer prototype is converted to a distributed lowpass prototype by inverting the even sections, according to Eq. (A.28) and Fig. A.13. Quarterwavelength of the sections corresponds to the center frequency of the quarterwave transformer, which is the center frequency of the lowpass prototype's stopband.

$$\begin{aligned} Z_k &= Z_0 G_k^{-1} \quad k \text{ odd} \quad k = 1, \ldots, N \\ Z_k &= Z_0 G_k \quad\;\; k \text{ even} \\ Z_s &= Z_0 \qquad Z_\ell = \begin{cases} Z_0 & N \text{ odd} \\ Z_0 \frac{\sqrt{1+\varepsilon^2}+\varepsilon}{\sqrt{1+\varepsilon^2}-\varepsilon} & N \text{ even} \end{cases} \end{aligned} \tag{A.28}$$

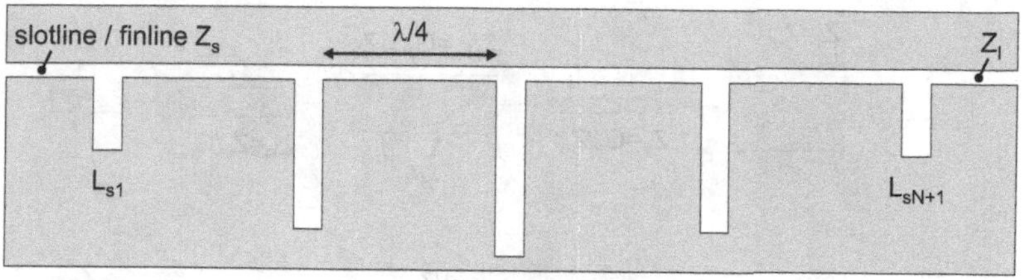

Figure A.14: Top view of exemplary slotline / finline distributed LPF with $(N+1)$ series inductors as inverters, separated by $N=4$ quarterwave interconnection lines.

Fig. A.14 depicts the top view of an exemplary slotline / finline distributed LPF[4] with $(N+1)$ series inductors as inverters, separated by $N = 4$ quarterwave interconnection lines [13, 14]. The LPF in Fig. A.14 is based on the quarterwave transformer and / or the corresponding distributed lowpass prototype. To determine the series inductor values L_s, all odd elements of the distributed LPF are inverted. The effect is removed by introduction of K inverters before and after each odd element. The $(N+1)$ K inverter values are chosen to allow for equalized impedances of every quarterwave section $Z_k = Z_0$, $k = 1, \ldots, N$. Fig. A.15 illustrates the process [5]. The inverter values are known from Eq. (A.29), and normalized to the geometric average of the adjacent impedances according to Eq. (A.30). Inserting the normalized inverter values into Eq. (A.12 to A.15) leads to the corresponding lumped element values (e.g. series inductances). It is necessary to calculate the individual phase shift ϕ_k, incorporated with each lumped inverter. These values have to be used to correct the quarterwave lengths ℓ_k, $k = 1, \ldots, N$.

[4] In case of finlines, there is a waveguide cut-off frequency greater than DC and the filter actually shows bandpass characteristics. Furthermore, due to even filter order source and load impedance differ. In Fig. A.14 the same finline is used at the input and output, which is often acceptable as the occurring reflections are low.

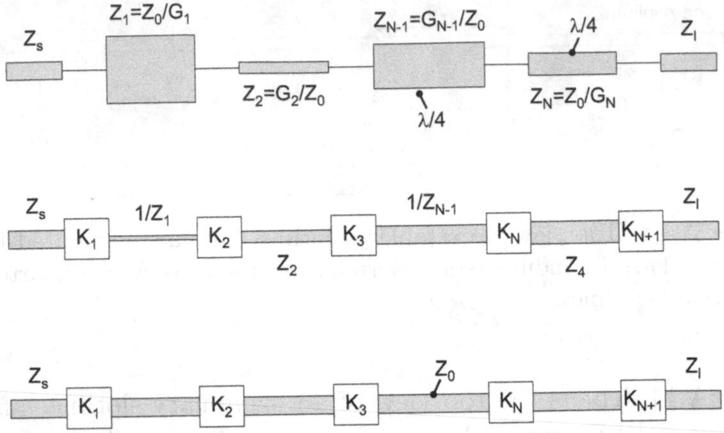

Figure A.15: Derivation of distributed LPF (bottom) from distributed LPF prototype (top) by introduction of K inverters.

$$K_k = \frac{\sqrt{\eta^2 + \sin^2\left((k-1)\frac{\pi}{N}\right)}}{\eta} \qquad k = 1, \ldots, N+1 \tag{A.29}$$

$$\frac{K_k}{\sqrt{Z_k^{-1} Z_{k+1}^{-1}}} \qquad k = 1, \ldots, N+1 \tag{A.30}$$

References

[1] G. L. Matthaei, L. Young, and E. M. T. Jones. *Microwave Filters, Impedance Matching Networks, and Coupling Structures.* New York: McGraw-Hill, 1964. ISBN 0890060991.

[2] J. D. Rhodes. *Theory of Electrical Filters.* New York: John Wiley & Sons, 1976. ISBN 0471718068.

[3] B. J. Minnis. *Designing Microwave Circuits by Exact Synthesis.* Boston: Artech House, 1996. ISBN 0890067414.

[4] I. Hunter. *Theory and Design of Microwave Filters.* IET Electromagnetic Waves Series, No. 48, 2001. ISBN 9780852967775.

[5] R. J. Cameron, C. M. Kudsia, and R. R. Mansour. *Microwave Filters for Communication Systems.* New York: John Wiley & Sons, 2007. ISBN 0471450227.

[6] P. Jarry and J. Beneat. *Advanced Design Techniques and Realizations of Microwave and RF Filters.* New York: John Wiley & Sons, 2008. ISBN 9780470183106.

[7] J.-S. Hong and M. J. Lancaster. *Microstrip Filters for RF / Microwave Applications.* New York: John Wiley & Sons, 2001. ISBN 0471388777.

[8] J. D. Rhodes. *Design Formulas for Stepped Impedance Distributed and Digital Wave Maximally Flat and Chebyshev Low-Pass Prototype Filters.* IEEE Transactions on Circuits and Systems, Vol. 22, No. 11, pp. 866-874, 1975.

[9] L. Young. *Tables for Cascaded Homogeneous Quarter-Wave Transformers.* IRE Transactions on Microwave Theory & Techniques, Vol. 7, No. 2, pp. 233-237, 1959.

[10] L. Young. *Tables for Cascaded Homogeneous Quarter-Wave Transformers (Correction).* IRE Transactions on Microwave Theory & Techniques, Vol. 8, No. 2, pp. 243-244, 1960.

[11] L. Young. *The Quarter-Wave Transformer Prototype Circuit.* IRE Transactions on Microwave Theory & Techniques, Vol. 8, No. 5, pp. 483-489, 1960.

[12] R. Levy. *Tables of Element Values for the Distributed Low-Pass Prototype Filter.* IEEE Transactions on Microwave Theory & Techniques, Vol. 13, No. 5, pp. 514-536, 1965.

[13] C. Nguyen and K. Chang. *Millimeter-Wave Low-Loss Finline Lowpass Filters.* Electronic Letters, Vol. 20, No. 24, pp. 1010-1011, 1984.

[14] C. Nguyen and K. Chang. *On the Design and Performance of Printed-Circuit Filters and Diplexers for Millimeter-Wave Integrated Circuits.* Int. Journal of Infrared and Millimeter Waves, Vol. 7, No. 7, pp. 971-998, 1986.

Publications

Within the scope of this thesis, the following scientific contributions have been published.

German Conference Contributions

[G1] M. Sterns, M. Hrobak, S. Martius, and L.-P. Schmidt. *Magnetically Tunable Filter from 72 GHz to 95 GHz*. German Microwave Conference, 2010.

[G2] M. Hrobak, M. Sterns, W. Stein, M. Schramm, and L.-P. Schmidt. *An Octave Bandwidth Finline Frequency Tripler*. German Microwave Conference, 2011.

[G3] M. Schramm, M. Hrobak, J. Schür, L.-P. Schmidt, and M. Konrad. *Synthetischer Netzwerkanalysator: Attraktive Messplatzerweiterung für bestehende RF-PXI-Systeme*. ATE-Technologietag, 2011.

[G4] M. Hrobak, M. Sterns, E. Seler, T. Schrauder, M. Schramm, and L.-P. Schmidt. *Broadband, Nonuniform Stripline Directional Couplers for use in VNA Testsets*. German Microwave Conference, 2012.

International Conference Contributions

[I1] M. Schramm, M. Hrobak, J. Schür, L.-P. Schmidt, and A. Lechner. *A Simplified Method for Measuring Complex S-Parameters Dedicated to Synthetic Instrumentation.* European Microwave Conference, 2010.

[I2] M. Schramm, M. Hrobak, J. Schür, L.-P. Schmidt, and M. Konrad. *MOS-16: A New Method for In-Fixture Calibration and Fixture Characterization.* Automatic RF Techniques Group Measurement Conference, 2011.

[I3] M. Schramm, M. Hrobak, J. Schür, L.-P. Schmidt, and M. Konard. *Impact of Different Post-Processing Methods on a One Receiver 2-Port Synthetic VNA Architecture.* European Microwave Conference, 2011.

[I4] M. Hrobak, W. Stein, M. Sterns, M. Schramm, and L.-P. Schmidt. *A Hybrid Broadband Microwave Triple Balanced Mixer Based on Silicon Crossed Quad Diodes.* European Microwave Conference, 2012.

[I5] M. Schramm, M. Hrobak, J. Schür, and L.-P. Schmidt. *A SOLR Calibration Procedure for the 16-Term Error Model.* European Microwave Conference, 2012.

[I6] M. Hrobak, M. Sterns, M. Schramm, W. Stein, and L.-P. Schmidt. *Planar Varistor Mode Schottky Diode Frequency Tripler Covering 60 GHz to 110 GHz.* International Microwave Symposium, 2013.

[I7] M. Hrobak, M. Sterns, M. Schramm, W. Stein, F. Poprawa, A. Zirof, and L.-P. Schmidt. *Planar D-Band Frequency Doubler and Y-Band Tripler on PTFE Laminates.* International Conference on Infrared, Millimeter, and Terahertz Waves, 2013.

[I8] M. Hrobak, M. Sterns, M. Schramm, W. Stein, and L.-P. Schmidt. *Planar Zero Bias Schottky Diode Detector Operating in the E- and W-Band.* European Microwave Conference, 2013.

[I9] M. Schramm, M. Hrobak, J. Schür, and L.-P. Schmidt. *A New Switch Correction Method for a Single-Receiver VNA.* European Microwave Conference, 2013.

Journal Papers

[J1] M. Hrobak, M. Sterns, E. Seler, M. Schramm, and L.-P. Schmidt. *Design and Construction of an Ultrawideband Backward Wave Directional Coupler.* IET Microwaves, Antennas, and Propagation, Vol. 6, No. 9, pp. 1048-1055, 2012.

[J2] M. Hrobak, M. Sterns, M. Schramm, W. Stein, and L.-P. Schmidt. *Design and Fabrication of Broadband Hybrid GaAs Schottky Diode Frequency Multipliers.* IEEE Transactions on Microwave Theory & Techniques, Vol. 61, No. 12, pp. 4442- 4460, 2013.

Books and Book Chapters

[B1] M. Hrobak. *Modeling of Interconnect Structures in Semiconductor Packages.* Berlin: Mensch und Buch Verlag, 2007. ISBN 3-86664-210-5.

[B2] M. Hrobak, M. Schramm, L.-P. Schmidt, and A. Lechner. *Untersuchung einer Streuparameter-Testumgebung aus modularen RF PXI Instrumenten.* Virtuelle Instrumente in der Praxis. Heidelberg: Huethig, 2009. ISBN 3-77852-976-5.

About the Author

Michael Hrobak received the Diploma degree in electrical engineering from the University of Applied Sciences (now OTH), Regensburg, Germany, in 2007. His Diploma thesis *Modeling of Interconnect Structures in Semiconductor Packages* [B1] concerned signal integrity issues and planar integration of microwave components within ball-grid-array packages of the Infineon Technologies AG, Munich, Germany. The thesis has been awarded the VDI-prize 2007 by the Association of German Engineers (VDI). From 2007 to 2013, he was with the Institute of Microwaves and Photonics (LHFT), Friedrich-Alexander University of Erlangen-Nuremberg, Germany, as a Research Associate, dealing with the development of millimeter-wave front end modules for synthetic semiconductor test systems. Since 2014 he is with the Microwave Department of the Ferdinand-Braun-Institut (FBH), Berlin, Germany, where he is involved in the development and measurement of monolithic integrated circuits using indium phosphide (InP) double heterojunction bipolar transistors (DHBT). His research interests are in the field of planar realizations of linear and nonlinear broadband components, including directional couplers and filters, balanced mixers, frequency multipliers and power detectors. Dr. Hrobak is a member of the IEEE association.

Printed in the United States
By Bookmasters